Wind Turbine Noise

Edited by
Dick Bowdler &
Geoff Leventhall

ISBN 978-1-907132-30-8

5 Wates Way, Brentwood, Essex CM15 9TB, United Kingdom

Contents

Introduction

In 1996 Springer-Verlag published Wind Turbine Noise, authored by Siegfried Wagner, Rainer Bareiss and Gianfranco Guidati. Although that was only 15 years ago, many things have moved on. When Wagner et al published their book, the installed capacity in Europe was about 3GW and it was projected to increase to 30GW by 2010 and to 100GW by 2030. In fact 30GW was reached in 2003 and it seems likely that the 2030 target will be achieved sometime in 2011 or 2012. At present, Europe is home to around half of the world's capacity but that proportion is gradually reducing as the other nations of the world take wind power on board.

Perhaps the most striking development, and one which is common to most countries, is the strength of the objections to wind turbine noise amongst a proportion of the population. A survey in the Netherlands [F. van den Berg, E. Pedersen, J. Bouma, R. Bakker, WINDFARM perception: Visual and acoustic impact of wind turbine farms on residents - Final report, 2008.] found that noise was often cited as the most annoying aspect of wind turbines. The reason for this is unclear, though it has been known for some time that the perception of noise is significantly influenced by psychological factors and, in particular, that non-acoustic factors can moderate the level of annoyance associated with a noise source, making its noise more or less annoying..

We now have better knowledge of how turbine noise is produced and how it is propagated than we did 15 years ago. There have been investigations into sound near and in buildings and into the subjective responses to turbine noise, techniques of measurement have improved and different regulatory criteria for acceptable turbine noise have been developed world wide. Nevertheless, it is important for both the developers of wind farms and for the neighbouring residents that knowledge of wind turbine noise and its effect on people is under continuous development. Only in this way can we know what noise levels are acceptable and, when noise levels are found not to be acceptable, how we can mitigate them. The knowledge may also allow us to design and build new, quieter turbines.

The turbines we discuss in this book are modern large upwind turbines with three aerofoil shaped blades and a rotor diameter of around 60m or more with pitch or stall control, though much of the book applies to a wider range of turbines. The book aims to set out the facts, as we know them at present, as accurately as possible. Whilst the book is at a good technical level, the authors have avoided excessive theoretical treatments but, with a subject such as this, it is inevitable that some mathematics has been included. There is an introductory chapter on the acoustical background. Thereafter the book is arranged logically, starting with how turbine noise is produced, how it is propagated, what happens when it arrives at the receiver and what effect the noise has on the people receiving it. There are then two technical chapters and, finally, appendices covering a number of peripheral subjects in brief.

This book originates from an idea by Geoff Leventhall. Geoff is an internationally known consultant in noise, vibration and acoustics. He holds a BSc in Physics and both an MSc and PhD in Acoustics. He has been an academic – at Chelsea College, University of London as Reader in Acoustics and Professor of Environmental Engineering at London South Bank University – and a commercial consultant. He has written numerous papers on acoustics and sits on several technical committees and is the organiser of two series of International Conferences: on Low Frequency Noise and, particularly, Wind Turbine Noise.

In the spring of 2010 Dick Bowdler became involved in the project. He has been noise consultant for over forty years first in London and then Scotland. In the last ten years he has specialised in wind farm noise and represented local authorities and objectors groups in planning procedures. He believes that it is most important to make sure decisions are based on the best knowledge we have of the scientific facts and that these facts should be developed where they are lacking. He has been an enthusiastic organiser, contributor and supporter of wind farm noise meetings and conferences.

In Chapter 1 Geoff Leventhall describes, in non-technical terms, the basics of acoustics for those readers who are not familiar with the subject.

Chapter 2, written by Stefan Oerlemans from the Netherlands, looks at how turbine noise is produced and at its characteristics . Appendix 1 relates to this chapter and is also written by him. Stefan has a degree in applied physics from Eindhoven University of Technology, a Master's thesis on the physical properties of free steam jets in air and a PhD thesis on 'Detection of aeroacoustic sound sources on aircraft and wind turbines'. He works as an aeroacoustics research engineer and scientist at the National Aerospace Laboratory, NLR, in The Netherlands. He has carried out several national and international research projects on flow-induced sound from aircraft, wind turbines, and helicopters and has a long list of papers on these subjects to his credit.

Chapter 3 looks at the propagation of turbine noise in the open air and how it differs in some respects from propagation of other noise sources. It is written by Andrew Bullmore and Andrew Peplow. Andrew Bullmore has a BSc in Engineering Acoustics and Vibration and a PhD in Acoustics both from the University of Southampton. He has worked full time in acoustics, for the first seven years in research and for the past nineteen years in general acoustical consultancy and applied research. He is Managing Partner of Hoare Lea Acoustics, the specialist noise and vibration consultancy division of Hoare Lea & Partners. He was one of the authors of the "ETSU" method of assessment used in the UK and some other countries. He is active in preparing noise assessments for wind farms. Since gaining his PhD in the modelling of the acoustic performance of roadside barriers, Andrew Peplow has been active in full time research with his main interest being in computational acoustics. His research includes the use of boundary integral equation formulations for sound propagation above non-flat terrains and within layered media. His research has been variously undertaken at the University of Bradford, ISVR at the University of Southampton and the Royal Technical University in Stockholm. He joined Hoare Lea Acoustics in 2008 where he has been applying advanced numerical techniques to solving practical problems, including the modelling of environmental noise from wind farms.

Chapter 4 is on the effects of the receiving environment. This is written by Bo Søndergaard who has an M.Sc. in Physics from the University of Aarhus. Bo worked for DELTA – a Danish independent testing company for 24 years and produced reports into the measurement of noise from wind turbines, in particular on measurement methodology and low frequency noise. Since 2010 he has been a consultant working for Grontmij, Acoustica specialising in noise from wind turbines. He is the Danish representative and convenor of the IEC Maintenance Team 11 working on a revision of IEC 61400-11, the standard for measurement of noise from wind turbines. He has been a prolific presenter of papers on wind turbine noise at meetings and conferences.

The swish of wind turbines that is heard close to them does not normally travel over distances more than a few hundred metres as is explained in Chapter 2. Nevertheless amplitude modulated noise from turbines is heard by people and has been measured at distances of at least 2km. Why this happens is poorly understood and Chapter 5 covers what is known of the subject at this time. It is written by Frits van den Berg (see below) and Dick Bowdler.

Chapter 6 deals with the effects of noise on people and is written by Frits van den Berg. Frits graduated in applied physics, then worked as a teacher and as a measurement and control specialist. From 1985, he worked in the Science Shop for Physics at the University of Groningen, supporting residential and environmental groups with problems in acoustics and radiation. His main topics were low frequency noise complaints, quiet areas and wind turbine sound (in which he obtained a PhD in 2006). He is now a senior advisor in environmental health at the Amsterdam Public Health Service.

The measurement of wind turbine noise is more complex than most noise measurements, largely because it varies with wind speed and temperature gradient. Chapter 7 deals with this. It is written by David Hessler. David holds a Bachelor of Arts degree from the University of Hartford CT and a Bachelor of Science degree in mechanical engineering from the University of Maryland. He is a

Principal at the acoustical consulting firm Hessler Associates, Inc. located near Washington, DC. He has carried out pre-construction noise assessments or operational tests at more than 60 wind projects throughout the United States and Canada. He has published various technical papers on wind turbine noise issues and conducted research into low frequency microphone self-noise in windy conditions.

The final chapter is a summary of international noise criteria for the regulation of wind farm noise. It is written by Mark Bastasch. Criteria are continuously evolving, so we can only present a snapshot at a given time. Mark holds a Bachelor of Science degree from the California Polytechnic State University at San Luis Obispo and a Master of Science degree from William Marsh Rice University. He is a registered professional acoustical engineer with CH2M HILL in the United States. His wind turbine experience includes some of the first major wind developments in the United States including the Stateline project in Oregon and Washington, which when built in 2001 was the largest in the world. Mark also assisted the American and Canadian Wind Energy Associations in the establishment of an independent international expert panel and provided technical assistance to the expert panel which published "Wind Turbine Sound and Health Effects" in 2009.

With such a diverse range of authors we have a diverse range of writing styles. We have not tried to impose any structure on individual authors but we have tried to keep a logical flow and avoid too much duplication. We hope this book will help more people to better understand the science of wind turbine noise and that it will encourage further work to develop knowledge in the future.

Dick Bowdler [dick@dickbowdler.co.uk]
Geoff Leventhall [geoff@activenoise.co.uk]
October 2011

Chapter 1. Basic Acoustics

Geoff Leventhall, (UK) Consultant.
Geoff@activenoise.co.uk

This chapter reviews the basic acoustics which is required to appreciate the subsequent chapters.

1. Sound pressure. Sound is produced by rapid pressure fluctuations in the air. A fluctuating pressure of one hundred thousandth (10^{-5}) of the static atmospheric pressure produces a very loud sound. As the atmospheric pressure is approximately 10^5 pascals, the loud sound results from 1Pa of pressure, which will be shown later to be 94dB re 20μPa. The frequency of the fluctuations may be between 20 times a second (20Hz), and up to 20,000 times a second (20,000Hz) for "audible" noise. Frequencies below 20Hz are commonly called "infrasound", although there is a very fuzzy boundary between infrasound and low frequency noise, whilst infrasound at high levels is audible. Low frequency noise might be from about 10Hz to about 200Hz

In addition to frequency, *f*, in hertz (Hz) , the quantities which define a sound wave include

— pressure, *p* in pascals
— wavelength, λ in meters
— velocity, *c* in meters per second. c= 340m/s approx, depending on temperature
— velocity = wavelength x frequency, c = λf

Thus, we can relate frequency and wavelength by velocity. For example:

Frequency Hz:	16	31.5	63	125	250	500	1000	2000	4000
Wavelength m:	22	11	5.4	2.7	1.4	0.68	0.34	0.17	0.085

It is useful to develop an appreciation of frequencies and related wavelengths, since this helps an understanding of noise propagation and control.

2. Sound pressure. The sound pressure in a wave is force per unit of area for the wave and has units of newtons per square meter (N/m^2), which is also called pascals (Pa). The sound pressure fluctuates above and below atmospheric pressure by a small amount and the time average is approximately zero. The sound pressure is therefore quantified by the square root of the square of the fluctuations, giving the root mean square (rms) value. (Squaring the pressure fluctuations makes all values positive).

3. Sound intensity is a measure of the rate of flow of sound energy in watts per square metre (W/m^2).

4. The sound power is a characteristic of the source, its rate of production of sound energy, expressed in watts. The sound power is the fundamental property of the source. Intensity and pressure at a measurement location depend on the transmission path from source to receiver.

5. Noise levels are measured with a Sound Level Meter or equivalent instrumentation, calibrated to read decibels

6. Noise frequencies (spectrum) are obtained from a frequency analyser which, in the case of simple analysis requirements, can be incorporated in the Sound Level Meter.

7. Noise Levels – the 'decibel'. The decibel is the logarithm of the ratio between two values of a quantity such as power, pressure or intensity, with a multiplying constant to give a convenient numerical result. Logarithms are useful for compressing a wide range of quantities into a smaller range. For example

$$\log_{10}(10/1) = 1$$
$$\log_{10}(100/1) = 2$$
$$\log_{10}(1000/1) = 3$$

and the ratio of 1000:1 is compressed into a value of 3. (Note that all logarithms in this chapter are to base 10)

This approach is advantageous for handling sound levels, where the ratio of the highest to the lowest sound which we are likely to encounter, is as high as 1,000,000:1. It is useful to take the ratios with respect to the quietest sound which we can hear. This is the threshold of hearing, which is at about 20μPa (2x10^{-5}Pa) of pressure, or 10^{-12} W/m^2 of intensity, for an "average" person.

When the word "level" is added to the word for a physical quantity, decibel levels are implied, denoted by L_X, where X is the symbol for the quantity.

Intensity levels, L_I, are expressed as

$$L_I = 10 \log \left[\frac{I}{I_0} \right] \qquad [1]$$

where I is the measured intensity and I_o is a reference intensity of 10^{-12} W/m^2.

However, it can be shown that intensity is proportional to the square of pressure, giving

$$\text{pressure level, } L_p = 10 \log \left[\frac{p}{p_o} \right]^2 \text{ dB}$$

$$= 20 \log \left[\frac{p}{p_0} \right] \text{ dB} \qquad [2]$$

where p is the measured pressure and p_0 is the reference pressure of 2x10^{-5} Pa.

Equ. 2 is the expression most commonly used. Substituting the numerical value of p_0 (2x10^{-5} Pa) and taking logarithms, leads to the decibel level as:

$$L_p = 20 \log p + 94 \text{ dB} \qquad [3]$$

Thus, if p = 1Pa, the sound pressure level is 94dB.

Similarly, Equ.1 leads to

$$L_I = 10 \log I + 120 \text{ dB} \qquad [4]$$

Note that if the sound pressure is doubled, that is $p \Rightarrow 2p$, L_p increases by 6dB.

If the sound intensity is doubled, that is $I \Rightarrow 2I$, L_I increases by 3dB. This is because $I \propto p^2$

The relation between sound pressure and decibels is shown in Figure 1 over the range of 14dB to 114dB, in which the pressure axis is a logarithmic scale. Examples from Figure 1 are:

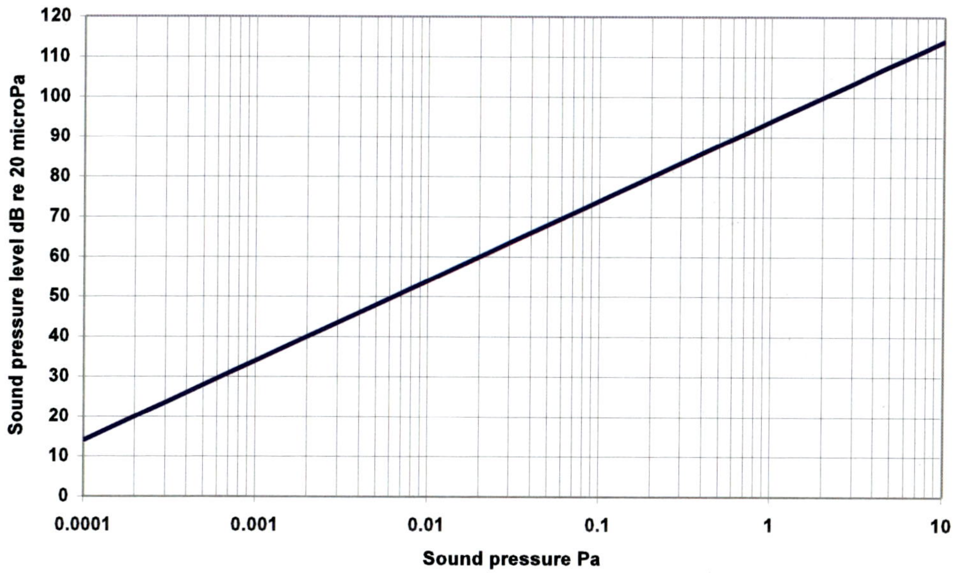

Figure 1. Sound pressure level and sound pressure

2 Pa = 100dB. 1 Pa = 94dB. 0.2Pa = 80dB. 0.02Pa = 60dB. 0.002Pa = 40dB etc.

Figure 1 does not depend on frequency, Thus, 60dB will be a pressure of 0.02 Pa, whether it occurs at 10Hz, 100Hz or 1000Hz

8. Sound power. Although the sound pressure is the quantity most frequently measured, it is not the most fundamental property of the source, since both intensity and pressure only describe what is occurring at a region in space. The more fundamental quantity is the sound power of the source, expressed in watts (joules/sec). Sound represents a flow of energy propagating from the source. The source acts as a reservoir of sound power: the rate of production of energy, joules/sec or watts. At a distance from the source we detect the flow of energy at our location. This flow is the acoustic intensity, the energy flow in watts per square meter.

The intensity can be measured directly, but we normally measure sound pressure. The reason for this is that measuring instruments, such as microphones, employ a diaphragm which, deflecting under the fluctuating force of the sound wave, converts its deflection to an electrical signal. Pressure = force/area, and it is the sound pressure which applies a force to the diaphragm. A good quality condenser microphone might have an output of about 50mV for 1 Pa of pressure

The reference levels for pressure and intensity are close to normal thresholds of hearing, so that the threshold is at approximately 0dB. An advantage of the choices of reference level of 2×10^{-5} Pa for pressure and 10^{-12} W/m^2 for intensity is that the decibel values of sound pressure level and sound intensity level are the same in a plane wave to within about 0.5dB.

Acoustic powers, P, are very low wattage and given in decibels as

$$L_W = 10\ log(P/P_0) \quad \text{dB} \qquad [5]$$

Where the reference power is $P_0 = 10^{-12}$ watts

$$\text{Then } L_W = 10logP + 120 \quad \text{dB} \qquad [6]$$

Equ. 6 shows that an acoustic power of 1 W is a sound power level of a source of 120dB.

Note that if the sound power is doubled, that is P doubles, L_W increases by 3dB. A wind turbine of sound power 100dBA is a source power of 10^{-2}W, A-weighted. Wind turbine source powers are typically 95 to 110dBA, depending on the size and design of the turbine

9. Addition and subtraction of decibels. As decibels are logarithmic ratios they do not add arithmetically. It is necessary to convert back to the original physical units, add or subtract intensity or $(\text{pressure})^2$ and then convert the result back to decibels. This means first converting the decibel levels, N, via antilogs, $10^{N/10}$, to the $(\text{pressure})^2$ or intensity, summing and then taking 10log (sum).

$$\text{If } N = 20\log(p/p_o) = 10\log(p/p_o)^2 \text{ dB} \quad [7]$$

$$\text{Then } (p/p_o)^2 = 10^{N/10} \quad [8]$$

Example
Add 55dB and 57dB sound pressure levels.

$55\text{dB} = 20\log(p_1/p_0) = 10\log(p_1/p_0)^2$
Then $(p_1/p_0)^2 = \text{antilog } 5.5 = 10^{5.5}$
similarly $(p_2/p_0)^2 = 10^{5.7}$

The sum of the sound pressure levels is $10^{5.5} + 10^{5.7} = 316228 + 501187 = 817415$

10log(817415) = 59.1dB, which is the addition of the two, and would normally be rounded to 59dB.

Subtract 60dB from 61dB.
The difference is
$10^{6.1} - 10^6 = 1258925 - 1000000 = 258925$.
10log(258925) = 54dB.

This shows that if a noise of level 60dB has one of level 54dB added to it, the resultant is 61dB.
We have to work back to the original quantities, $(\text{sound pressure})^2$, which is a measure of the energy in the sound wave, add them, and then work forward to the total decibel level. Figure 2 can be used to simplify addition and subtraction. For addition, the scales are used as shown in the Figure. For subtraction, enter the difference between the two levels on the upper scale. The amount to be subtracted from the smaller decibel level is given on the bottom scale.

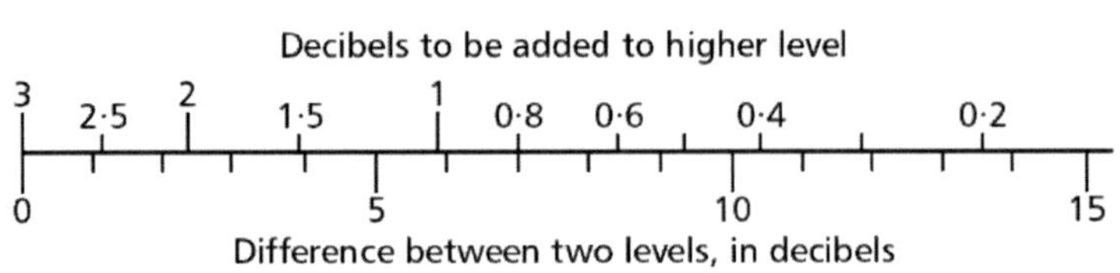

Figure 2. Addition and subtraction of decibels

Addition: Add 55 and 57dB.
Bottom scale. Difference between levels is 2dB.
Top scale. Add 2.1 to the higher level.
Sum is 59.1dB

Subtraction: Subtract 60dB from 61dB
Top scale. Difference between levels is 1dB
Bottom scale: Subtract 6dB from the smaller level.
Difference is 54dB

A general relation for adding N equal sound levels is that the resultant is greater than each of the components by an amount 10log(N). Thus, five equal sources of 60dB each add to 60 + 10log(5) = 67dB.

10. The Point source If the source is small compared with the wavelength, it approximates to a point source, which can be visualised as a small bubble whose surface vibrates in and out, so that the bubble expands and contracts in three dimensions at the frequency of the radiated sound. Here, small means small compared to the wavelength of the sound. As the sound spreads out in a sphere, at distance r from the centre of the bubble, the area through which the radiated sound travels is $4\pi r^2$, or the surface area of the sphere, and the inverse square law of radiation applies, similar to sources of light. At a distance, r, from a source of sound power W, the intensity, which is defined as the rate of flow of energy per unit area is

$$I = W/4\pi r^2 \text{ watts/m}^2 \qquad [9]$$

As the distance doubles, the spherical area through which the sound passes increases as r^2, which is a four times increase in area, or a 6dB fall in level for a doubling of distance.

By taking logarithms of Equ. 9, and noting that log 4π = 11, intensity level is expressed in decibels as

$$\text{Intensity level, } L_I = L_w - 20logr - 11 \text{ dB.} \qquad [10]$$

But as the decibel sound pressure levels and sound intensity levels are numerically very similar, we can immediately write this in the more familiar pressure terms as

$$\text{Sound pressure level, } L_p = L_w - 20logr - 11 \text{ dB} \qquad [11]$$

This is the basic equation for determining the sound level at a distance from a source and is the starting point for predictions of wind turbine noise at a receiver.

If the sound is constrained into a hemisphere by, for example, a reflecting surface below and close to it, the surface area through which the radiation passes changes from $4\pi r^2$ to $2\pi r^2$ and the sound pressure level becomes

$$L_P = L_w - 20logr - 8 \text{ dB} \qquad [12]$$

The $20logr$ term for free propagation, either spherical or hemispherical, means that, if distance r is doubled or halved, the pressure level change is 6dB (20log 2=6).

Equations 11 and 12 are the basis of predictions for wind turbine noise, but practical calculations require additional propagation and absorption factors to be taken into account, as explained in later Chapters. For example, a simple prediction of A-weighted wind turbine noise includes air attenuation of 0.005dB/m, leading to

$$L_P = L_W - 20logr - 11 - 0.005r \; dBA \qquad [13]$$

11. Directivity of a source. Directivity refers to how a source transmits sound preferentially in one or more directions. The directivity compares the transmission of sound in one direction with that which would be received in that direction if the source was radiating uniformly. That is, directivity ratio D is

$$D = I_0/I_{av}$$

Where I_0 is the intensity measured at a distance, r, from the source and I_{av} is the averaged value of intensity over a sphere of radius r. The Directivity Index is defined as $DI = 10 \log D$. However, for large sources such as wind turbines, it is not possible to measure the average of the radiations over a surrounding surface, and the practical measurement is to determine the sound pressure levels at a number of points at the same height around a wind turbine, which is equivalent to measurements around a circle surrounding the wind turbine.

12. The line source An ideal line source can be visualised as an infinite cylinder with a surface which expands and contracts in a similar manner to the surface of the bubble for the point source, except that the cylinder expands and contracts in only two dimensions. In a coherent line source, all parts of the surface move in phase and the source possesses strong directional characteristics. A coherent line source can be approximated by a series of closely spaced, in-phase point sources, where the spacing between sources is small compared with the wavelength of the radiated sound. The source characteristics approximate to those of an ideal line source when the line is several wavelengths long. An incoherent line source is produced by a series of closely spaced sources which radiate in random phase.

The area through which unit length of an infinite line source radiates changes by $2\pi r$ as r changes. Consequently, the intensity per unit length is is given by $I = W/2\pi r$ watts/m^2, where W is the sound power per unit length. The fall off in level from a line source is at 3dB per doubling of distance, as shown in Equ 14.

$$L_P = L_W - 10\log r - 11\ dB \qquad [14]$$

A row of wind turbines is sometimes referred to as a "line source", but does not satisfy the requirements for a genuine line source, as typical source separation distances are much greater than a wavelength. One must distinguish between a line source and a line of sources. In the latter case, each component is considered as a point source and the sounds received at a receptor point are summed.

13. Summation of the effects of a number of sources This will be illustrated by using a line of sources, which might be wind turbines, although the same principles can be applied to other sources and other configurations.

Consider Figure 3, which represents a line of 11 wind turbine sources, each separated by distance d, and an observation point P in line with the central source The distance of any source from P is given by

$$r = \sqrt{x^2 + (nd)^2}$$

where n is the source number and x is the distance of P from the central source. Application of Equ.13 permits the contribution of each source to be determined at point P and these are summed logarithmically to determine the total effect at P as $S_P = 10\log\left[\sum_{i=-n}^{i=+n} 10^{\frac{S_i}{10}}\right]$ where S_i is the contribution at P from source i.

We are normally interested in how the sound level changes with distance from the source and will use a change in distance x to illustrate this. The distance dependent parts of propagation equation,

Equ.13, are *-20logr – 0.005r* dB, which is plotted in Figure 4, both with and without air attenuation, for a source spacing of d = 400m.

Figure 4 shows that for the configuration of Figure 3, with d=400m, an observer is mainly influenced by the nearest turbine up to a distance of about 100m, and the fall off in level up to that point is 6dB per doubling of distance. The dashed line in Figure 4 is at 6dB per doubling of distance, as would result from a single source. At greater distances, the adjacent turbines +1 and –1 begin to influence the observer and add to that from turbine 0, increasing the overall level and so reducing the

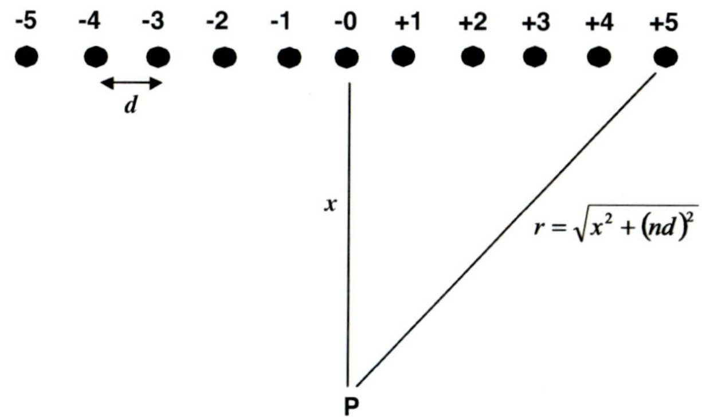

Figure 3. A line of sources each separated by distance d m

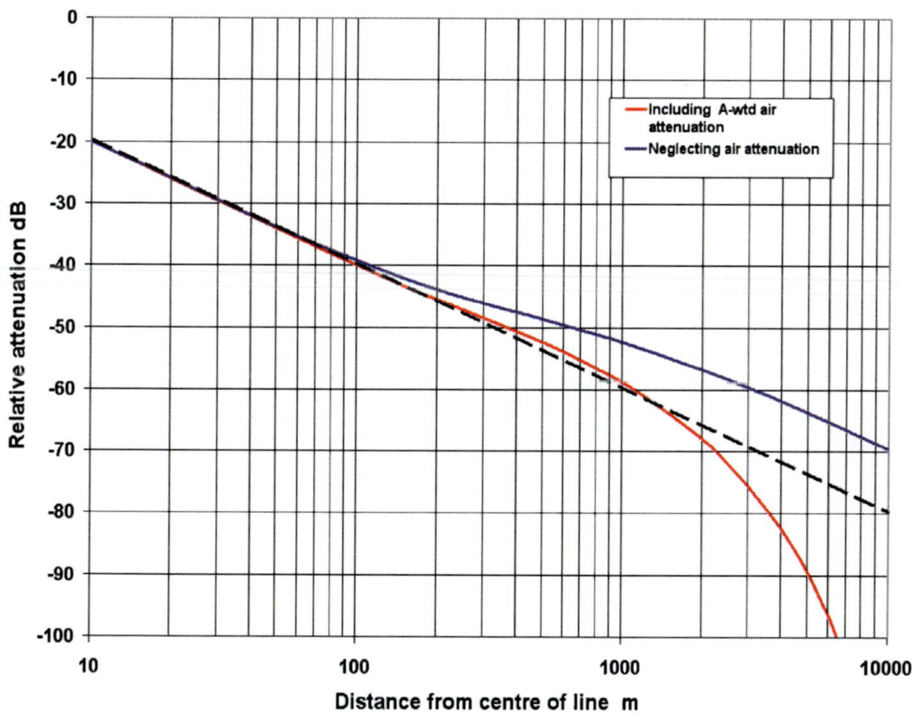

Figure 4. Fall-off of in level with distance from a line of 11 wind turbines, spaced 400m apart. Line length = 4000m The dashed line falls at 6dB per doubling of distance

rate of fall-off to less than 6dB. This continues as the distance from the line increases and the distances r from the turbines becomes closer to distance x. At large distances , all values of r and x are similar to each other and the whole line acts as a point source, giving 6dB fall off for doubling of distance. However, as there are 11 sources, the level of the sound is 10log(11) = 10.4dB greater than that of a single source, as shown on the right side of the plot. Some differences occur if the observation point is not central to the line of turbines, but the trends are similar.

In a real situation there are additional variables to take in to account as described in Chapter 3, whilst distances are normally obtained from entering the source and receiver coordinates into the calculation.

14. Measuring noise. A sound level meter, or equivalent equipment, is the standard way of measuring noise. Environmental noise is normally assessed by the A-weighting. Although hand-held instruments appear to be easy to use, lack of understanding of their operation and limitations, and the meaning of the varied measurements which they can give, may result in misleading readings.

The operation of a sound level meter is indicated in Figure 5 The microphone is normally a capacitor (condenser) type, whose output is amplified and passed to a digital processor and then to filter networks before displaying the result. The filter networks include the A, C and Linear (Z) weighting plus octave and one third octave filters, depending on the complexity of the instrument. Narrow band analysis is also available on some instruments.

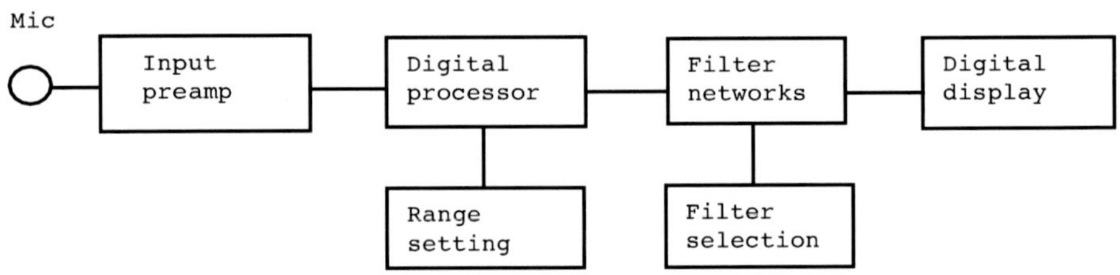

Figure 5. Operation of digital sound level meter

The weighting filter networks are an important part of the sound level meter, as they give an indication of the frequency components of the sound. The filters are as follows:

- A-weighting: on all meters
- C-weighting: on most meters
- linear (Z-weighting): on many meters
- octave filters: on some meters
- third octave filters: on some meters
- Narrow band: on a few meters

The main classes of sound level meter are Type 1 (precision) or Type 2 (general purpose), of which Type 1 is the more accurate. There is also Type 3 for survey work and preliminary investigations and Type 0 (high precision), mainly for laboratory measurements.

15. Limits for sound level meters. The lowest frequency which can be measured is determined by the microphone and the input electronics. The lowest level is determined by self noise in the instrument.

Many sound level meters give third octave analysis down to 6.3Hz., which must be matched by the microphone. All microphones include pressure equalisation which connects the back of the diaphragm to the atmospheric pressure, which acts on the front of the diaphragm. Without this pressure equalisation the microphone would become a barometer. Pressure equalisation can be slowed down by tight sealing, in order to enable the microphone to operate down to around 1Hz, but this low frequency must also be matched by the preamplifier and the sound level meter electronics.

16. Sound level meter weighting networks. These are shown in Figure 6. Originally, the A-weighting was intended for low levels of noise. C-weighting was intended for higher levels of noise. The weighting networks were based on human hearing contours at low and high levels and it was hoped that their use would mimic the response of the ear. This concept, which did not work out in practice, has now lapsed and A- and C-weighting are used at all levels. Most of the noise levels quoted for turbines and for background noise in this book are A-weighted, as this is widely used for legislation. Linear weighting is used to detect low frequencies. A specialist G-weighting is used for infra-sound below 20 Hz.

The A-weighting depresses the levels of the low frequencies, as the ear is less sensitive to these. This is acceptable if there is largely middle and high frequency noise present, but if there are high tonal components in the noise at low frequencies, the A-weighting may not give a valid measure. A-weighting is adequate for placing noises of similar spectrum (frequency components) in order from worst to best but, if the spectra are very different, the A-weighting is not reliable as an indicator of subjective response. Wind turbine noise spectra generally follow a similar form.

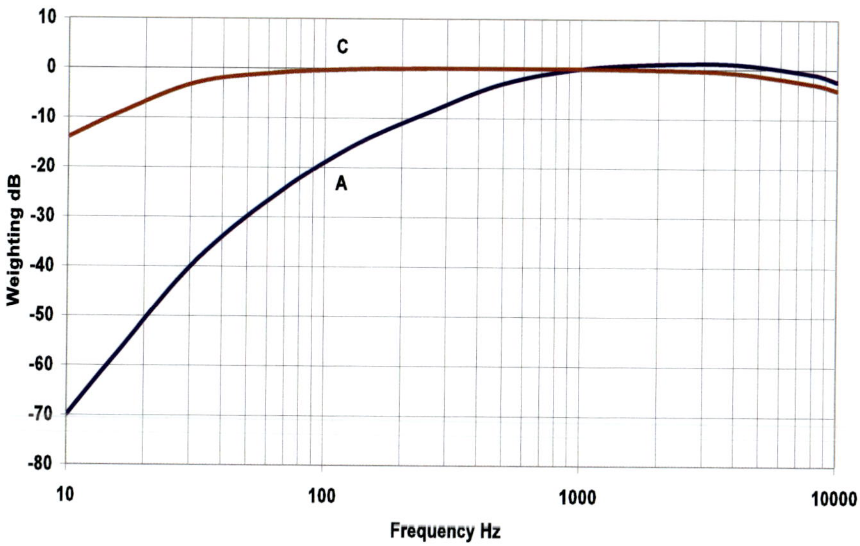

Figure 6. Weighting networks

17. Equivalent level ($L_{eq,T}$) This is the steady level over a period of time, T, having the same energy as that of the fluctuating level actually occurring during that time. A-weighted equivalent level, designated $L_{Aeq,T}$, is used for many legislative purposes, including for assessment of wind turbine noise.

Mathematically,

$$L_{Aeq,T} = 10\log\left\{\frac{1}{T_2 - T_1}\int_{T_1}^{T_2}\left[\frac{p_A(t)}{p_0}\right]^2 dt\right\} \qquad [15]$$

where T_1 is the start time of the noise, T_2 is the end time of the noise $T=T_2-T_1$, p_0 is the reference pressure of 20 μPa, $p_A(t)$ is the A-weighted instantaneous sound pressure (Pa). Here we are averaging the fluctuations in the pressure of the noise and converting the average to decibels.

18. Percentiles (L_N) These are a statistical measure of the fluctuations in overall noise level, i.e. in the envelope of the noise, which is usually sampled a number of times per second, typically ten times. The most useful percentiles are L_{90} and L_{10}. The L_{90} is the level exceeded for 90% of the time and represents a low level in the noise. It is often used to assess background noise. The L_{10} is the level exceeded for 10% of the time and is a measure of the higher levels in a noise. Modern computing sound level meters give a range of percentiles. Note that the percentile is a statistical measure over a specified time interval. Percentiles cannot be combined like decibel levels, without knowledge of the statistics of the noise.

19. Frequency analysis This gives more detail of the frequency components of a noise. Frequency analysis normally uses one of three approaches: octave band, one-third octave band or narrow band.

The filters can be incorporated in sound level meters or be externally connected components. For an octave band filter, the higher limit is twice the lower limit. For a $^1/_3$-octave band filter the higher limit is about 1.25 times the lower limit. A narrow band analysis can be in fractional octaves, such as $^1/_{12}$-octave or in constant bandwidth analysis using, say, filters of constant bandwidth (e.g. 4 Hz) over the whole frequency range. The general relations for a 1/n octave filter are:

Centre frequency $f_c = (f_1\ f_2)^{1/2}$ and $f_2 = 2^{\ 1/n}\ f_1$ where f_1 is the lower band limit and f_2 is the upper band limit. These give $f_c = 2^{1/2n}\ f_1$.

For example, an octave filter(n=1) at f_c =1000Hz has f_1 = 700Hz and f_2 = 1400Hz, with a bandwidth equal to 70% of the centre frequency

A third octave filter (n=3) at 1000Hz has f_1= 890Hz and f_2 = 1120Hz, whilst its bandwidth is about 23% of the centre frequency

Commonly used centre frequencies for octave band filters are1Hz, 2Hz, 4Hz, 8Hz, 16Hz, 31.5 Hz, 63 Hz, 125 Hz, 250 Hz, 500 Hz, 1 kHz, 2 kHz, 4 kHz and 8 kHz but lower and higher bands are also defined.

Third octave band filters are in the sequence **1**, 1.25, 1.6, **2,** 2.5, 3.15, **4** ,5, 6.3. **8**, 10,12.5, **16**, 20, 25, **31.5**, 40, 50, **63**, 80,100, **125**, 160Hz etc. There are three one-third octave bands to each octave band. The lowest frequencies are usually used for vibration investigations, not acoustic measurements.

Narrow band analysis is most useful for complex tonal noises. It could be used, for example, to determine a fan tone frequency, to find the frequencies of vibration transmission from machinery or to detect system resonances. All analyses require an averaging over time, so that the detail of those fluctuations in the noise levels, which might occur during the averaging, is normally lost.

Criteria for assessment of noise are based on dBA, octave bands or $^1/_3$-octave band measurements. These measures give increasingly detailed information about the noise. Additionally, narrow band analysis gives even more information. T H Pedersen (2010) has compared an analysis of the same wind turbine sound by the different methods, with results shown in Figure 7. As the bandwidth becomes smaller, the measured level reduces, since there is less energy in the narrower band. An exception is if the band is dominated by a tone, in which case the level is the same for all bandwidths. In order to make analyses comparable, the level is sometimes given as that in a band 1Hz wide. This is known as the spectrum level, whilst the result of analysis in any other bandwidth is called the band level.

Different bandwidths can be related. For example, for a uniform spectrum, as there are three one-third octave bands in an octave band, each third octave must be 10log(3)dB lower than the octave band, giving 4.8dB, which is usually taken as 5dB. When the level of the spectrum is changing rapidly with frequency an average level for each band must be used in the addition. The octave band level is then given by

$$BL_{OCT} = 10log\left(10^{\frac{BL1}{10}} + 10^{\frac{BL2}{10}} + 10^{\frac{BL3}{10}}\right) \quad [16]$$

where BL1, BL2 and BL3 are the third octave band levels.

Any number of narrow bands can be combined into a wider band in this way.

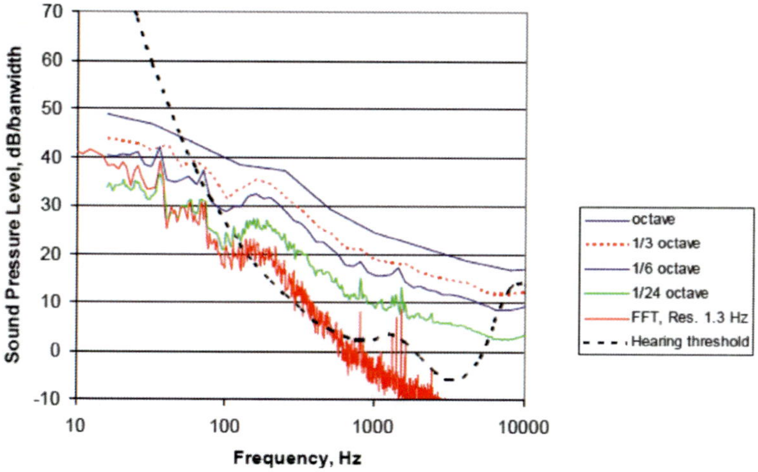

Figure 7. Different analyses of the same wind turbine sound of 33dBA

CONCLUDING REMARKS

Most predictions of wind turbine noise are carried out using commercial software incorporating the data for attenuation mechanisms and other factors, which it is necessary to include in a full calculation. However, an understanding of basic acoustics is necessary to be able to interpret the results of the calculations.

REFERENCE

Pedersen T H 2010 Low frequency noise from large wind turbines.EFP06 project report http://www.madebydelta.com/imported/images/A401929-Danish-Energy-Authority-EFP-06-project-Final-report-for-LF-noise-from-large-wind-turbines-av127210.pdf

Chapter 2. Primary Noise Sources

Stefan Oerlemans (Netherlands), National Aerospace Laboratory, NLR.
Stefan.Oerlemans@nlr.nl

INTRODUCTION

In order to progress the development of wind power and reduce noise annoyance, it is essential to reduce the sound of wind turbines. However, before we can reduce the sound, we first have to know and understand the sources of the noise. This chapter deals with the primary noise sources of modern large wind turbines, and is divided into the following sections:

- Introduction
- Potential sources of wind turbine noise
- Characteristics of airfoil noise
- Field measurements of wind turbine noise
- Prediction of wind turbine noise
- Reduction of wind turbine noise
- Summary

After this introduction we will first discuss the different potential source mechanisms and their characteristics. Then, on the basis of field measurements, we will show that aerodynamic noise from the blades is generally dominant. Moreover, using the characteristics of trailing edge noise, we will be able to understand the swishing character of wind turbine noise. Next, we will show that wind turbine noise can be accurately predicted using a semi-analytical, semi-empirical prediction method, which only needs the blade geometry and operating conditions as input. Finally, we will discuss the possible ways of *reducing* wind turbine noise, and describe validation experiments for different noise reduction concepts.

Throughout this chapter we will see that directional acoustic measurement methods play an important role in experimental wind turbine noise research. This so-called *phased array* technique is explained separately in Appendix 1 at the end of the book. Since we are concerned with *primary* noise sources, we will not consider secondary effects due to sound propagation or wind shear*. These effects are addressed in other chapters. The present chapter is mainly based on experience gained at the Dutch National Aerospace Laboratory NLR during the last 10-15 years. As a consequence, most of the examples presented here are taken from projects in which NLR was involved. A limited number of references to relevant work from other groups is given in this chapter. A more complete overview of the available literature can be found in [1].

POTENTIAL SOURCES OF WIND TURBINE NOISE

In this chapter we focus on modern large wind turbines, which typically have a horizontal axis and an upwind rotor. The sound from such a turbine can be divided in mechanical noise and aerodynamic

*Wind shear is considered a secondary effect because – as long as trailing edge noise is dominant – it has practically no effect on wind turbine overall sound power generation. In some cases periods of increased amplitude modulation are reported, which may be explained by dynamic stall due to extreme wind shear. Note that, although the effect of wind shear on the noise *generation* is generally small, wind shear can affect the propagation and perception of wind turbine noise.

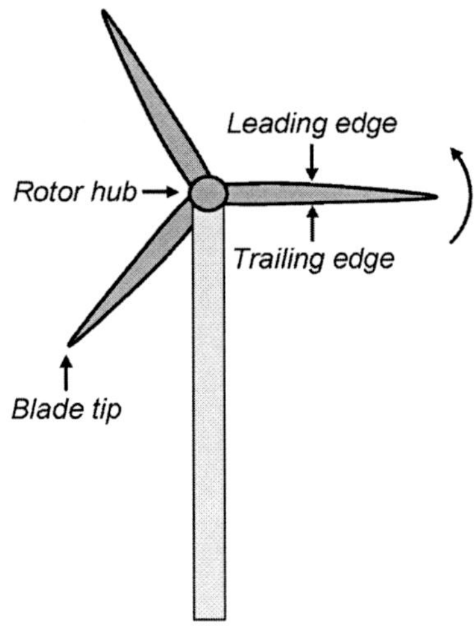

Figure 1. Wind turbine as seen from a downwind position

noise [2]. Mechanical noise can be generated by different components in the hub, such as the gearbox or the generator, and may contain several tones. However, provided that mechanical noise is adequately treated, aerodynamic noise from the blades is generally the dominant noise source. Therefore, in this section we will briefly discuss the flow around a wind turbine blade, followed by a description of potential aerodynamic source mechanisms.

Flow Around a Wind Turbine Blade

In order to understand the different mechanisms of aerodynamic noise, we first have to look at the air flow around the blade. If we look at a wind turbine from a downwind observer position (i.e., we look in the *upwind* direction), for most modern turbines the blades rotate in anti-clockwise direction (Figure 1). As the upper blade cuts through the air, it experiences an air flow from left to right (Figure 2). As illustrated in Figure 3, the effective flow speed U perceived by the blade is composed of the wind speed U_w and the rotational flow speed U_r. For modern wind turbines, the rotational speed at the blade tip is

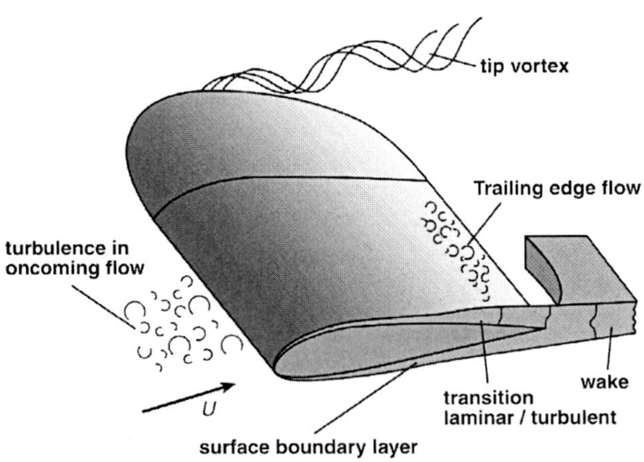

Figure 2. Flow around the outer part of a wind turbine blade [2]

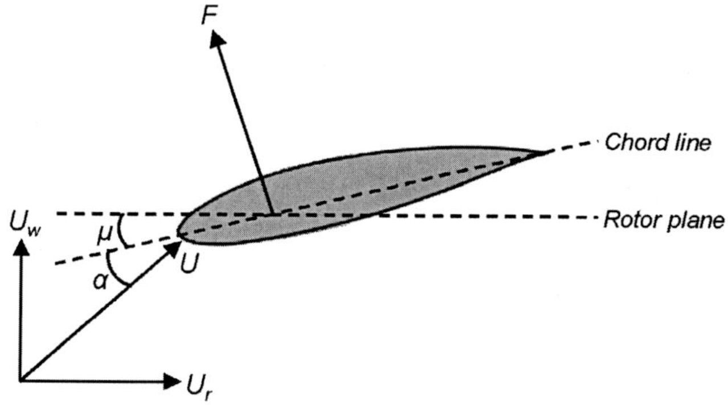

Figure 3. Airfoil with definition of flow angles

typically 75 m/s, while the wind speed at rotor height is of the order of 10 m/s. The cross section of the blade (the *airfoil*) is designed such that it diverts the incoming flow downward (i.e., towards the rotor plane), thus producing a reaction force $\boldsymbol{F}$. The component of this force in the rotor plane is what makes the rotor blades turn. This is similar to the lift force produced by aircraft wings to keep airplanes in the air. Since the blades divert the flow towards the rotor plane, the rotor effectively slows down the incoming wind, thus extracting energy from it. The lower surface of the airfoil (i.e., the upwind side), where the relatively low flow speed results in a high pressure, is called the *pressure side*. The upper surface, with a high flow speed and low pressure, is called the *suction side*.

The angle between the effective flow direction and the airfoil chord line is called the angle of *attack* α. The angle between the chord line and the rotor plane, μ, is the sum of the (fixed) local blade twist and the (adjustable) blade pitch angle. Thus, Figure 3 illustrates that the angle of attack can be increased by *reducing* the pitch angle. Similarly, a higher wind speed results in an increase in angle of attack. Normally, a higher angle of attack means a higher reaction force F and therefore a higher torque.

As the air flows past the blade, a thin layer of air develops which, due to viscosity, partially 'sticks' to the blade surface (Figure 2). This layer, which is usually less than a few centimetres thick, is called the *boundary layer*. At the blade surface, the air velocity (relative to the blade) is zero, while at the edge of the boundary layer the velocity is equal to the undisturbed flow speed. The structure of the boundary layer depends on the Reynolds number Re = UC/ν, with the blade chord and ν the kinematic viscosity (about $15 \cdot 10^{-6}$ m^{2}s^{-1} in air). For the outer part of a large wind turbine blade (where practically all the noise is produced), is typically a few millions or more. For these Reynolds numbers, the boundary layer, which is initially laminar, will generally become turbulent somewhere along the chord. Whereas the laminar boundary layer is organized in layers, the turbulent boundary layer is more chaotic and contains turbulent eddies (vortices).

The thickness of the boundary layer can be characterized using $\delta_{0.99}$, the distance from the surface where the mean velocity equals 99% of the undisturbed flow speed. However, this distance is rather ill-defined due to the asymptotic behaviour of the velocity profile at the edge of the boundary layer. Therefore, the thickness of the boundary layer is often characterized using the *displacement thickness* δ^*. This is the distance by which the surface would have to be moved in a hypothetical non-viscous (uniform) flow to maintain the same volumetric flow as in the case with boundary layer. The displacement thickness is typically 1/8 to 1/5 of $\delta_{0.99}$, depending on the shape of the velocity profile. For a turbulent boundary layer over a flat plate, the displacement thickness at distance C from the leading edge can be estimated using the relation $\delta^*/C \approx 0.05\text{Re}^{-1/5}$ [3]. For an airfoil, an increase in angle of attack typically results in a thicker boundary layer on the suction side, while the pressure side boundary layer becomes thinner.

Aerodynamic Noise Sources

The different aerodynamic noise source mechanisms are illustrated in Figure 2. First, turbulence may be present in the oncoming flow, for example generated by the atmospheric boundary layer or by the wake from upwind turbines in a wind farm. The interaction between the incident eddies and the blade surface may cause *inflow turbulence noise*. This noise source strongly depends on site-specific and time-dependent atmospheric conditions, and it is an open issue to what extent inflow turbulence noise contributes to the overall sound level of a wind turbine. The second source mechanism is related to the tip vortex, which is formed due to the pressure difference between the pressure and suction side of the blade. Turbulence is convected over the tip edge, and separated flow may occur. The interaction between the turbulent flow and the tip surface may cause *tip noise*. This noise source depends on the strength of the tip vortex and on the planform of the blade tip. Finally, noise may be generated by the interaction between the turbulent flow over the 2D airfoil and the trailing edge of the blade. Since this sound, like tip noise, is generated in an undisturbed inflow, it is referred to as airfoil *self noise* [4].

Depending on the flow conditions, several self noise mechanisms may occur (Figure 4). On the outer part of a large wind turbine blade (**Re** > 1 million), typically a turbulent boundary layer develops along the blade surface, which remains attached up to the trailing edge. As the turbulent eddies are convected past the trailing edge, their sound is scattered at the trailing edge, causing broadband noise. This turbulent-boundary-layer-trailing-edge noise or briefly *trailing edge noise* (Figure 4a) usually defines the lower bound of wind turbine noise, and is considered to be the most important noise source for modern large wind turbines.

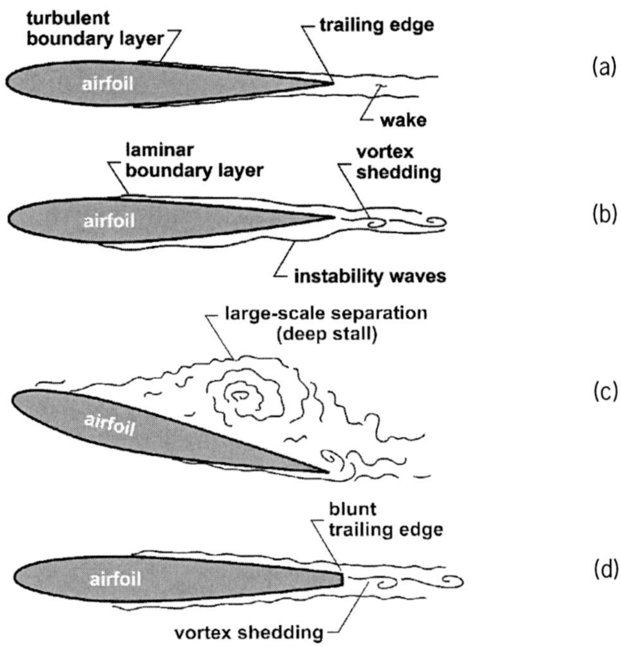

Figure 4. Airfoil self-noise mechanisms [4]

If the Reynolds number is smaller than about 1 million, the boundary layer on either side of the airfoil may remain laminar up to the trailing edge. Upstream radiating noise from the trailing edge may then trigger laminar-turbulent transition or boundary layer instabilities (Tollmien-Schlichting waves), which in turn radiate as trailing edge noise. If such a feedback loop occurs, high levels of tonal noise may be generated. This type of whistling noise is called *laminar-boundary-layer-vortex-shedding-noise* (Figure 4b), and can be prevented by tripping the boundary layer, which induces transition from laminar to turbulent flow. This noise source is considered to be only relevant for small wind turbines, which have relatively small blades with Reynolds numbers below 1 million.

As the angle of attack increases, at some point the flow will separate from the suction side of the airfoil. This corresponds to so-called stall. Stall causes a substantial level of unsteady flow around the airfoil, which may lead to a significant increase in noise [4,5,6,7]. For mildly separated flow this *separation-stall noise* (Figure 4c) appears to be radiated from the trailing edge, whereas deep stall causes low-frequency radiation from the airfoil as a whole. Stall noise is considered to be of minor importance for modern pitch-controlled wind turbines.

Blunt trailing edge noise (Figure 4d) occurs when the trailing edge thickness is increased above a critical value. Periodic Von Karman type vortex shedding from the trailing edge may then result in tonal noise. Normally blunt trailing edge noise can be prevented by proper design of the blades, i.e., a sufficiently small thickness of the trailing edge.

In summary, the most relevant potential noise sources for a modern wind turbine are trailing edge noise, inflow turbulence noise and tip noise. In the next section the characteristics of the different airfoil noise sources will be discussed in more detail.

CHARACTERISTICS OF AIRFOIL NOISE

In this section we will first address some general theoretical aspects of the interaction between turbulence and an airfoil surface. Next, we will focus on the theoretical and experimental characteristics of trailing edge noise and other airfoil noise sources. Finally, we will consider the effect of source motion on the frequency and level of the radiated sound.

Interaction Between Turbulence and an Airfoil Surface

For the flow around the outer part of a wind turbine blade, the Mach number $M = U/c$, with c the speed of sound (about 340 m/s), is of the order of 0.2. At these low Mach numbers, *free* turbulence (away from the blade surface) is a very inefficient noise source. Therefore, the radiated aerodynamic sound from an airfoil will be dominated by the *interaction* between the turbulence and the airfoil surface. The acoustic effect of the turbulent flow depends on the length scale Λ of the turbulent eddies [8]. For trailing edge noise this turbulent length scale is the boundary layer displacement thickness δ^* at the trailing edge, for inflow turbulence noise it is the scale of the incident eddies. For a mean flow velocity U, the disturbances occur at a frequency $f \approx U/\Lambda$, which is equal to the frequency of the emitted sound $f \approx U/\lambda$, with λ the acoustic wavelength. Thus, if the incident eddies are much larger than the airfoil chord C, the acoustic wavelength will also be much larger than the chord, so that the airfoil is acoustically *compact*. The eddies will then cause a fluctuating force on the complete airfoil, resulting in low-frequency sound which radiates as a *compact dipole*. It can be shown analytically that in this case the acoustic power, which is proportional to the square of the acoustic pressure p, scales with the sixth power of the flow speed: $p^2 \sim U^6$. Furthermore, the radiation pattern or *directivity* is given by $p^2 \sim \sin^2\theta$, where θ is the angle with respect to the flow direction. This directivity pattern implies that maximum radiation occurs in the direction perpendicular to the mean flow (Figure 5).

If the eddies are much *smaller* than the airfoil chord, they will induce only local pressure fluctuations which do not affect the global aerodynamic force. The sound of the turbulent eddies is scattered at the leading edge (for inflow turbulence noise) or at the trailing edge (for trailing edge noise), resulting in high-frequency *edge noise*. Using a semi-infinite flat plate approximation, it can be shown analytically that in this case the acoustic power scales with the *fifth* power of the flow speed: $p^2 \sim U^5$. Moreover, the directivity is now given by $p^2 \sim \sin^2(\theta/2)$, which means that maximum radiation occurs in the upstream direction for trailing edge noise (i.e., in the direction of the leading edge), and in the downstream direction for leading edge noise (Figure 5).

Note that in Figure 5 the quantity on the x-axis is U. This is because the compactness factor $C/\lambda \sim MC/\Lambda$. Thus, for a given chord length C and turbulent length scale Λ, with increasing flow speed the

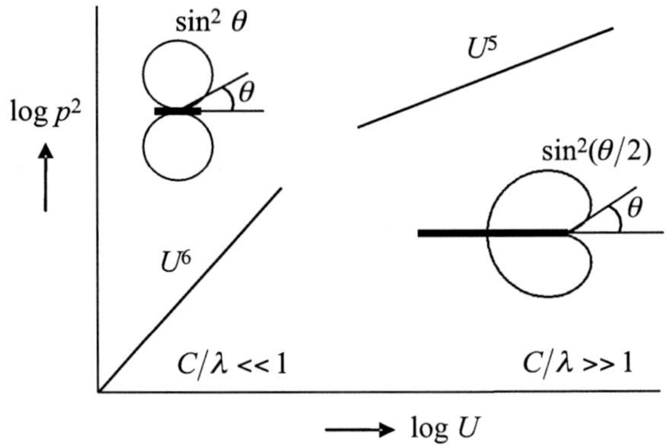

Figure 5. Directivity and speed dependence of airfoil noise [8]

source behaviour changes from a compact dipole to an edge noise source. Since for trailing edge noise the turbulence length scale is usually much smaller than the airfoil chord ($\delta^*/C \ll 1$, see previous section), the transition between the two acoustic regimes will occur at relatively low subsonic speeds. Therefore, we expect trailing edge noise to exhibit non-compact behaviour. Later in this chapter we will indeed see that the non-compact trailing edge noise characteristics match the experimental results best, even at low frequencies, where the acoustic wavelength is of the same order as the blade chord. Moreover, it will be shown that the $\sin^2(\theta/2)$ directivity largely explains the swishing character of wind turbine noise.

Theoretical Relations for Trailing Edge Noise

The trailing edge noise from a wind turbine blade is composed of the contributions from all radial segments of the blade. The characteristics of the noise from a given segment depend on the local flow parameters (Figure 6). As mentioned above, the *frequency* of trailing edge noise is proportional to U/δ^*, and is characterized by the Strouhal number $St = f\delta^*/U$: the peak of the broadband spectrum occurs at a Strouhal number of order 0.1. The *level* of the (non-compact) trailing edge noise from each radial segment can be estimated using the following expression, which was analytically derived using the semi-infinite flat plate approximation [9]:

$$p^2 \sim U^5 \frac{L\delta^*}{r^2} \cos^3\gamma \ \sin^2(\theta/2)\sin\varphi \qquad (1)$$

In this equation, L is the span of the blade section, r is the distance between the source and the observer ($r \gg L$), and the angles are defined in Figure 7. The boundary layer displacement thickness at the trailing edge, δ^*, represents a characteristic turbulence length scale. The dependence of p^2 on

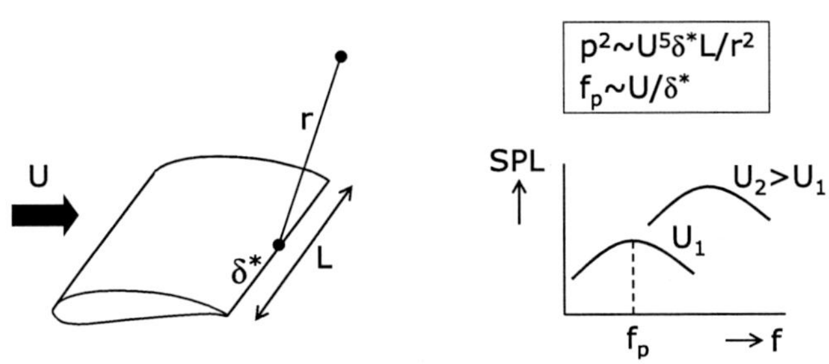

Figure 6. Trailing edge noise from a blade segment. SPL is Sound Pressure Level

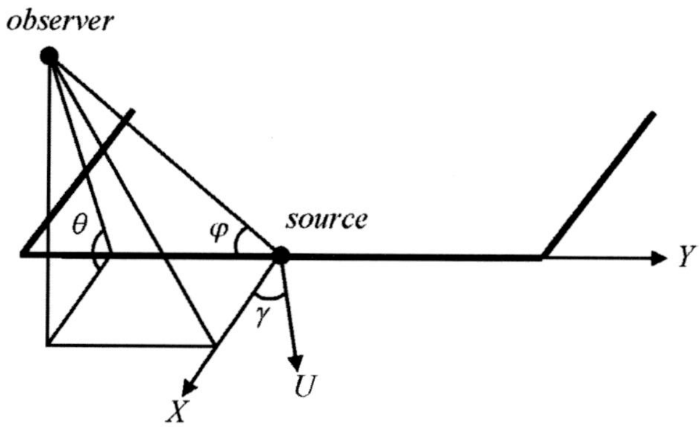

Figure 7. Definition of angles for trailing edge noise from a semi-infinite flat plate

L/r^2 is straightforward: doubling the span results in a doubling of acoustic energy (+3 dB), whereas a doubling of the distance results in a 75% reduction in acoustic energy (-6 dB).

Eq. (1) illustrates a number of interesting properties of trailing edge noise. First, we recognize the U^5 speed dependence of non-compact edge noise, as discussed in the previous section. Thus, as illustrated in Figure 6, an increase in flow speed will typically result in an increase in the level (and frequency) of the trailing edge noise spectrum (since δ^* depends only weakly on U, see previous section). This strong speed dependence of the sound level is the main reason why practically all turbine noise is produced by the outer part of the blades: doubling the flow speed (all else being equal) results in an increase in source strength of 15 dB!

The $\sin^2(\theta/2)$ directivity was also discussed in the previous section: the sound radiation is symmetrical about the chord line and maximum radiation occurs in the direction of the leading edge, where $\theta = \pi$ (Figure 8). It should be noted that, even though the pressure and suction side boundary layers contribute independently to the total noise level, the sound radiation is symmetrical. Due to scattering at the trailing edge the sound generated by the pressure side boundary layer will be radiated symmetrically (in anti-phase) to *both* sides, and the same is true for the suction side boundary layer. Since the suction side boundary layer thickness at the trailing edge, δ_s^*, is normally larger than δ_p^*, the sound produced by the suction side boundary layer usually has a higher level and lower frequency than the noise generated on the pressure side (Figure 8). The factor $\sin\phi$ in Eq. 1 describes the directivity in the lateral direction, and shows that the sound radiation goes to zero in the direction of the trailing edge.

The factor $\cos^3\gamma$, finally, shows that the trailing edge scatters sound most effectively when the path of the turbulent eddies is perpendicular to the trailing edge. Thus, the sound may be reduced by sweeping the edge with respect to the flow direction (like a swept aircraft wing or a swept propeller blade). Theoretically, a sweep angle of 45° already gives a noise reduction of almost 5 dB! This idea

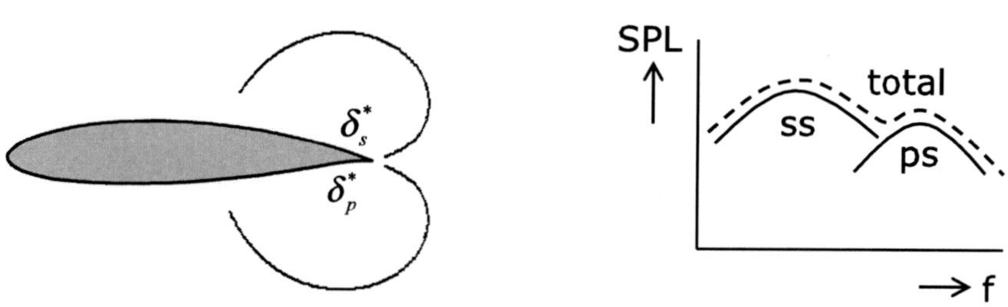

Figure 8. Directivity of trailing edge noise (left) and contributions of airfoil suction side (ss) and pressure side (ps) to total trailing edge noise spectrum (right)

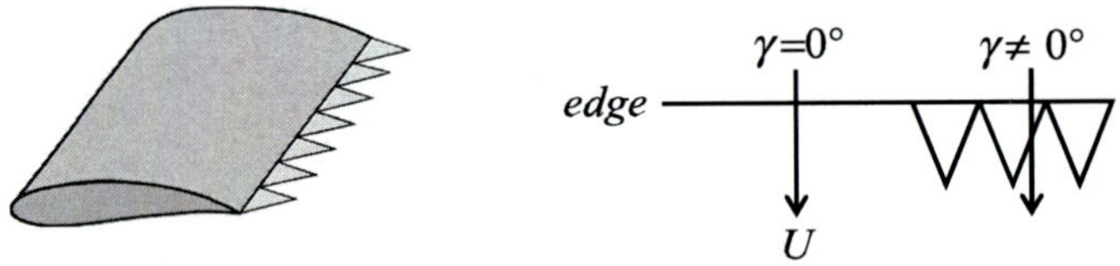

Figure 9. Trailing edge serrations

can be extended to the use of trailing edge *serrations* (Figure 9). For a serrated trailing edge the angle between the eddy path and the edge is smaller than 90°, so that the sound is scattered less effectively [10]. This noise reduction concept will be discussed later in this chapter.

The relations described above are based on the interaction between turbulence and the trailing edge of a semi-infinite plane. As an alternative theoretical approach, trailing edge noise can also be analyzed in terms of convecting surface pressure patterns or 'gusts' [11]. This fluctuating pressure field can be regarded as the 'footprint' of the boundary layer turbulence, and its temporal and spatial transform - the wave number-frequency spectrum - can be used to compute trailing edge noise [12]. In the limits for the compact dipole and for non-compact edge noise, this theory yields the same directivity patterns as shown in Figure 5. In the transition region between these limits, where the airfoil chord is of the order of the acoustic wavelength, the directivity pattern is found to exhibit multiple lobes and a reduced intensity in the direction of the leading edge. This modified directivity pattern should be taken into account when predicting the noise from a wind turbine blade. The different trailing edge noise theories were reviewed and unified by Howe [13].

Measurements of Trailing Edge Noise from Two-Dimensional Airfoils

In order to verify the theoretical relations described above, in the last decades various groups have performed aeroacoustic wind tunnel tests on two-dimensional blade sections (Figure 10). Typically, these measurements are carried out in open jet wind tunnels (Figure 11). The airfoil model is mounted between two (acoustically lined) endplates and the test section is surrounded by an anechoic chamber to prevent acoustic reflections. The microphones are placed outside the tunnel flow to prevent flow-induced noise from the microphones. In some cases a phased microphone array is used to localize the noise sources on the airfoil (see Appendix 1). Several parameters may be varied, such as the wind tunnel speed, the airfoil angle of attack, and the shape of the airfoil. Furthermore, a spanwise roughness strip or *trip* may be applied to the airfoil surface, to enforce boundary layer transition from laminar to turbulent. In this way higher Reynolds numbers (where this transition occurs naturally) can be simulated. Trips may also be used to investigate the effect of surface roughness on the airfoil noise. On a wind turbine blade roughness may occur due to, for example, dirt or insects.

Figure 10. Two-dimensional airfoil sections for wind tunnel measurements

Figure 11. NLR's small anechoic wind tunnel KAT with white airfoil model between endplates and out-of-flow microphone array

A typical example [14] of an acoustic source map produced using a microphone array is shown in Figure 12. This map shows the noise sources on an airfoil in a clean tunnel flow (i.e., no inflow turbulence). Clearly, all noise is radiated from the trailing edge of the model. Using these maps airfoil noise spectra can be determined for different conditions. As an example, Figure 13 shows the trailing edge noise spectra for a NACA0012 airfoil at 4° effective angle of attack. Note that in open jet wind

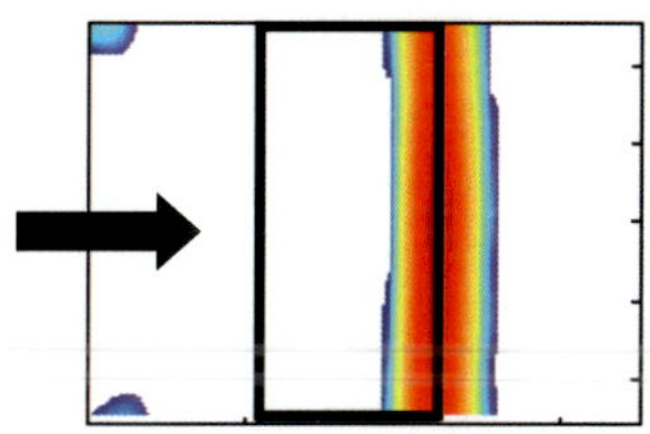

Figure 12. Acoustic source map showing trailing edge noise radiation from an airfoil in a clean flow. The black rectangle indicates the airfoil contour. Flow is from left to right

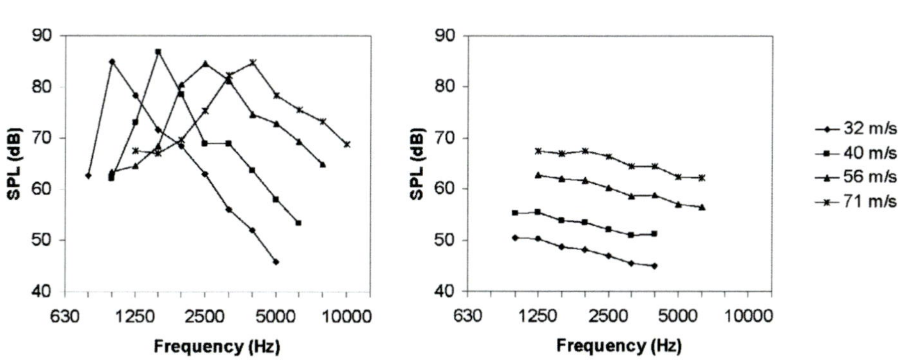

Figure 13. Airfoil noise spectra for clean (left) and tripped (right) NACA0012 airfoil at $\alpha_{eff}=4°$

tunnels the *effective* angle of attack is usually smaller than the *geometrical* angle. This is due to the fact that the tunnel flow is deflected by the airfoil, resulting in a smaller lift force than in an infinite flow. The model chord is 0.23 m (9 inch), yielding Reynolds numbers between 0.5 and 1.1 million. It can be seen that for the clean airfoil clear spectral peaks occur, the frequency of which increases with wind speed. These spectra are indicative of laminar-boundary-layer-vortex-shedding-noise. After application of 0.25 mm trips at 5% chord on both sides of the airfoil, broadband trailing edge noise is generated, resulting in a dramatic decrease in sound levels.

If we plot these trailing edge noise spectra in a normalized manner, with $SPL_{norm} = SPL - 10\log(\delta^* U^5)$ as a function of the Strouhal number, we see that all spectra collapse on a single 'universal' curve (Figure 14). For this figure the boundary layer thickness was not measured, but roughly estimated using $\delta^*/C \approx 0.05\mathrm{Re}^{-1/5}$. However, similar results were also obtained using *measured* values of the displacement thickness, in a comprehensive experimental study on NACA0012 airfoils with varying chord lengths [4]. Acoustic results for other airfoil shapes and flat plates show a similar behaviour [15]. It should be noted that in some studies slightly different scaling approaches are used to obtain the best collapse of the experimental data. Nevertheless, the global dependence of the trailing edge noise spectrum on flow speed appears to be well described by the relations given in the previous section.

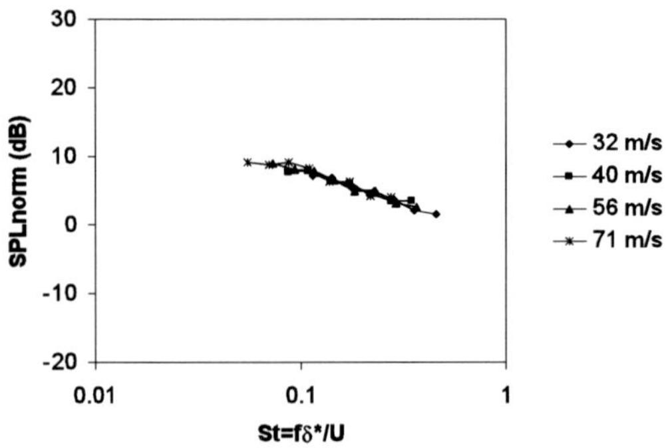

Figure 14. Normalized trailing edge noise spectra as a function of Strouhal number for tripped NACA0012 airfoil at α_{eff}=4°

The *directivity* of trailing edge noise from two-dimensional airfoils has also been investigated in wind tunnel measurements. In general, the experimental results are in accordance with the flat plate directivity function from Eq. 1. For example, Figure 15 shows that the sound radiation is symmetrical with respect to the chord line, even though the flow is asymmetrical due to the nonzero angle of attack. However, due to geometrical constraints in the wind tunnel, experimental validation of the directivity function has been limited to the plane normal to the trailing edge, for directions not too close to the plane of the airfoil [16]. Thus, the directivity function has only been validated for ϕ=90° and approximately $45° < |\theta| < 135°$. We will see later that the characteristics of wind turbine noise can be used to assess the complete trailing edge noise directivity function.

Blunt Trailing Edge Noise

Blunt trailing edge noise occurs when the trailing edge thickness h is increased above a critical value. If the ratio h/δ^* is larger than about 0.3, periodic vortex shedding may occur, resulting in a Von Karman type vortex street. The alternating vortices in the near wake produce surface pressure fluctuations close to the trailing edge, leading to tonal noise. The frequency of the tone is determined by the Strouhal number $St = fh/U \sim 0.1$. Thus, sharpening of the trailing edge will shift the peak towards the ultrasound region. As h increases, the frequency and the bandwidth of the tone decrease, but the broadband part

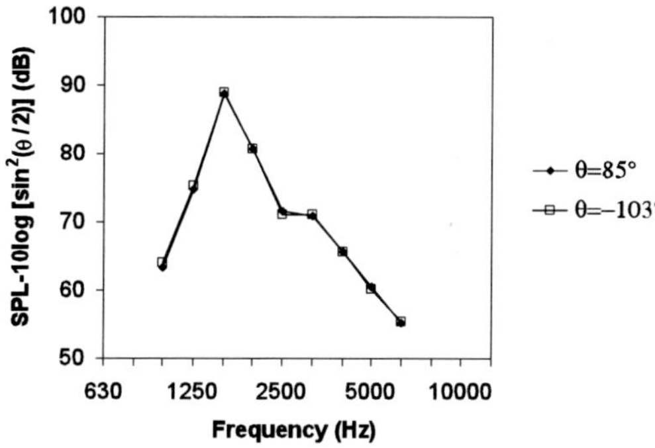

Figure 15. Airfoil noise spectra for clean NACA0012 airfoil at 40 m/s and α=4°.

of the spectrum remains the same as for a sharp edge. Since for a wind turbine blade the flow speed varies along the radius, blunt trailing edge noise may not result in a single tone or hump in the spectrum, but may rather appear as a broadband spectral increase.

The directivity and speed dependence of blunt trailing edge noise are considered to be the same as for broadband trailing edge noise: for $C/\lambda<1$ the source behaves like a compact dipole with $\sin^2\theta$ directivity and U^6 speed dependence, while for $C/\lambda>1$ it is an edge noise source with $\sin^2(\theta/2)$ directivity and U^5 speed dependence. The amplitude of blunt trailing edge noise also depends strongly on the trailing edge geometry (Figure 16). A round shape or a 60°-90° wedge may produce amplitudes two or three times higher than a rectangular edge, while a wedge angle smaller than 45° or a bevel angle less than 60° can give much lower amplitudes [8].

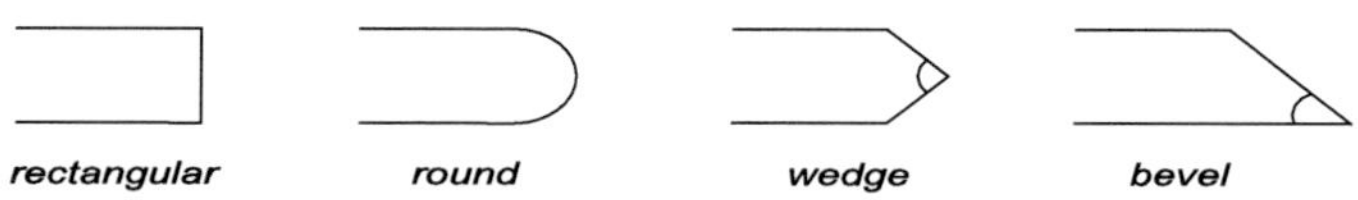

Figure 16. Blunt trailing edge geometries

Inflow Turbulence Noise

The characteristics of inflow turbulence noise depend strongly on the properties of the incoming turbulence. As mentioned above, turbulent eddies with a length scale Λ much larger than the airfoil chord will lead to compact dipole noise with a $\sin^2 o$ directivity and a U^6 speed dependence. Small eddies will induce local pressure fluctuations, causing leading edge noise with a U^5 speed dependence. On the basis of the semi-infinite flat plate approximation, a $\sin^2((\pi-\theta)/2)$ directivity pattern is expected, with maximum radiation in the direction of the trailing edge (i.e., the pattern is inverted with respect to trailing edge noise). However, the actual directivity pattern will depend on the details of the airfoil geometry. The frequency of the radiated sound is given by $f\sim U/\Lambda$. Thus, the spectrum of inflow turbulence noise is directly related to the spectrum of the incoming turbulence. Natural atmospheric turbulence is expected to cause broadband noise for frequencies up to 1000 Hz. Other sources of inflow turbulence may be the tower wake (for downwind rotors) or the turbulent wake from upwind turbines in a wind farm. It is currently still an open issue to what extent inflow turbulence noise contributes to the overall sound level of a large wind turbine [17]. For small wind turbines inflow turbulence noise may be more important, because these are often placed in a turbulent environment (e.g. rooftops). Moreover, small-scale turbulence produces higher frequencies, which are closer to the center of the audible range, while for small turbines the dominant trailing edge noise frequencies are beyond this range.

The characteristics of inflow turbulence noise were investigated experimentally [14] in the same wind tunnel as described above for trailing edge noise (Figure 11). Inflow turbulence was generated using a turbulence grid inside the nozzle. The grid had diagonally oriented 12-mm bars and a mesh width of 60 mm. The turbulence in a flow can be quantified using the *turbulence intensity*, which is the ratio of the root-mean-square of the velocity fluctuations and the mean flow speed. The present grid produced turbulence with an intensity of about 11% at the leading edge of the airfoil (without grid the turbulence intensity was 1% or less). It should be noted that this level of turbulence is much higher than that which is usually experienced in the atmosphere at typical rotor speeds.

The source map in Figure 17 shows that in the presence of severe inflow turbulence leading edge noise is dominant. The corresponding spectra in Figure 18 illustrate the dependence on wind speed. First, we see that tripping does not affect inflow turbulence noise. This is not surprising, since the leading edge noise depends on the turbulence in the inflow, and not on the boundary layer. Interestingly, the normalized plot shows that the best data collapse is obtained if the data is scaled using the *sixth* power of the tunnel speed. Thus, in contrast to the trailing edge noise discussed above, inflow turbulence noise appears to behave like a compact dipole, at least in this experiment. Similar to the trailing edge noise, the inflow turbulence noise radiation was found to be symmetrical with respect to the chord line.

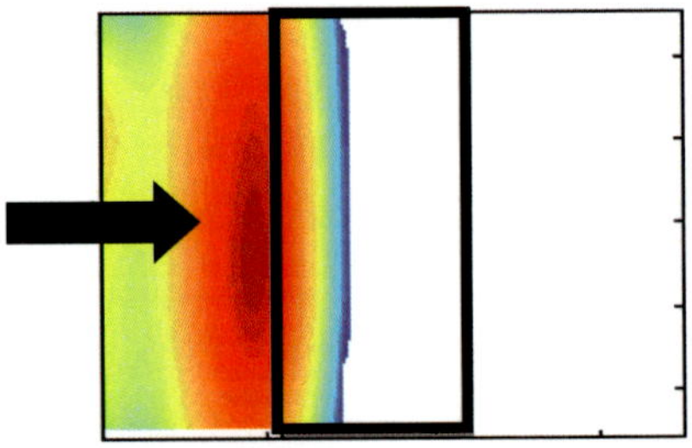

Figure 17. Acoustic source map showing leading edge noise radiation from an airfoil in a turbulent flow. The black rectangle indicates the airfoil contour. Flow is from left to right

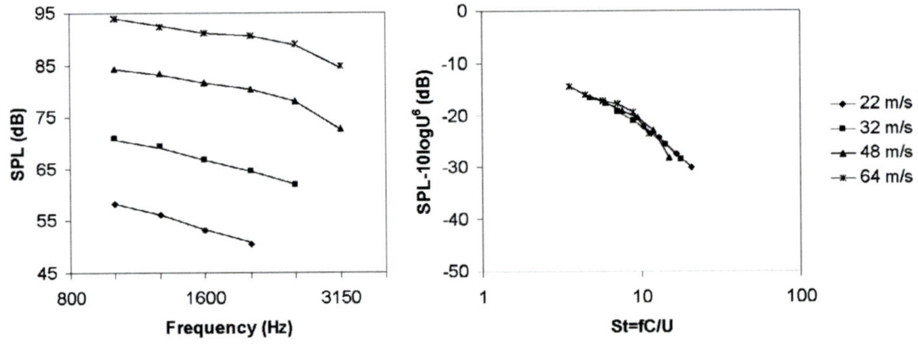

Figure 18. Inflow turbulence noise spectra for S822 airfoil at α_{eff}=8°. Left: measured spectra with trip (markers) and without trip (lines). Right: normalized spectra

The dependence of inflow turbulence noise on airfoil shape was also investigated. Figure 19 shows the leading edge noise spectra together with the airfoil shapes, in order of decreasing sound level. It can be clearly seen that the inflow turbulence noise level increases with increasing 'sharpness' of the leading edge (or decreasing airfoil thickness). This trend is well reproduced by semi-empirical prediction codes [18].

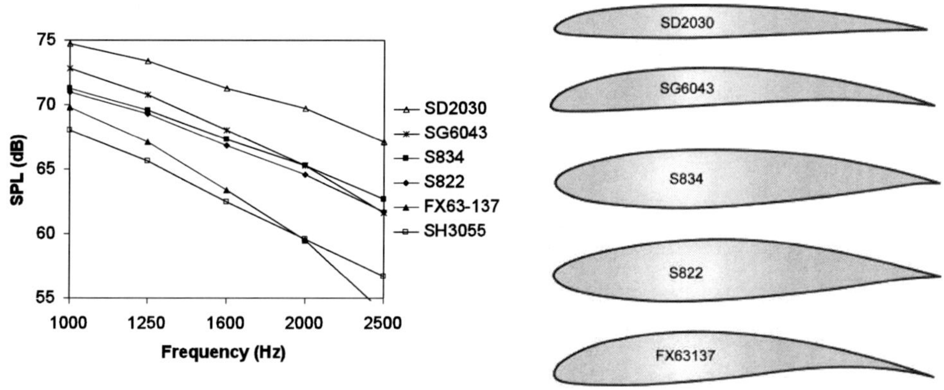

Figure 19. Inflow turbulence noise spectra for different airfoils at 32 m/s and α_{eff}=8°. The airfoil shapes are plotted in order of decreasing sound level

Tip Noise

Tip noise is caused by the three-dimensional flow at the very tip of the blade. Due to the pressure difference between the pressure and suction side of the blade, a tip vortex is formed. Turbulence is convected over the tip edge and separated flow may occur. The mechanism of noise production can be considered to be the same as for trailing edge noise [4]. Thus, we expect the frequency and level of tip noise to depend on the (spanwise) length scale of the tip vortex and the convection velocity over the edge. These parameters are again related to the blade load distribution, the strength of the tip vortex, and the planform shape of the tip [2]. Tip noise typically has a broadband, high frequency character and is not considered to be the most important noise source for modern large wind turbines. Nevertheless, it can make a significant contribution to the total sound level in specific cases. Blade tip noise is related to flap side-edge noise from aircraft wings, although a flap tip usually has a rectangular planform and a flat side-edge. Noise reduction concepts for flap edge noise include fences [19] (similar to winglets) and porous edges [20].

Moving Sources

For rotating wind turbine blades the sound source is moving with respect to the (stationary) observer. In this section we will discuss the effect of source motion on the frequency and level of the perceived sound. Consider a sound source moving at speed U in a stationary medium and radiating sound at frequency f. The frequency perceived by a stationary observer is then given by

$$f' = \frac{f}{1 - M\cos\vartheta}, \tag{2}$$

with $M = U/c$ and ϑ the angle between the source velocity and the source-observer line (at emission time). This is the well-known *Doppler frequency shift*. If the source moves towards the observer the perceived frequency increases, and vice versa. For sound from a rotating wind turbine blade with a tip Mach number of 0.2, a Doppler factor of 1.25 is found for an observer in the rotor plane. This means that during the blade revolution the perceived frequency of a 1 kHz signal varies between 830 Hz and 1250 Hz!

Besides the frequency shift, source motion also affects the *amplitude* of the perceived sound. This effect is called Doppler or convective amplification, and its magnitude depends on the nature of the source. For the low Mach number aerodynamic noise sources on wind turbine blades, the amplitude p of the perceived sound is modified by the factor

$$\frac{1}{\left(1 - M\cos\vartheta\right)^2}.$$

This means that the sound level increases when the source moves towards the observer and vice versa. For a wind turbine blade rotating at a tip Mach number of 0.2, this factor amounts to 3-4 dB for an observer in the rotor plane. So even if the source has a constant strength and a uniform directivity, the observer in the rotor plane will perceive a variation in sound level of 7 dB during the revolution of the blade! This example illustrates that we have to account for convective effects in the analysis or prediction of wind turbine noise.

It should be noted that, while the frequency shift is determined by the velocity of the source relative to the observer, the amplification factor depends on the speed with respect to the medium. Thus, in the acoustic wind tunnel tests described above, we have to account for convective amplification, although there is no frequency shift.

FIELD MEASUREMENTS OF WIND TURBINE NOISE

This section deals with acoustic field measurements of wind turbine noise [21,22]. First, we will discuss some features of wind turbine noise that can be measured with a single microphone. Next, source localization measurements with a phased microphone array are considered. These experiments will be used to explain the swishing character of the noise and to show that trailing edge noise from the blades is the dominant source mechanism.

Single-Microphone Measurements

Several interesting features of wind turbine noise can be measured with a single microphone. Figure 20 shows a picture of the turbine that will be used throughout this chapter to illustrate the different aspects of the sound. The three-bladed rotor has a diameter of 94 m and the tower height is 100 m. The three blades are not cleaned, tripped or treated otherwise. The turbine is pitch-controlled and has a rated power of 2.3 MW.

Figure 20. A typical modern large wind turbine. The noise source distribution in the rotor plane (averaged over many revolutions) is projected on the picture

The average spectrum of this turbine is shown in Figure 21. It can be seen that the turbine produces broadband noise, with the highest levels between 315 Hz and 1250 Hz. The absence of spectral peaks or humps suggests that mechanical noise is not important for this turbine, and that blade noise is dominant. By integrating this spectrum we can calculate the overall sound pressure level OASPL of the turbine, and investigate how it depends on wind speed (Figure 22). Note that rather than wind speed, in this figure the sound levels are plotted as a function of power, because this quantity was measured more accurately. It can be seen that the noise increases only gradually with increasing wind speed. This can be intuitively understood from the fact that the effective flow speed perceived by the blades increases only slowly with wind speed. It should be noted that wind turbine noise is most annoying at low wind speeds, when background noise levels are low. This is reflected in the IEC norm for wind turbine noise measurements [23], which demands measurements between wind speeds of 6 m/s and 10 m/s (at 10 m height).

The measurements discussed above were taken with a microphone on a ground plate, about one rotor diameter upwind of the turbine. By carrying out measurements at different locations, the *directivity* of the wind turbine noise can be determined. This will be discussed in more detail later, in relation to the prediction of wind turbine noise.

Rather than considering *averaged* sound levels, we can also look at the sound level as a function of time (Figure 23). In order to show the dependence on blade position more clearly, the sound level

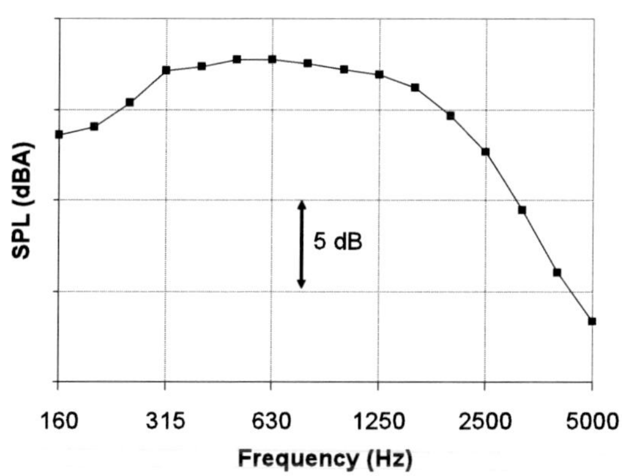

Figure 21. Wind turbine noise spectrum

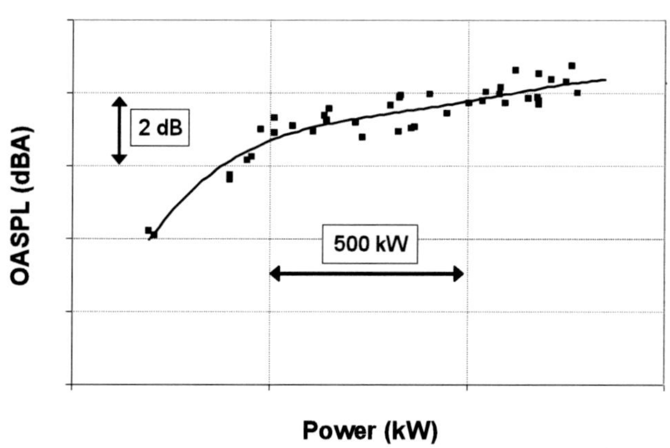

Figure 22. Wind speed dependence of turbine noise (measured values and trend line)

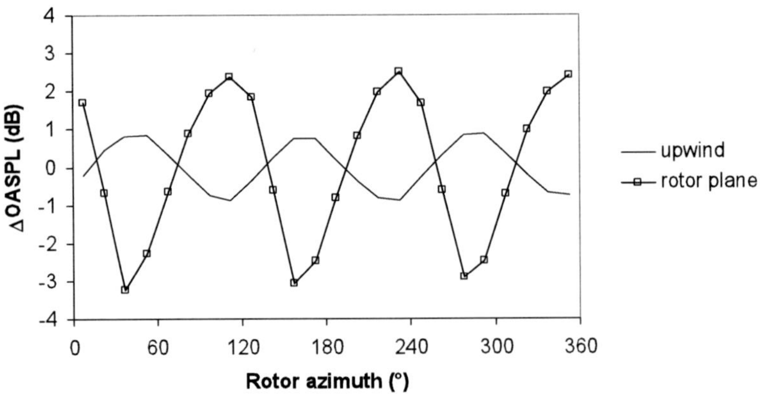

Figure 23. Sound level variation as a function of rotor azimuth, for two measurement locations at about one rotor diameter distance from the turbine

variation is shown as a function of rotor azimuth. For both measurement locations we can clearly identify three humps, corresponding to the passage of the blades. This amplitude modulation of the broadband aerodynamic blade noise at the blade passing frequency is often referred to as *swish.* It can be seen that for an observer close to the rotor plane the swish is much more pronounced than at an upwind position. Furthermore, the level variations for these two positions are in anti-phase. As expected, the noise levels from the three different blades are practically the same.

The above results illustrate that single-microphone measurements are useful for quantifying and characterizing the noise from wind turbines. However, in order to understand the measured trends, and to be able to predict and reduce wind turbine noise, we would like to know what the dominant *source* is. It may be trailing edge noise, but also for instance inflow turbulence noise or tip noise. In order to get more information about the source mechanisms we have to revert to source localization techniques. This is the subject of the next section.

Noise Source Distribution in the Rotor Plane

Source localization measurements were done using a 148-microphone acoustic array, mounted on a horizontal wooden platform of about 16x18 m^2 (Figure 20 and Figure 24). The distance between the platform and the tower was roughly the same as the tower height, resulting in a 'view angle' of about 45°. By placing the scan plane of the array in the rotor plane, the noise source distribution in the rotor plane could be determined (Figure 25). It is important to realize that no assumptions or adjustable settings were used to produce these results. The source maps simply show from what direction the sound is coming, for an observer at the platform position. Also note that the source distribution is

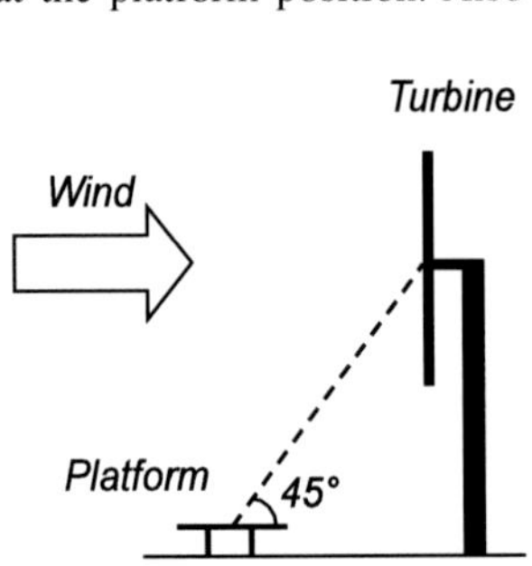

Figure 24. Test set-up for array measurements on a wind turbine

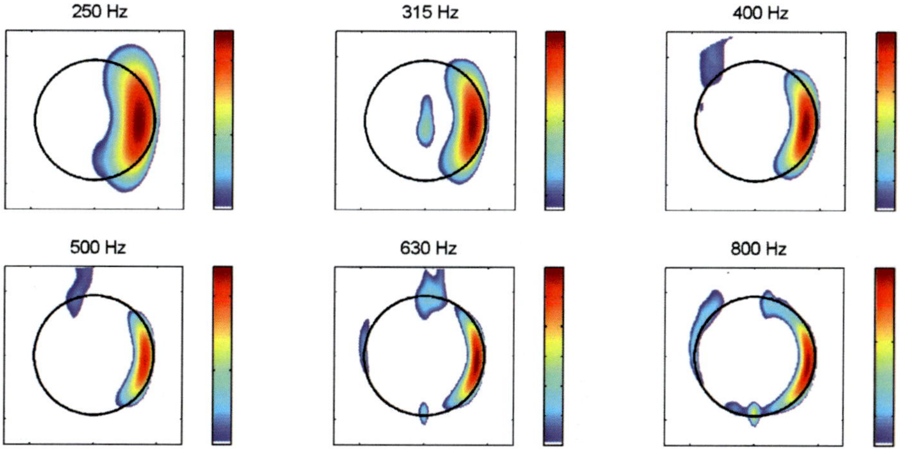

Figure 25. Average noise source distribution in the rotor plane as a function of frequency. The turbine rotates in clockwise direction and the tip trajectory is indicated by the black circle. The range of the colour scale is 12 dB

averaged over a large number of revolutions (more than 20 minutes of measurement time), covering the wind speed range between about 6 m/s and 10 m/s.

Several interesting observations can be made in Figure 25. First, we see that practically all noise is produced by the blades, although the nacelle also generates some noise at 315 Hz. Note that the 'sources' between 10 and 12 o'clock for 400 Hz - 630 Hz are an artefact of the array technique. Second, most of the noise is radiated from the outer part of the blades, but not the very tip. This rules out tip noise as the dominant noise source. The dominance of the outer part of the blades can be understood from the fact that aerodynamic noise scales with the 5th or 6th power of the flow speed, as discussed before. We can also observe a minor high-frequency source where the blades pass the tower. This may originate from (1) reflection of blade noise on the tower, (2) impingement of blade tip vortices on the tower, and/or (3) the upstream influence of the tower on the flow field around the blade. However, the most striking feature in the source maps is that practically all noise is produced during the downward movement of the blades. This asymmetric source pattern explains the swishing character of the noise perceived on the ground (Figure 23): each time a blade passes the 3 o'clock position, a swish is radiated to the observer at the array position. Note that by the time the sound arrives at the observer, the blade has already travelled further. As a result, the swish sometimes *seems* to be generated when the blade is close to the tower.

Now the remaining question is *why* the source distribution in the rotor plane is asymmetric. Assuming that the effects of wind shear and rotor tilt angle are small, the flow conditions on the blade, and consequently the acoustic source strength, are independent of rotor azimuth. In other words, the blades produce a constant sound level during the complete revolution. Therefore, we expect the strong asymmetry to be due to convective and directivity effects. The trailing edge noise directivity function in Eq. 1 exhibits maximum radiation in the direction of the leading edge. Thus, if the blade moves downward, the trailing edge noise is mainly radiated towards the ground, while if the blade moves upward the noise radiates into the sky. As a result, an observer on the ground will hear the trailing edge noise mainly when the blades move downward. Furthermore, the convective factor in Eq. 3 shows that the amplitude of the perceived sound increases when the source moves towards the observer, and decreases when it moves away from the observer. Again, this means that the descending blade will be noisier than the ascending blade for an observer on the ground. Thus, the characteristics of trailing edge noise seem to be qualitatively in line with the experimental results.

To see if trailing edge noise directivity and convective amplification can *quantitatively* explain the observed asymmetry, the magnitude of both effects, given by the factor

$$\frac{\sin^2(\theta/2)\sin\phi}{\left(1-M\cos\vartheta\right)^4} \tag{4}$$

is calculated for an observer on the ground, one rotor diameter upwind from the turbine. The source is assumed to be located at the tip of the blade, which rotates at a Mach number of 0.2. For simplicity the rotor tilt angle and the blade pitch and twist are assumed to be zero. Figure 26 shows the magnitude of the directivity and convective factors in Eq. 4 as a function of rotor azimuth (0° corresponds to the 12 o'clock position). As expected both effects increase the perceived sound level during the downward movement of the blades, and decrease the level during the upward movement. The directivity factor has the largest effect. The total asymmetry is more than 12 dB and the maximum occurs between 3 and 4 o'clock, which is in good agreement with the acoustic source maps in Figure 25. Thus, the asymmetry in the source maps can be well explained using the characteristics of trailing edge noise.

Now the question is what the source distribution would look like if the dominant source mechanism were inflow turbulence noise or compact dipole noise. The convective amplification factor for these

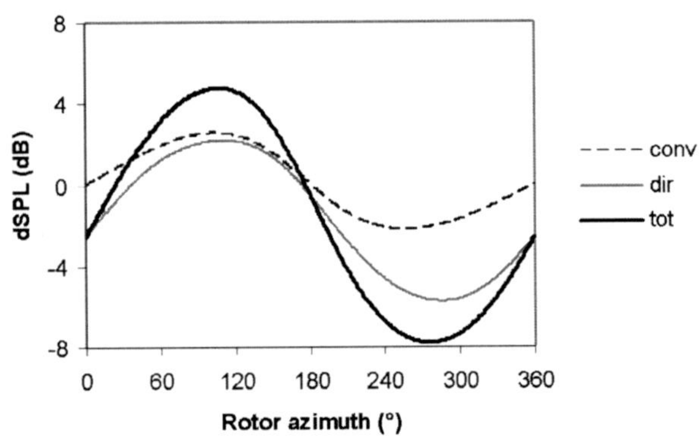

Figure 26. Calculated convective amplification and directivity factors for non-compact trailing edge noise

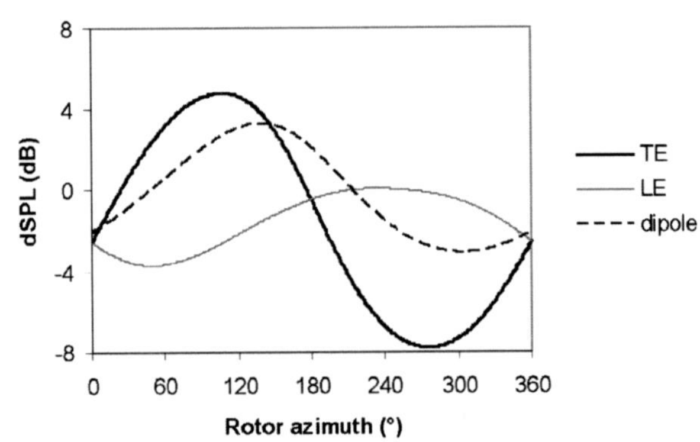

Figure 27. Calculated effect of convective amplification and directivity for non-compact trailing edge noise, non-compact leading edge noise, and compact dipole noise

sources is the same as for trailing edge noise, but as discussed before the directivity factor is different: for inflow turbulence (leading edge) noise, maximum radiation occurs in the direction of the trailing edge, while a compact dipole radiates strongest in the direction perpendicular to the plane of the blade. The magnitude of these directivity factors (including convective amplification) is shown in Figure 27, again for an observer on the ground, one rotor diameter upwind from the turbine. As a reference the curve for non-compact trailing edge noise is shown as well.

It can be seen that for leading edge noise the maximum level would be radiated during the *upward* movement of the blades, while for a compact dipole the asymmetry would be less than 8 dB. Thus, the experimental results can *only* be explained using the directivity function for non-compact trailing edge noise. This is true even at the low frequencies, where the acoustic wavelength is of the same order as the blade chord. Thus, the blades produce a constant sound level during the complete revolution, but the trailing edge noise radiates mainly in the direction in which the blade is moving. As a result, an observer on the ground hears the blades most clearly during their downward movement, which causes the swishing character of the sound. It should be noted that the asymmetry in the source maps is purely an effect of observer position: an observer above the turbine would mainly hear the ascending blades, while an observer on the rotor axis would perceive a circular source pattern, without swish. We will further discuss the swish amplitude for different observer locations, as shown in Figure 23, in the section about the prediction of wind turbine noise.

We can also use the trailing edge noise characteristics to explain the fact that in Figure 25 the source radius increases with increasing frequency. As mentioned before, the peak frequency of trailing edge noise is proportional to U/δ^*. For increasing radius the flow speed increases, and the boundary layer thickness typically decreases due to the smaller blade chord. Thus, the frequency of the emitted sound increases with increasing radius.

Noise Sources on the Blades

In addition to the source distribution in the rotor plane, array measurements can also be used to identify the noise sources on the individual blades. This is done by defining three rotating scan planes which move along with the blades. To investigate the effect of different blade treatments, measurements were done on an 850 kW wind turbine with one cleaned blade, one tripped blade, and one untreated blade. The rotor diameter was 58 m. Tripping was done using zigzag tape of 0.4 mm thickness over the complete radius, at 5% chord on the suction and pressure side of the blade. Like the turbine discussed above, practically all perceived noise was produced by the outer part of the blades, during the downward part of the revolution.

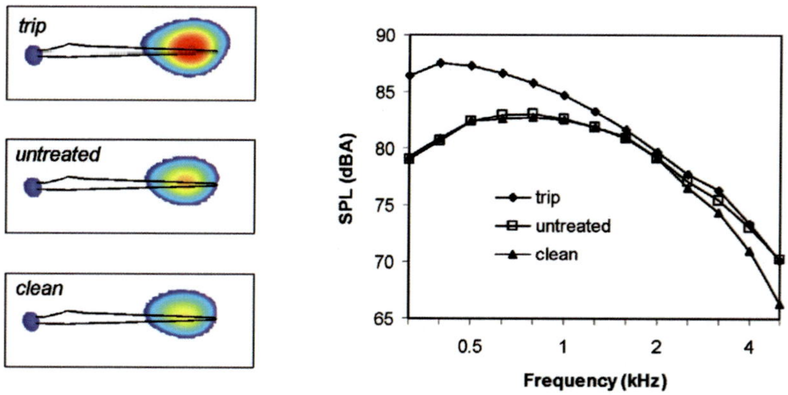

Figure 28. Acoustic source maps (800 Hz) for individual blades and corresponding noise spectra. The black line indicates the blade contour (leading edge on lower side). The range of the colour scale is 12 dB and the scale is the same for the three blades

The source maps for the individual blades, shown in Figure 28, are in line with this. Moreover, the corresponding noise spectra show that the tripped blade is much noisier than the other two, especially at low frequencies. This can be explained by the characteristics of trailing edge noise: tripping leads to a thicker boundary layer, which in turn yields lower frequencies and higher sound levels. Since tripping was previously found to have no effect on inflow turbulence noise (Figure 18), these results are another indication that trailing edge noise is the dominant mechanism (at least after tripping). This was further confirmed by the fact that the blade noise spectra are proportional to the fifth power of the local flow speed. Moreover, analysis of the narrowband noise spectra for individual radial stations of the blade shows that tonal blunt trailing edge noise is not important for these turbines.

All in all, the experimental results for both turbines provide strong evidence that trailing edge noise is the dominant noise source for typical modern wind turbines. This fact will be used in the next section to develop a wind turbine noise prediction method based on trailing edge noise characteristics.

PREDICTION OF WIND TURBINE NOISE

The ability to predict wind turbine noise is essential for the design of quiet turbines and for assessing possible noise pollution around wind farms. There are several different approaches to calculating the noise from a wind turbine, with different degrees of sophistication. On one side of the spectrum we have the 'rule-of-thumb' methods, which only need the rotor diameter or rated power as input. Although these empirical methods can provide an order-of-magnitude estimate of the sound level, they do not enable us to assess design modifications or noise characteristics. On the other side of the spectrum, we have complete numerical methods, which combine Computational Fluid Dynamics (CFD) and Computational AeroAcoustics (CAA) to calculate the sound of an airfoil section from first principles. These methods are computationally rather expensive and at present do not allow us to calculate the sound from a complete three-dimensional wind turbine rotor with high Reynolds number flow around the blades.

Somewhere in between these two extremes we have the semi-analytical, semi-empirical methods. These models contain scaling laws derived from theoretical analyses, and are complemented with results from aerodynamic and acoustic measurements. The different potential source mechanisms are considered separately. This type of method represents the current state-of-the art and will also be applied in the present section [24]. The models are based on physical relations and therefore allow us to investigate the influence of several design and operation parameters, such as the blade planform, blade twist, airfoil shape, pitch angle, RPM, and wind speed. At the same time they are fast enough to allow a quick assessment of different design modifications. Moreover, we will see below that these methods are capable of producing *accurate* predictions.

In the previous section we saw that trailing edge noise is the dominant noise source for typical modern wind turbines. In the present section we will use this knowledge to develop a semi-analytical, semi-empirical prediction method for wind turbine noise. First, the prediction method is described in detail. Then, the predictions will be validated against experimental results. Finally, the prediction method will be applied to calculate the noise footprint of a typical wind turbine.

Prediction Method

Since the field measurements show that broadband trailing edge noise is the dominant noise source for both turbines, only this noise source is incorporated in the prediction method. This means that inflow-turbulence noise, blunt trailing edge noise, and tip noise are discarded. The prediction method only needs the blade geometry and turbine operating conditions as input, and contains no fitting or tuning parameters. The calculation can be divided into three steps: (1) blade aerodynamics, (2) trailing edge noise source strength, and (3) directivity and convective amplification. These steps will be described in the following.

First, the blade is divided into a number of radial segments (21 for the present calculations). For each segment, the local flow parameters (Reynolds number, angle of attack) are calculated using a blade element momentum method. An airfoil design and analysis code is then used to calculate the boundary layer displacement thicknesses on the pressure and suction side. The effects of atmospheric turbulence, wind shear, and yaw are neglected in the calculations, i.e., stationary and axisymmetric conditions are assumed.

In the second step, the source spectrum for each radial segment is calculated using a semi-empirical, semi-analytical prediction code for trailing edge noise from two-dimensional airfoil sections [4]. This code is based on comprehensive acoustic and aerodynamic measurements on NACA0012 airfoils with varying chord lengths, at various flow speeds and angles of attack. The total trailing edge source strength due to the turbulent boundary layer is the sum of three contributions of the following form:

$$SPL_i = 10\log\left(\frac{\delta_i^* M^5 L}{r^2}\right) + A_i\left(\frac{St_i}{Sr_i}\right) + K_i\,, \tag{5}$$

where δ^* is the displacement thickness, M the Mach number, L the span of the blade segment, r the distance to the observer, and K an empirical constant which depends on the Mach and Reynolds numbers. The function A describes the spectral shape as a function of the ratio between the Strouhal number $St = f\delta^*/U$ (with U the local flow speed) and the empirical peak Strouhal number Sr. The three contributions (here denoted by the index i) are the pressure side boundary layer, the suction side boundary layer, and an additional contribution to account for nonzero angle of attack. Note that Eq. 5 is essentially a combination of the analytical relation in Eq. 1 (without the directivity factors), and an empirically found 'universal' normalized trailing edge noise spectrum, similar to Figure 14.

In the third step, directivity and convective effects are included to obtain the effective source strength perceived by an observer at a specified position. These effects are accounted for using a modified version of Eq. 4. The numerator in this equation describes the directivity for trailing edge noise from a semi-infinite flat plate, and has maximum radiation in the direction of the leading edge. However, as mentioned before, the directivity function for a finite-chord airfoil exhibits a reduced intensity in the direction of the leading edge. Therefore, the numerator of Eq. 4 is modified to reflect this finite-chord effect. The corresponding directivity functions are shown in Figure 29. In these plots, the source position at the trailing edge is located at the centre of the sphere, the chord line runs along the x-axis and the trailing edge runs along the y-axis. Note that this modification only affects the sound at observer positions close to the rotor plane, which means that the results in Figure 26 and Figure 27 (for an upwind observer) are still valid.

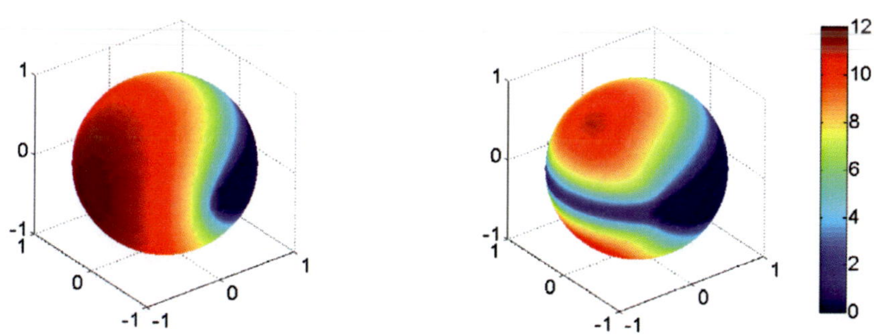

Figure 29. Trailing edge noise directivity function for semi-infinite flat plate (left) and finite-chord airfoil (right). The range of the colour scale is 12 dB

Validation Against Experiment

In order to validate the prediction method, calculations were carried out for both turbines mentioned in the field measurements section. In the present section we will show some representative comparisons between prediction and experiment for the 2.3 MW turbine shown in Figure 20. First, Figure 30 shows the measured and calculated overall sound level as a function of rotor power. It can be seen that the

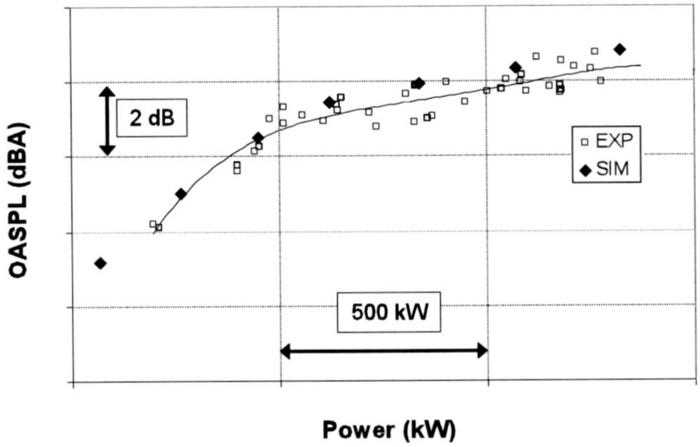

Figure 30. Measured and predicted rotor noise levels as a function of rotor power

trend in the experiments is well represented by the predictions: the difference between measurement and prediction is less than 1-2 dB, which is smaller than the scatter in the experimental data. For the other turbine a similar agreement was found. This accuracy is considered satisfactory for the present semi-empirical prediction model. Besides the overall sound level, also the *shape* of the predicted rotor noise spectrum (not shown here) was found to agree well with the experiments.

In order to assess the directivity of the turbine noise, in the experiments eight ground microphones were placed on a 240-m circle around the turbine, at 45° intervals (Figure 31). In this figure, β indicates the misalignment angle between turbine and array, and ξ indicates the position of a ground microphone with respect to the downwind direction. The measured and calculated directivity are shown together in Figure 32. The experimental directivity pattern shows two dips in the crosswind directions, i.e., for measurement positions close to the rotor plane (ξ=90° and ξ=270°). The predicted curve follows the measured curve within 1-2 dB, with two 6 dB dips in the cross-wind direction. These dips can be understood from the reduced levels of the trailing edge noise directivity function close to the plane of blade (Figure 29).

The calculated source distribution in the rotor plane was also used to produce simulated source maps (Figure 33). Comparison with the measured source maps in Figure 25 shows good agreement: as in the experiments, the simulated source maps show dominant noise radiation from the outer part of

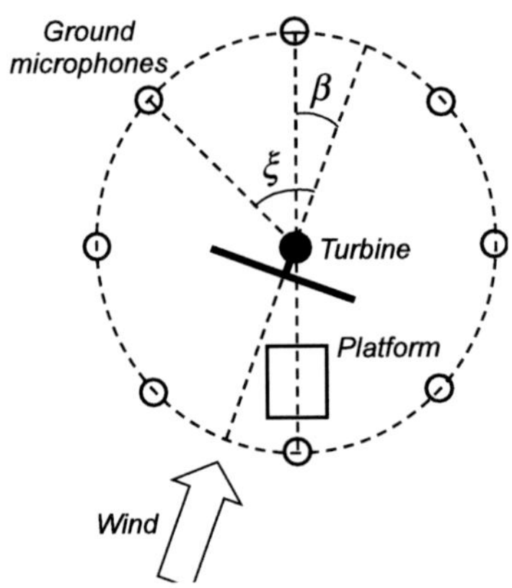

Figure 31. Top view of test set-up for directivity measurements

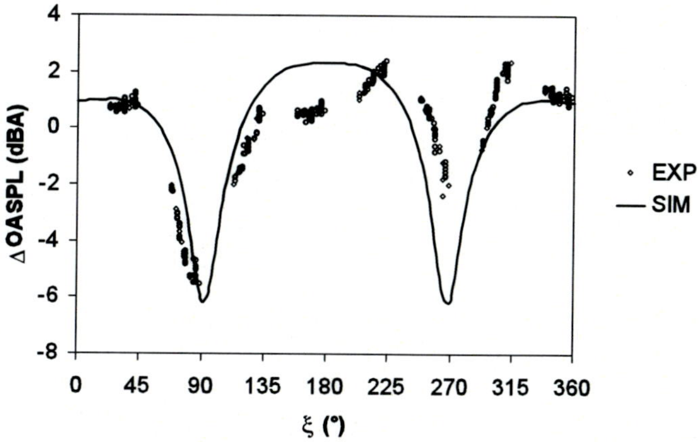

Figure 32. Measured and predicted turbine noise directivity

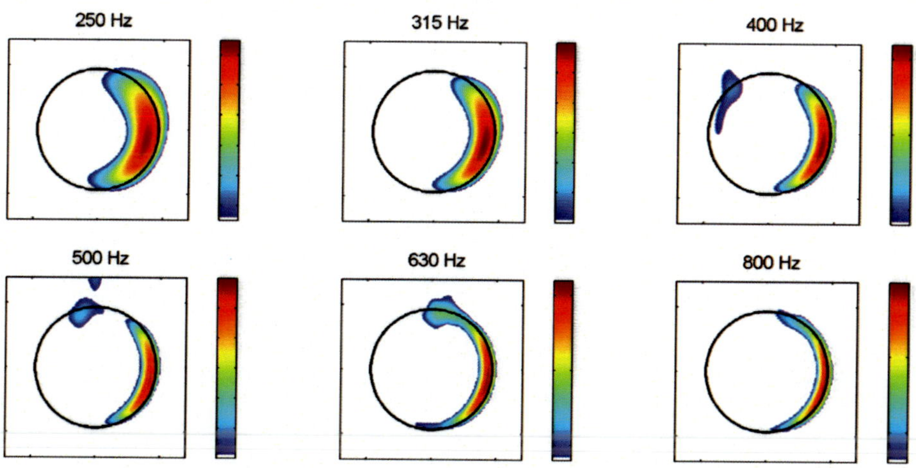

Figure 33. Predicted source maps (compare to measured maps in Figure 25)

the blades, during their downward movement. Furthermore, the blade sources move outward with increasing frequency.

Whereas the previous source maps concern misalignment angles β around 0°, experimental data are also available for large misalignment angles. The measured and predicted source maps for these angles are shown in Figure 34. It can be seen that the location of the source region shifts upward or downward when the right or left hand side of the rotor plane is turned towards the array, respectively. This can be qualitatively explained by the change in the component of the blade velocity in the direction of the array, which results in a change in convective amplification. In short: you mainly hear the blade when it is coming towards you. At high misalignment angles the array resolution decreases due to the oblique view angle. Again a good qualitative agreement between simulation and experiment is found, indicating that the changes in source pattern are well captured by the trailing edge noise prediction method.

Since the simulated source maps successfully reproduce the asymmetric source distribution found in the experiments, we expect that the variation in perceived sound level (swish) is also captured by the predictions. Figure 23 shows the variation in sound level for two measurement positions. From plots like these we can determine the (peak-to-trough) swish amplitude as a function of observer

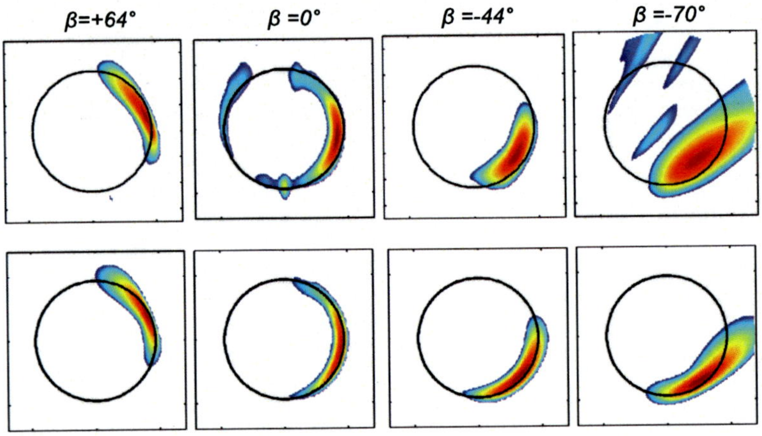

Figure 34. Measured (upper row) and predicted (lower row) source maps at different misalignment angles (800 Hz)

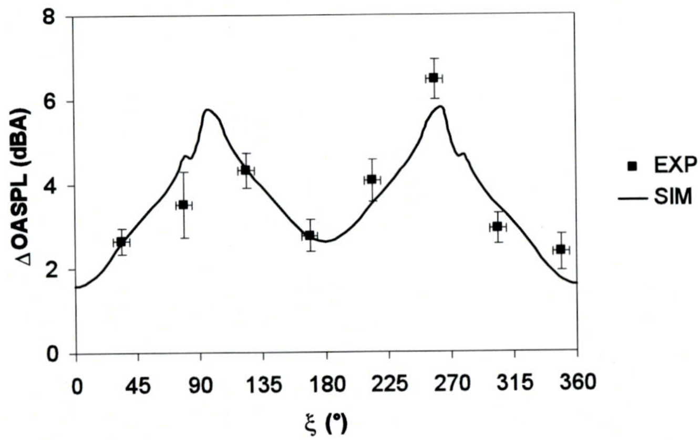

Figure 35. Measured and predicted swish amplitude as a function of far field position

position. The experimental swish amplitudes for the eight ground microphones (Figure 31) are shown together with the predicted values in Figure 35. It can be seen that the swish amplitude is predicted within 1 dB for all directions. The highest swish is observed for microphones close to the rotor plane, where it can reach values of more than 6 dB.

In summary, the results in this section show that the characteristics of wind turbine noise can be well predicted using a semi-analytical, semi-empirical prediction method. Good agreement is obtained between prediction and experiment, not only in terms of source spectra and overall sound levels (as a function of rotor power), but also in terms of the noise source distribution in the rotor plane (as a function of frequency and observer position). Moreover, the turbine noise directivity and swish amplitude in different directions are accurately predicted. This allows us to apply the prediction method to calculate the noise at larger distances, which is the subject of the following section.

Prediction of Noise and Swish Footprints

The prediction method is now applied to calculate time-dependent noise footprints of the 94-m diameter turbine shown in Figure 20, which is considered to be representative of a modern large wind turbine. Figure 36 shows instantaneous turbine noise footprints (top view) for four different rotor azimuth angles, up to a distance of ten times the rotor diameter. The turbine is located at the centre of the footprint, and the wind goes from left to right. The rotor azimuth at observer time is indicated in the upper right corner of each footprint. In order to limit the range of the dB scale, the levels are normalized using the horizontal distance to the turbine.

The footprints show two waves of increased sound level, one in each cross-wind direction, which start close to the turbine and propagate outward with the speed of sound. The wave to the side of the descending blade is generated when the blade is at 1 o'clock, while the wave to the side of the ascending blade is generated when the blade is at 6 o'clock. Or, as mentioned before: you mainly hear the blade when it is coming towards you. After $\psi = 180°$ the cycle repeats and both waves can be seen to propagate further to the edge of the footprints. The distance between two successive waves is about 5 rotor diameters, which is consistent with the time period of 1.33 s between the passage of two blades (the RPM is 15) and a speed of sound of 340 m/s.

Due to the passage of these sound waves from the blades, the noise levels in the crosswind directions vary significantly, while in the upwind and downwind directions the levels are quite constant at large distances. This is illustrated in Figure 37, which shows footprints of the average sound level (averaged over a complete revolution) and the swish (variation in sound level during a complete revolution). It can be seen that both footprints do not change significantly beyond a distance of a few rotor diameters. For both cross-wind directions, the average level is lower than in the upwind and downwind directions, but the *variation* in level is larger. Even at a large distance, trailing edge noise directivity and convective amplification may cause swish amplitudes up to more than 5 dB in the cross-wind directions. This may be an explanation for the 'Amplitude Modulation' phenomenon reported by many observers. Also note that at small distance to the turbine (about one rotor diameter) substantial swish is observed in all directions, which is consistent with the experimental results described above (Figure 35).

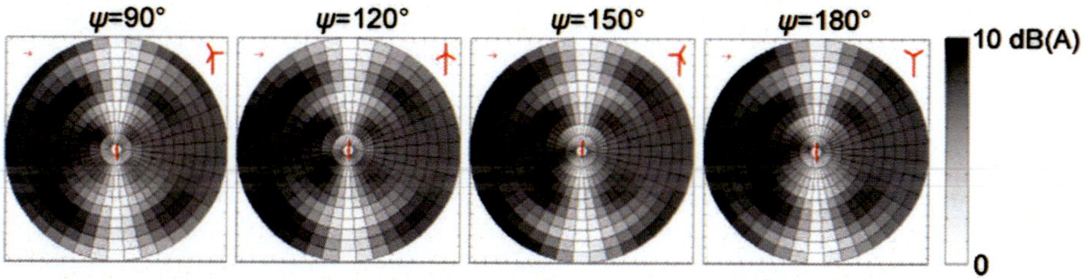

Figure 36. Predicted instantaneous noise footprints for increasing rotor azimuth angle. The wind goes from left to right

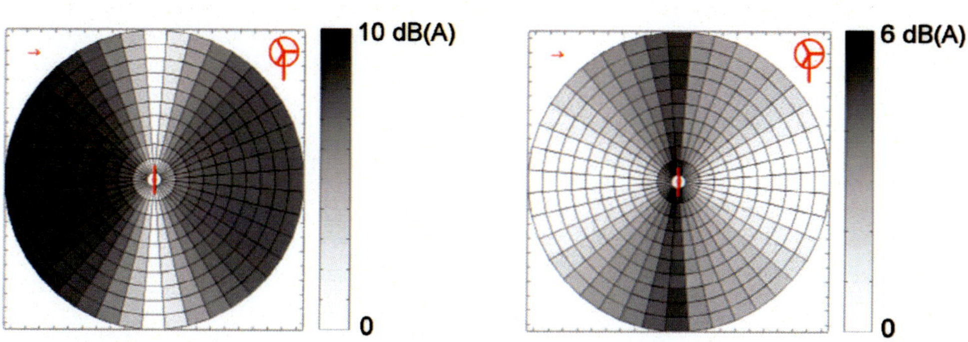

Figure 37. Predicted average footprint (left) and swish footprint (right) for a complete revolution

REDUCTION OF WIND TURBINE NOISE

In the preceding sections we have shown that trailing edge noise is the dominant noise source for typical modern wind turbines, and that we can predict the characteristics using a semi-analytical, semi-empirical method. In the present section we will address the question how we can reduce wind turbine noise. We will here focus on trailing edge noise; suggestions for reducing other sources (e.g. inflow turbulence noise or blunt trailing edge noise) can be obtained from the characteristics described earlier in this chapter. A reduction in turbine noise can be exploited in several ways. First, it can prevent turbines from having to operate at reduced power during the night, due to noise constraints. Second, it enables us to erect more turbines on a given site, within the same overall noise limits. A noise reduction of 3 dB potentially doubles the number of turbines! Third, the distance to settlements can be reduced, so that more sites become available. Finally, quiet blade designs enable a larger rotor diameter and/or a higher RPM, thus increasing the energy yield for a given turbine. In this section we will describe a number of noise reduction concepts and discuss their performance in wind tunnel and field tests.

Reduction Concepts

As mentioned before, trailing edge noise usually defines the lower bound of wind turbine noise. If all other potential noise sources are removed, trailing edge noise is what remains. In order to *further* reduce the noise from a wind turbine blade we should therefore reconsider Eq. 1, which describes how the trailing edge noise level depends on the different parameters. Since the sound scales with the fifth power of the flow speed, an obvious noise reduction technique is to lower the turbine RPM. In fact this 'low noise operation mode' is sometimes used to reduce sound emission from a turbine during the night. Theoretically, a 20% reduction in RPM already gives a noise reduction of 5 dB. However, the big disadvantage is that a reduction in RPM also reduces the power production. The same holds for an increase in blade pitch angle μ, which reduces the local angle of attack α (see Figure 3), yielding a smaller displacement thickness δ^* and as a result less noise. However, due to the reduced lift the power output is also reduced. Therefore, our objective will be to reduce trailing edge noise *without energy loss*.

One way to achieve a noise reduction without energy loss is to modify the airfoil shape such that the displacement thickness is reduced, while the aerodynamic performance is maintained [25]. The idea is schematically illustrated in Figure 38 (note that these are not the actual airfoil shapes). The sound from the reference airfoil is dominated by the low-frequency peak due to the suction side boundary layer. To reduce this peak, the optimized airfoil has less loading on the suction side, so that the suction side boundary layer thickness is reduced. As a result, the suction side peak has a lower level and higher frequency. In order to maintain the performance, the pressure side loading of the optimized airfoil is somewhat increased, which increases the pressure side boundary layer thickness. Nevertheless, the total sound level is reduced with respect to the reference airfoil. The optimized airfoil concept has been assessed using two-dimensional airfoil sections in small wind tunnels like the KAT (Figure 11), but also on a 4.5-m diameter model rotor in the larger DNW-LLF wind tunnel (Figure 39). Depending on the reference airfoil, average overall noise reductions of up to about 4 dB were obtained, without a significant degradation in aerodynamic performance. The assessment of this concept on a full-scale wind turbine will be discussed in the next section.

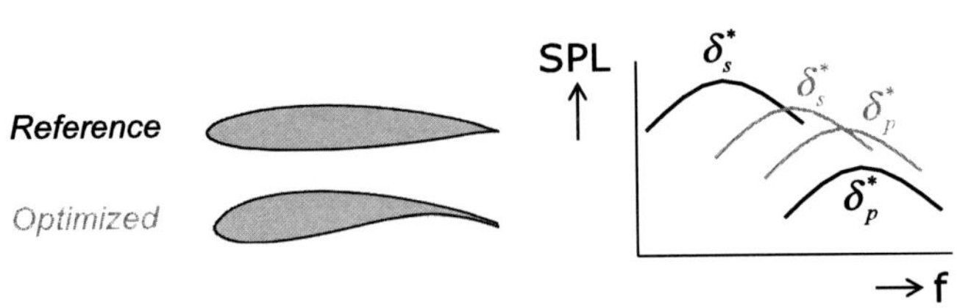

Figure 38. Reduction of trailing edge noise by modifying the boundary airfoil shape

Besides changing the flow parameters, we can also try to reduce noise by decreasing the acoustic radiation efficiency of the trailing edge. An example of this are the trailing edge serrations discussed before (Figure 9). By changing the angle γ between the path of the turbulent eddies and the normal to the trailing edge, the sound is scattered less effectively by the trailing edge. Serrations are intended to affect only the acoustic scattering at the trailing edge, and their aerodynamic effect should be small. Thus, the effects of an optimized airfoil shape and serrations are expected to supplement each other. The serration concept was investigated in a number of experimental studies on two-dimensional airfoils and model scale rotors. Typically, overall noise reductions of up to about 4 dB were achieved. In order to prevent excess high-frequency noise, it was found to be critical to align the plane of the serrations with the trailing edge flow (Figure 40). For a real turbine this may be complicated, since the trailing edge flow angle depends on the wind speed. The assessment of the serration concept on a full-scale wind turbine will be discussed in the next section.

Another method to reduce the acoustic radiation efficiency is to add trailing edge brushes, which smooth the streamwise discontinuity at the trailing edge (Figure 41). This concept is inspired by the silent flight of the owl, which is attributed to the fluffy feathers at the trailing edge of its wings. Flexible brushes align automatically with the trailing edge flow and have shown significant noise reduction potential in wind tunnel tests on flat plates and two-dimensional airfoils [26]. However, the successful application of the brush concept on a full-scale wind turbine has not been reported yet.

Figure 39. Model scale wind turbine (rotor diameter 4.5 m) in DNW-LLF wind tunnel

Figure 40. Alignment of serrations with trailing edge flow

Assessment of Noise Reduction Concepts on a Full-Scale Wind Turbine

Two of the above-mentioned reduction concepts, the optimized airfoil shape and the trailing edge serrations, were tested on the 2.3 MW, 94-m diameter wind turbine shown in Figure 20. In order to compare the blade performance for identical weather and operation conditions, the rotor consisted of

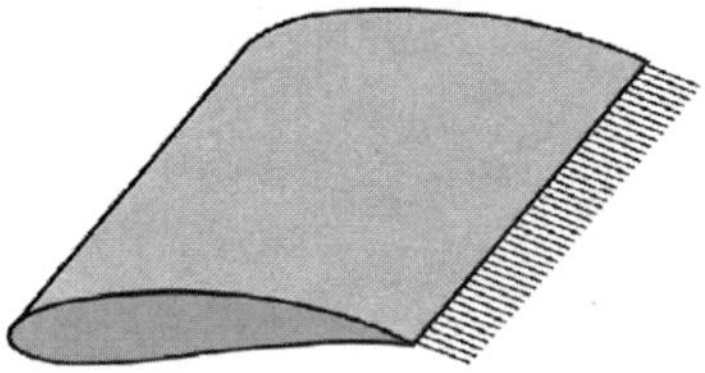

Figure 41. Trailing edge brushes

one standard (baseline) blade, one standard blade with trailing edge serrations, and one blade with an optimized airfoil shape. Since all the noise is produced by the outer part of the blades (Figure 25), only this part was modified. The optimized blade was nominally identical to the baseline blade, except for the outer ~30%, where it had a new airfoil. The serrated blade had the same nominal geometry as the baseline blade. The serrations were mounted to the outer 12.5 m of the blade (Figure 42). Their length was about 20% of the local chord, and varied as a function of radius. The plane of the serrations was aligned with the flow direction at the blade trailing edge (as determined from flow calculations). By aligning the serrations with the flow (at least for a certain wind speed range) it was attempted to minimize their aerodynamic impact and prevent increased high frequency noise by flow through the teeth.

From power and load measurements on the baseline and modified rotor, it was found that the in-plane and out-of-plane blade loads on the serrated blade (and, to a lesser extent, also on the optimized blade) were slightly higher than on the baseline blade, causing the aerodynamic performance of the modified blades to be similar or slightly better than the baseline blade. During the measurements on the baseline rotor, it was also verified that the sound produced by the three unmodified blades was practically the same: for the two blades used again in the present test campaign the difference was less than 0.05 dB.

Figure 42. Climber removing trips from serrated blade

In order to assess the differences in sound level between the three blades, we *could* use a single microphone to produce graphs like the one in Figure 23. However, this would not give us the complete picture, because it only shows the swish produced by each blade during a small part of the revolution. In reality each blade may produce noise during the complete revolution (as will be seen below). Therefore, more detailed acoustic measurements were carried out, using a large horizontal microphone array positioned about one rotor diameter upwind from the turbine (Figure 24). Using three rotating scan planes, the noise sources on the individual blades (averaged over many revolutions) could be localized and quantified (Figure 43). This figure shows that for low frequencies both modified blades are significantly quieter than the baseline blade, especially the serrated blade. For high frequencies however, both modified blades are noisier than the baseline blade, especially the optimized blade. These trends are illustrated in Figure 45, which shows the integrated spectra for the three blades, averaged over all wind speeds between 6 m/s and 10 m/s (at 10 m height). These spectra confirm the low-frequency noise reduction and high-frequency noise increase for the modified blades.

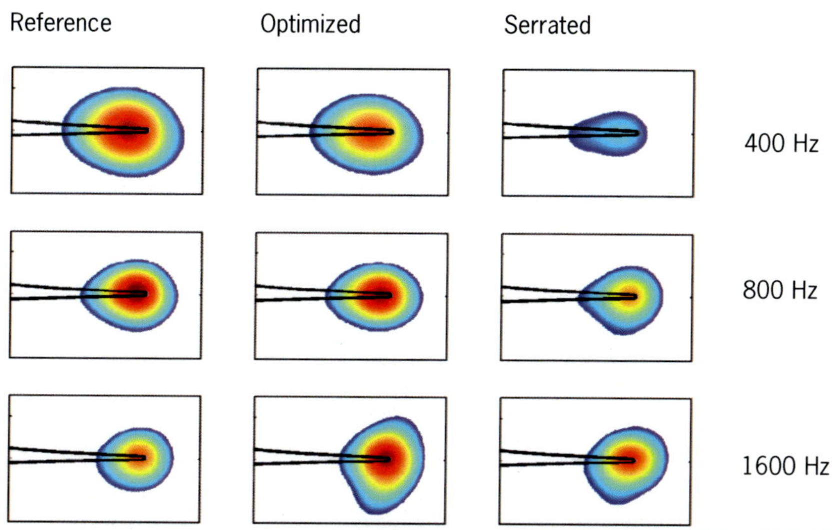

Figure 43. Acoustic source maps for baseline blade and two modified blades, averaged over all wind speeds. The black contour indicates the outer 32 m of the blade (trailing edge on upper side). The range of the dB scale is 12 dB

Since such a noise increase was not observed in the preceding two-dimensional wind tunnel tests, a more detailed analysis of the field results was carried out. This analysis showed that the noise increase at high frequencies is due to tip noise, which is mainly radiated during the upward part of the revolution. This is illustrated in Figure 44, which shows the dependence of the blade noise on rotor azimuth. The excess tip noise is found to be most important at low wind speeds, where the tip loading is high (Figure 46). This suggests that the tip noise is caused by the slightly higher blade loads on the modified blades, which may have led to a stronger tip vortex. As a result of the increased tip noise at low wind speeds, the overall noise reduction increases with increasing wind speed. Based on the spectra in Figure 45, *average* overall noise reductions of 0.5 dB and 3.2 dB are obtained for the optimized blade and the serrated blade, respectively. This 3 dB reduction demonstrates that wind turbine noise can be halved without adverse effects on the aerodynamic performance! Note that in this context 'halve' refers to the acoustic energy, i.e., two wind turbines *with* serrations would produce the same amount of noise as one turbine without serrations. It does *not* imply that the subjectively perceived loudness is reduced by 50% if a turbine is equipped with serrations. For this a noise reduction on the order of 10 dB would be required.

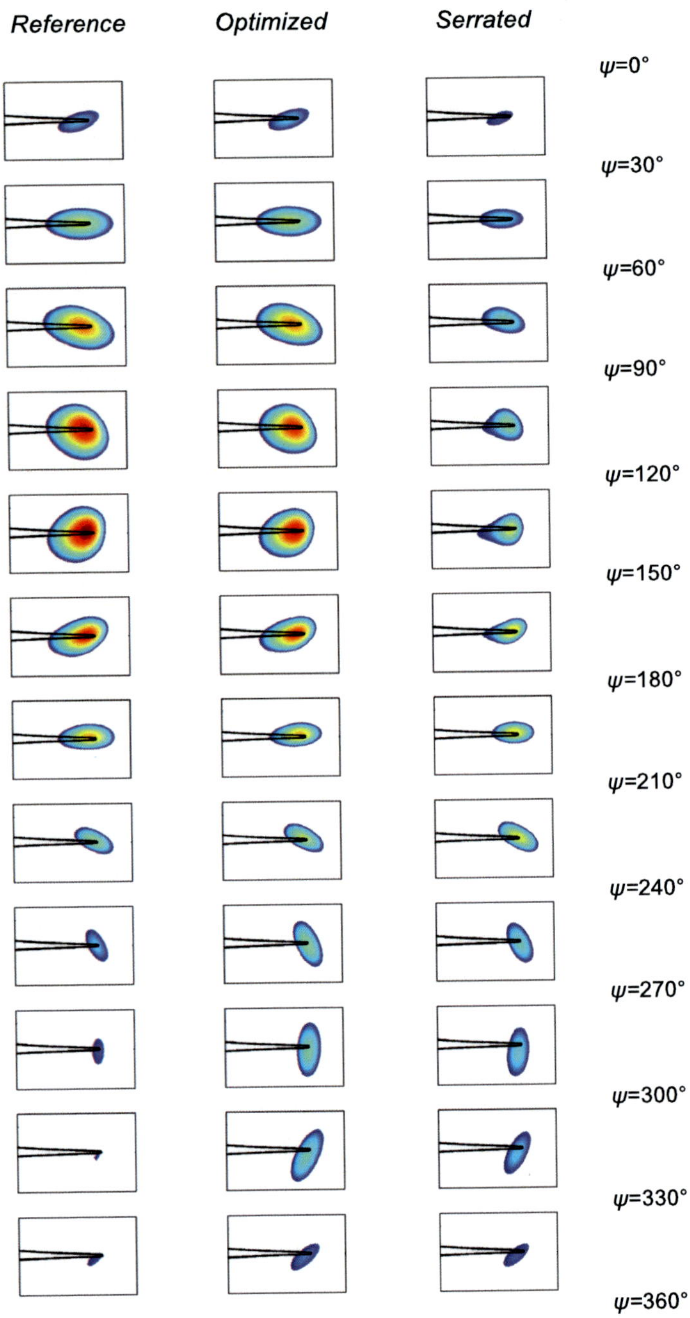

Figure 44. Dependence of blade noise on rotor azimuth (ψ=0° corresponds to 12 o'clock). The maps are summed over all frequencies and averaged over all wind speeds. The range of the colour scale is 12 dB and the maximum is the same for all maps

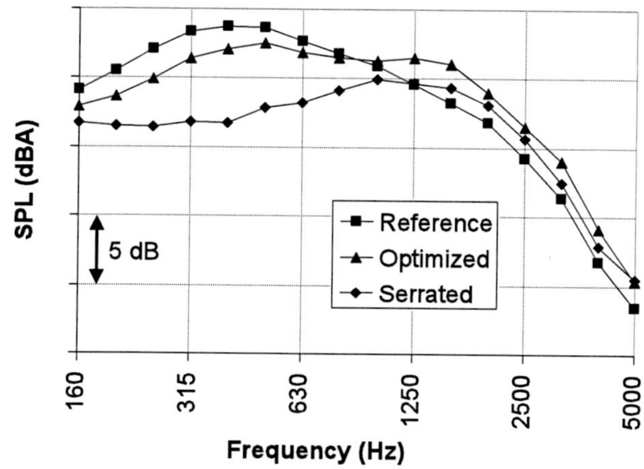

Figure 45. Integrated spectra for baseline blade and two modified blades, averaged over all wind speeds

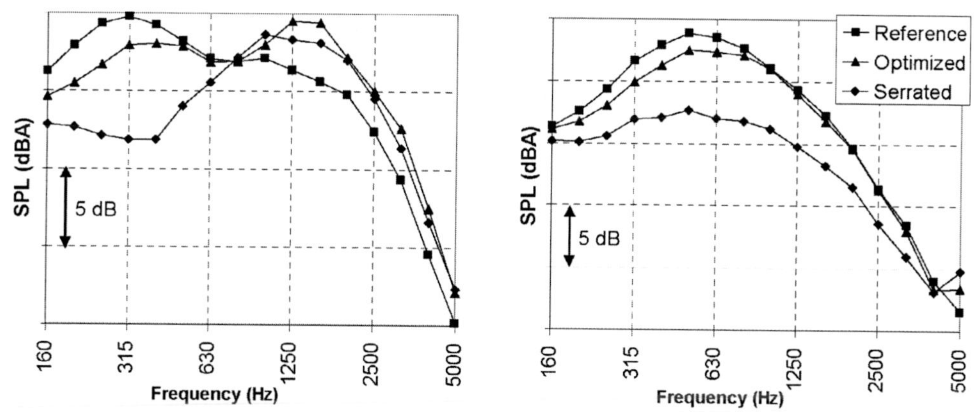

Figure 46. Integrated blade noise spectra at wind speeds of 7 m/s (left) and 10 m/s (right)

CONCLUSION

In this chapter we have addressed the primary sources of wind turbine noise. First, the different potential source mechanisms were described and the theoretical characteristics of flow-induced sound from wind turbine blades were explained. It was shown that the sound power scales with at least the 5th power of the flow speed, and that the directivity is generally not uniform.

Then, the noise sources on wind turbines were characterized experimentally by means of acoustic array measurements in the field. The test results indicated that the dominant noise source for typical modern large wind turbines is broadband trailing edge noise from the outer part of the blades. The swishing character of the sound can be explained by trailing edge noise directivity and convective amplification: each time a blade moves towards the observer, a higher sound level is perceived.

Next, it was shown that a semi-analytical, semi-empirical prediction method can accurately predict the characteristics of wind turbine noise. Good agreement was found between predictions and experiment, not only in terms of sound levels and spectra, but also with regard to swish and turbine noise directivity. Calculated noise footprints show that swish amplitudes up to 5 dB can be expected for cross-wind directions, even at large distance.

Finally, different noise reduction concepts were described and two of them, an optimized airfoil shape and trailing edge serrations, were tested on a full-scale turbine. Both blade modifications yield a significant trailing edge noise reduction at low frequencies, but also exhibit increased tip noise at high frequencies. Nevertheless, the experimental results show that wind turbine noise can be halved by means of serrations, without adverse effects on the aerodynamic performance. In future, even larger reductions may be achieved using flexible trailing edge brushes.

REFERENCES

[1] S. Oerlemans, Detection of aeroacoustic sound sources on aircraft and wind turbines, PhD Thesis University of Twente, The Netherlands, 2009 (http://doc.utwente.nl/67363/). ISBN 978-90-806343-9-8. NLR Report NLR-TP-2009-511.

[2] S. Wagner, R. Bareiss, and G. Guidati, *Wind Turbine Noise*, Springer, 1996.

[3] H. Schlichting, *Boundary-Layer theory*, McGraw-Hill, 1979.

[4] T.F. Brooks, D.S. Pope, and M.A. Marcolini, *Airfoil self-noise and prediction*, NASA Reference Publication 1218, 1989.

[5] R. W. Paterson, R. K. Amiet, C. L. Munch, *Isolated airfoil - tip vortex interaction noise*, AIAA paper 74-194, 1974.

[6] M. R. Fink, D. A. Bailey, *Airframe noise reduction studies and clean-airframe noise investigation*, NASA-CR-159311, 1980.

[7] S. Moreau, M. Roger, J. Christophe, *Flow features and self-noise of airfoils near stall or in stall*, AIAA paper 2009-3198, 2009.

[8] W.K. Blake, *Mechanics of Flow-Induced Sound and Vibration*, Volumes I and II, Academic Press, 1986.

[9] J.E. Ffows Williams and L.H. Hall, Aerodynamic sound generation by turbulent flow in the vicinity of a scattering half plane, *Journal of Fluid Mechanics*, Vol. 40, part 4, 1970.

[10] M.S. Howe, Noise produced by a sawtooth trailing edge, *The Journal of the Acoustic Society of America* 90, 1991.

[11] R.K. Amiet, Noise due to turbulent flow past a trailing edge, *Journal of Sound and Vibration* 47, Vol. 3, 1976.

[12] T. F. Brooks and T. H. Hodgeson, Trailing edge noise prediction using measured surface pressures, Journal of Sound and Vibration 78, 1981.

[13] M. S. Howe, A review of the theory of trailing edge noise, *Journal of Sound and Vibration* 61, Vol. 3, 1978.

[14] P. Migliore and S. Oerlemans, Wind tunnel aeroacoustic tests of six airfoils for use on small wind turbines, *Journal of Solar Energy Engineering* Vol. 126, 2004 (Chapter 7 of Ref. 1).

[15] M. Herr, C. Appel, J. Dierke, R. Ewert, *Trailing-Edge Noise Data Quality Assessment for CAA Validation*, AIAA paper 2010-3887, 2010.

[16] F. V. Hutcheson, T. F. Brooks, *Effects of angle of attack and velocity on trailing edge noise*, AIAA paper 2004-1031, 2004.

[17] S. Moreau, M. Roger, Competing Broadband Noise Mechanisms in Low-Speed Axial Fans, *AIAA Journal*, Vol. 45, No. 1, 2007.

[18] P. J. Moriarty, G. Guidati, P. G. Migliore, *Recent improvement of a semi-empirical aeroacoustic prediction code for wind turbines*, AIAA paper 2004-3041, 2004.

[19] J. W. Slooff, W. B. de Wolf, H. M. M. van der Wal, J. E. J. Maseland, *Aerodynamic and Aero-acoustic Effects of Flap Tip Fences*, AIAA paper 2002-0848, 2002.

[20] W. Dobrzynski, Almost 40 Years of Airframe Noise Research: What Did We Achieve?, *Journal of Aircraft*, Vol. 47, No. 2, 2010.

[21] S. Oerlemans, P. Sijtsma, B. Méndez López, Location and quantification of noise sources on a wind turbine, *Journal of Sound and Vibration* 299, 869-883, 2007 (Chapter 5 of Ref. 1).

[22] S. Oerlemans, M. Fisher, T. Maeder, K. Kögler, Reduction of wind turbine noise using optimized airfoils and trailing-edge serrations, *AIAA Journal* Vol. 47, No. 6, June 2009 (Chapter 8 of Ref. 1).

[23] IEC norm 61400-11, *Wind turbine generator systems - Acoustic noise measurement techniques*, 2002.

[24] S. Oerlemans, J. G. Schepers, Prediction of wind turbine noise and validation against experiment, *International Journal of Aeroacoustics*, Vol. 8, No. 6, 2009 (Chapter 6 of Ref. 1).

[25] T. Lutz, A. Herrig, W. Würz, M. Kamruzzaman, E. Krämer, Design and wind tunnel verification of low noise airfoils for wind turbines, *AIAA Journal*, Vol 45, No. 4, 2007.

[26] Herr, M., *Design criteria for low-noise trailing edges*, AIAA paper 2007-3470, 2007.

Chapter 3. Sound Propagation from Wind Turbines

Andrew Bullmore (UK) and Andrew Peplow (UK). Hoare Lea Acoustics.
AndrewBullmore@hoarelea.com, AndrewPeplow@hoarelea.com.

CONTEXT

In an attempt to enable more informed and consistent decision making with regard to the potential acoustic impact of proposed wind farm installations, noise arising from operational wind turbines is frequently required to be predicted at the planning stage. The results of such predictions are then objectively rated in various ways, depending on the country of application, in order to assess the likely acceptability of the proposed development in noise terms. Any such noise impact assessment procedure thus requires the appropriate modelling of the environmental sound field that may arise around operational wind farm developments.

Environmental sound modelling describes the process of theoretically estimating sound levels from a source (or sources) of interest at a location (or locations) of interest under a prescribed set of environmental conditions. However, the physical environment is usually not fixed but is instead characterised by constantly varying conditions. These changing conditions generally cause the actual sound field to vary in both time and space. Thus the output of any environmental sound model will only represent a particular 'snapshot' in time and space of the range of actual environmental sound levels that could in practice occur.

The modelling of environmental sound fields around wind farms generally involves consideration of the relatively complex scenario of sound energy being emitted from multiple sources (wind turbines) propagating via multiple complex transmission paths to result in sound fields at multiple receivers (most usually residential dwellings). It is the predicted sound fields at these receivers that are then used to inform some assessment decision as to the likely acceptability of the proposed wind farm development.

Any environmental sound modelling exercise may be encumbered by the total uncertainty arising from potential uncertainty in all elements of the calculation chain. Any such uncertainty creates the risk that the sound model may produce 'inaccurate' results that incorrectly inform the assessment decision. Given the scale of social and financial costs that may be associated with wind turbine noise impact assessment outcomes, it is important that the consequence of any uncertainties are fully acknowledged and appropriately accounted for by all involved parties.

In order to provide a representative output for the purposes of an environmental impact assessment it may be necessary to define several different sets of conditions for which predictions are to be produced. For example, the assessment may be intended to evaluate, for any given location, the 'worst case' noise levels that may be encountered, or the noise levels that occur for more than a certain proportion of the time, or the noise levels that occur for a specific combination of environmental conditions or over a specified time period, an example of the latter being the requirement to evaluate the long term L_{Aeq} or L_{DEN} level over a whole year.

The challenge facing the acoustic practitioner is to develop a tailored assessment strategy that strikes the appropriate balance between the scale of resources required for the predictive study and the costs (social, financial or otherwise) resulting from an incorrect decision that may arise from a compromised assessment. To achieve such a balance frequently requires the assessment design to 'begin at the end'.

That is, prior to selecting an appropriate strategy, the nature and scale of the decision to be made should be considered together with how the predicted noise data will be used to inform the decision. This type of approach provides the opportunity to focus resources on the elements of a study that have most influence on the decision for which the assessment is intended to inform.

The key consideration here is that environmental sound modelling exercises should be used in a way that appropriately informs the decision to be made. It is worth emphasising that there are two main risks when considering the implication of an incorrect assessment outcome for wind turbine sound. The first risk is that the predictions fail to represent the full scale of sound levels that may occur in practice, thereby leading to a situation where environmental sound levels breach acceptable levels with potential social and financial consequences. The second risk is that a prediction study overestimates noise levels experienced in practice, potentially leading to unnecessary development costs including lost energy generation opportunities over the life of the wind farm.

INTRODUCTION

This chapter addresses the issue of sound propagation from wind turbines to locations distant from those turbines. The term 'distant' is used here to imply propagation distances typically of around 1 km or more although it may be appropriate to consider lesser distances in some instances.

There already exist many detailed publications on the subject of sound propagation outdoors, some of which are general in their approach and some of which relate to very specific sources including, for example, road traffic, trains, military weapons, aircraft and industrial installations. It is not the aim of this chapter to replicate a comprehensive treatise on the subject. Rather, the aim is to present a discussion of outdoor sound propagation in the specific context of its practical application to wind turbines. For a more complete overview of the basic principles of environmental sound propagation the interested reader is referred to any of the referenced texts, and in particular those by Salomons [1], Attenborough, Li and Horoshenkov [2]and Bies & Hansen [3], all of which deal in detail with the fundamental theories of environmental sound propagation and all of which have been relied upon to inform the present chapter content.

The chapter is divided into a number of quite distinct sections that are aimed at leading the reader through the topic. The technical sections commence with a discussion of the physical factors affecting the propagation of sound energy through the atmosphere. There then follows an overview of the means by which these physical factors have been incorporated into practical noise propagation calculation procedures with varying degrees of complexity, whilst the final section considers the 'accuracy' of these procedures when applied to real world wind farm installations. However, before commencing these sections, it is considered useful to present an overview of the propagation of sound as it relates specifically to wind turbines.

In common with the majority of situations in which sound in the outdoor environment is concerned, consideration of wind turbine sound can be conveniently separated into three constituent elements: sound energy of a certain level and character is emitted from the **source** to then be modified as it **propagates** through the atmosphere to result in sound immissions of a modified level and character at different **receiver** locations. Chapter 2 dealt with the source and chapter 4 will discuss the receiver locations. This chapter focuses on the propagation path element of the problem and how the sound energy emitted from wind turbines may be modified in level and character as it travels from source to receiver through the atmosphere. Notwithstanding this focus of attention on propagation effects, the issue of source modelling will be returned to later in this chapter when the 'accuracy' of model outputs is considered.

The effect of the propagation of sound through the atmosphere is to modify the amplitude and phase characteristics of sound energy as it travels between source and receiver. What is required is a model of the propagation path such that the sound pressure at the receiver can be predicted for a known sound energy input. There exist two main means by which the filtering effects associated with sound propagation can be modelled.

The first approach involves modelling the real physical mechanisms which cause the changes in amplitude and phase of the sound energy as it propagates from the source to the receiver. The success of this approach requires that the effects of the real physical parameters (for example atmospheric temperature, humidity, wind velocity, etc.) governing each mechanism can be identified and quantified. The second approach requires that the relationship between the system output and the system input be established by means of measurement. An empirical mathematical model can then be developed that best relates the output to the input by minimising some error function. It is important to realise that the mathematical model in this latter case will not necessarily mimic the system through modelling the separate physical mechanisms involved. Rather, the model is required to represent the effect of the system as a whole through the adoption of possibly abstract parameters, although in practice these parameters are likely to have some dependence on real physical quantities. Alternatively, some combination of the two foregoing approaches may be considered.

When making the decision as to the 'best' approach to modelling the propagation of sound outdoors it is important first to consider the main controlling factors. Sound waves spread geometrically through a typically non-homogeneous atmosphere above a ground which may also have variable acoustic properties. All of these factors can affect both the phase and amplitude of the sound as it propagates. This leads to a received frequency spectrum which is heavily influenced by the types of ground surfaces between the source and receiver. Source geometry and type determines the amount of attenuation due to the geometrical spreading. Source and receiver positions determine the attenuation due to atmospheric absorption and the phase difference between direct and reflected sound at the receiver. Atmospheric conditions also affect the extent of atmospheric absorption. Furthermore, the ground may not be flat, which can lead to shadow zones or multiple reflections. Obstructions may be encountered along the propagation path causing shadow zones, diffraction and scattering of the sound waves. Wind and temperature gradients can cause sound to refract either upward or downward, also leading to shadow zones or multiple reflections, respectively. Additionally, fluctuations of wind and temperature by atmospheric turbulence may cause sound to be scattered into acoustical shadow zones.

The potential effects of the foregoing factors will be discussed in more detail later in this chapter. For the present, however, it is sufficient to recognise that many of the factors mentioned above can vary both spatially and temporally, even over relatively short distances and timescales. Thus sounds heard at the typical separation distances considered here of around 1 km or more can, in practice, often be heard to change in both level and character in matters of seconds. Such subjectively discernible changes can occur even when the source is constant, so it is clear that the changes must be due to variations in the propagation path. Herein lies one of the key problems of physically modelling environmental sound propagation: whilst it may be possible to accurately model the effects of varying meteorological parameters on a propagating sound wave, the output of even the most advanced model will always be reliant on the precision to which these parameters are known along the whole of the propagation path. A sound wave will propagate across a distance of 1 km in approximately 3 seconds. It thus follows that, in order to precisely model the effects of changes in meteorological parameters, the values of all significant controlling parameters must be known at every point and moment in time as the sound wave travels from source to receiver. Based on current, or even foreseeable, measurement technology such detailed information is unlikely to be available. Whilst scanning LIDAR or SODAR meteorological measurement systems offer the potential of much more information over that obtained using traditional instrumented meteorological masts, they still come nowhere close to providing a full description of the propagation path. It should also be remembered that meteorological conditions are just one part of the equation and that variable ground conditions along the propagation path should also be similarly quantified.

Of course, environmental sound arising from wind turbines is not unique in its variability: most environmental sound fields are inherently variable. The solution to overcoming this variability in measured data is frequently to perform some form of averaging. In particular the use of temporally averaged measurement indices (for example the L_{DEN}, $L_{Aeq,T}$ or $L_{A90,T}$) is particularly common. However, this averaging approach has generally found less favour when calculating environmental sound levels where it has traditionally been more common to calculate a 'typical worst case' level based on a relevant combination of 'worst case' parameters. The difference with wind turbine sound in the environment is that, not only is the propagation path variable, but the source acoustic characteristics can also vary with changes in meteorological parameters. The propagation distances considered are also considerably greater than those generally considered for other sources of environmental sound. The consequence of this combination of effects is that the variability of wind turbine sound levels at typical receptor locations can be considerably greater than that associated with other sources of sound. There is also the additional consideration that wind turbine sound levels are frequently of a similar overall level to the general level of environmental sound in an affected area. The variation of wind turbine sound levels may thus result in the sound of the wind turbines ranging from being audible to inaudible depending on the prevailing conditions and time of day. Thus the adoption of a 'worst case' type scenario may result in the conclusion that the operation of the wind turbines would result in an unacceptable level of sound. However, more detailed analysis may reveal that this level of sound would only occur for a small percentage of the time and could, in any event, be mitigated by operational constraints. Thus a wind farm that may initially have been concluded to have a significant noise impact on the local sound environment based on a simple 'worst case' analysis could ultimately be concluded to have an insignificant impact and thus be perfectly viable.

The foregoing discussion has illustrated the potential problems of developing an 'accurate' physical model of propagation effects. However, the discussion also sheds some light on the problems of developing an empirical model based on observed data, whereby the sound pressure resulting from the operation of the wind farm is directly related to the sound energy emitted from the wind turbines. Such a model may be relatively simple to develop in cases where there is certainty concerning the input and output data, but in the present case there may be significant uncertainty associated with the available data. The output data comprises the sound measured at the receiver locations. It has already been identified that the levels of sound arising from the operation of the wind turbines may be comparable with, or even below, the naturally occurring levels of environmental sound experienced at the location of interest. It is therefore a strong possibility that the measured sound levels may contain both wind turbine sound and general environmental sound. The potential corruption of the data with sound that is not wind turbine related can seriously limit the certainty associated with any model that may be derived from that data. Furthermore, the input data, which comprises the sound energy emitted by the turbines, is equally subject to uncertainty. The standard assumption adopted for wind farm assessments is that the sound power output of a wind turbine will be of a given level for any given hub height wind speed. This assumption is usually then extended to all wind turbines seeing the same wind speed at any given moment in time. In practice this is not the case as, in a multiple wind turbine installation, the wind conditions experienced by each individual turbine will generally be different at any given moment in time. Thus the input is, in most practical cases, not a single turbine emitting a known sound power but instead a number of turbines each emitting different sound powers whose characteristics are less certain. There also exists the problem of choosing the physical variables to observe and subsequently use in developing the propagation model. These may be restricted to what are considered to be the 'key' meteorological factors such as wind speed and wind direction monitored at a fixed location and heights, but there is then some a priori assumption as to the significance of these selected variables and the location(s) where these variables should best be observed. Ultimately, what will result is a model that provides some guide as to the received sound level likely to occur for any given combination of observed variables. However, the output can never be a single sound level that will occur, but rather a range of sound levels together with an associated estimate as to the likelihood of where the actual sound level may fall within that range.

As a consequence of the foregoing, many approaches to environmental sound modelling have developed, each of which can lead to different numerical outcomes. This, in conjunction with the challenges of objectively rating inherently variable environmental sound fields, can create a significant risk of incorrect assessment outcome when using environmental sound predictions for decision making purposes. It is important that this risk is understood and accounted for when proposing and designing prediction based studies. All parties that commission or rely on the findings of environmental sound predictions for wind farm developments should have an awareness of the process of modelling if an appropriate balance is to be struck between the scale of the investigation and the potential risks of an incorrect assessment outcome.

MODELLING WIND TURBINE ENVIRONMENTAL SOUND FIELDS – ACCURACY, VARIABILITY, UNCERTAINTY, AND RISK

The success of environmental sound modelling is often judged solely by consideration of the 'accuracy' of the model. However, whilst accuracy is an important consideration, success should not be judged on this basis alone. Judging the reliability of modelling through statements of decibel differences between predicted and measured sound levels fails to recognise the varied applications and complexity of situations for which predictions are used. If made without consideration of the context in which the results are to be used then such comparisons, whilst generally useful, may undermine the utility of the modelling exercise when the output results are, in fact, quite fit for purpose. To illustrate this point it is recalled that a noise model only produces an estimated 'snapshot' of a sound field that, in reality, is inherently variable in both time and space. This variability introduces a fundamental difficulty when attempting to define a relationship between a constant predicted value and a potentially widely varying range of sound levels that could be measured in practice. The reliability of sound modelling must ultimately be judged on its merits as an objective reference in decision making, and this judgement should be made according to the specific application and situation under consideration. Providing that modelling studies are used with an awareness of the relative benefits and limitations of predictions when compared to other possible references upon which a decision could be made, predictions can ultimately provide a reliable basis for decision making purposes. This requirement of information being fit for purpose places an onus on practitioners to deliver the outputs of predictive studies with accompanying contextual information that enables decision makers to understand how such information can be used. This includes the need for any end user to have an appreciation of variability in environmental sound fields and, equally important, the potentially quite distinct differences between 'variability', 'uncertainty' and 'risk' in the context of the assessment being undertaken: variability of wind farm environmental sound fields does not necessarily imply uncertainty, and likewise uncertainty need not always mean increased risk.

In the UK and some other countries environmental sound assessment of operational wind turbine installations relies on comparisons between the wind turbine sound and the background sound (the underlying sound in the absence of influence from the installation in question). Thus, these forms of assessment are burdened by the challenges of dealing with variability in two distinct sound fields. In terms of both the background and wind turbine sound, changing source and propagation conditions will ultimately gives rise to changes in both predicted and actual sound fields. The reflection of these changes in predictions or measurements is thus a representation of real trends exhibited in the sound field. To aid understanding of the causes of variability, the following sets out examples of how background and specific sound levels can vary.

Component	Examples of Component Variations
Source	*Background sound* • Changing natural sound source contributions e.g. wind effects, diurnal and seasonal variations in the composition of natural sound sources such as streams and wind disturbed vegetation and human activity related sources such as transportation and agriculture.
	Wind turbine sound • Operational characteristics e.g. continuous or intermittent operation, wind effects, load settings, noise constrained operation • Diurnal and seasonal effects due to changing atmospheric conditions, including varying wind shears • Directionally varying sound characteristics. • Varying sound characteristics and features such as tones and blade swish
Transmission	• Seasonal effects such as varying ground cover conditions (e.g. grass or ice) that affect absorption of propagating sound, or open windows/doors in summer for indoor assessment locations. • Varying atmospheric conditions e.g. wind speed/direction/shear, temperature changes with increasing height above ground, pressure, and humidity.

Variability of environmental sound fields presents one of the most critical challenges in any attempt to rate the field by objective means, whether by measurement or prediction. The variability raises the question of how an objective rating of a sound field in one set of conditions may relate to those occurring in other valid assessment conditions. The use of well intended measures to suppress the degree of variability exhibited in objective ratings may in fact undermine the validity of an assessment finding by failing to recognise how actual sound variability may influence the decision making process. For example, two separate sound fields each characterised by identical average noise levels could invoke very different assessment outcomes if one of the fields was characterised by constant levels consistent with the average value, whilst the other field was characterised by widely varying levels with maximum levels substantially above the average at certain times.

Given the reliance on knowledge both of the background sound and the wind turbine sound for the assessment of wind turbine installations, it is important to recognise how each sound field may vary. Furthermore, some assessment outcomes require long term average values to be derived thus necessitating a quantitative knowledge of the distribution of the varying levels. Importantly, when evaluating appropriate conditions for which an assessment should take account, it is necessary to identify any relationships that may exist between the variability exhibited by the background and the wind turbine sound fields. The combined effects of variability in both sound fields can lead to critical regions where the ability to discern the causes of variability becomes significant. In assessments where the priority is to determine if the wind turbine sound exceeds the background sound, the critical region occurs when the potential range of specific and background sound levels overlap. The situation therefore results from the combination of two factors, namely the relative magnitudes of the sound levels and the extent to which each may vary. An important consequence is that, outside of the potential overlap region, the wind turbine and background sound levels may exhibit even higher degrees of variability but, due to the increased relative difference between background and wind turbine sound levels, this increased variability may have no impact on the assessment outcome. In short, outside of the critical overlap region, there is no risk of incorrect assessment outcome despite potentially high levels of variability, even if the sources of this variability are not known. However, within the critical overlap region the risk of incorrect assessment outcome becomes significant as long as the sources of variability remain unknown.

Whether the specific sound of the installation in question is being compared against a background related limit or fixed value limit, unidentified or misunderstood variability can create significant uncertainty as to whether an objective sound field rating produced for one set of conditions will provide a complete representation of the sound field. Subsequently, if unknown variability or uncertainty overlaps some threshold value at which different assessment outcomes are triggered, there is a consequent risk of an incorrect assessment outcome occurring.

FACTORS GOVERNING ENVIRONMENTAL SOUND PROPAGATION

The past few decades have seen a plethora of research papers and publications devoted to gaining a better understanding of the propagation of sound outdoors. These have varied from highly theoretical approaches to practical, measurement based approaches. Three more recent publications, however, have brought together much of this research into a more easily referenced form. These are the texts by E.M. Salomons [1], K. Attenborough, K. Ming Li and K. Horoshenkov [2] and the various outcome reports of the EU funded Harmonoise/Imagine projects [8] to [14]. The interested reader is strongly recommended to refer to these publications for a more in-depth discussion of the topics that are necessarily discussed only in summary form in the present chapter, as well as to access a much more comprehensive bibliography on the subject.

In many practical cases, and certainly for wind farms comprising multiple wind turbines, the situation under study requires the calculation of propagation from multiple sources to multiple receiver locations. However, the seeming additional complexity of such multiple path analyses reduces to the problem of analysing for each receiver location the contribution of each source separately, then summing the various source contributions. Thus it is generally valid to limit the present discussion to that of calculating the propagation from a single source to a single receiver. It is only in the final analysis that the consequences of summing multiple source contributions at a single receiver location need be considered.

Given the above, in order to understand the possible factors affecting sound propagation, it is helpful to consider the 'simple' case of the propagation of sound from a single point source having a fixed sound power output to a single point receiver location. Despite the apparent physical simplicity of this case, it nevertheless allows the investigation of all the key factors that may affect the propagation of outdoor sound. In this respect the highly structured approach adopted by Salomons [1] can be very effectively applied to the development of an understanding of the basic controlling elements of sound propagation outdoors. By adopting this approach it is possible to build up a model that steadily introduces an increasing number of complicating effects into the propagation model. The following images, after Salomons [1], indicate schematically the manner in which these different factors are gradually brought into the discussion throughout the following sections of this chapter.

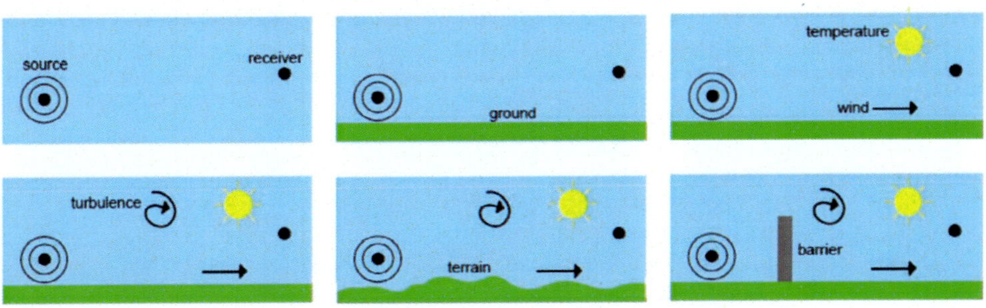

Figure 1. Diagrammatic representations, after Salomons [1], of the manner in which the basic factors affecting the propagation of sound outdoors have been isolated for individual consideration in the present document, and how these factors are gradually combined to build up a complete model of all the key factors required to be accounted for in order to make up a comprehensive sound propagation model

Reading from left to right, the individual diagrams in Figure 1 represent the effects on sound propagation of:

- geometric attenuation and atmospheric absorption in an unbounded homogeneous atmosphere;
- the additional effect of a bounding flat ground surface;
- the additional effect of in-homogeneities in the atmosphere through wind and temperature variations that affect sound speed profiles;
- the additional effect of further in-homogeneities in the atmosphere through turbulence;
- the additional effect of a non-flat ground surface;
- the effect of barriers.

Whilst is useful is to understand how each of these effects acting in isolation can modify the propagation of sound, it must always be remembered that in many instances the different effects can interact. Therefore, the cumulative effect under any given set of circumstances may not necessarily be the sum total of each effect considered in isolation. Good examples here include the potential reduction in the attenuation provided by an acoustic barrier when placed downwind of a source of sound, or the effect that an acoustic barrier can have on negating the effect the ground would otherwise have on the propagating sound.

Before looking at each of the situations illustrated in the Figure 1, it is useful to consider what is sought from the exercise of modelling sound propagation. In this respect it must be remembered that environmental sound fields are inherently variable, and part of this variability results from changes in the propagation paths from source to receiver. Such changes may occur over a wide range of timescales from seconds (due, for example, to local air turbulence or local wind and/or temperature effects) to seasons (due, for example, to changes in ground cover). The question must therefore be asked at the outset of any sound modelling exercise as to what quantity will best inform the desired outcome goal. For example, is the most informative quantity the instantaneous noise level expected under a specific set of conditions, or is it the worst case noise level expected under any foreseeable set of conditions, or is it possibly even the long term average noise level experienced over, say, a whole year? The other key determining factor when considering the appropriate choice of propagation model is the nature of the sound being considered. Is it important, for example, to assess the overall A-weighted noise level due to a source of noise that is broad band in character, or is it some tonal or much narrower band feature of the noise that is of interest: accurate modelling of the latter can present significantly greater challenges than the modelling of the former.

The task of representing in a simple, generalised form the propagation of sound from a source to some receptor location is usually approached by the use of a generic equation of the type shown in equation (1) below:

$$L_p(r) = L_w + \Sigma_i A_i \qquad [1]$$

where:

- $L_p(r)$ is the sound pressure level at a distance 'r' meters from the source;
- L_w is the sound power level of the source
- A_i is a series of modifying factors that either attenuate or enhance the transmission of the sound energy as it propagates from source to receiver

The modifying factors that are of relevance to wind turbine sound are categorised in the following discussion in five parts **A1 – A5.** It is the aim of this section to consider what the relevant modifying factors are, and then to identify the physical processes by which each of these factors may affect the propagation of sound energy. Only by developing this level of understanding is it then possible to determine the key physical parameters that each of the modifying factors depends upon and whether or not each factor is dependent on the frequency of the sound.

The Effect of Geometric Attenuation (A1)

The propagation of sound away from a simple point source radiating into an unbounded homogeneous atmosphere is subject to two attenuating effects. The first of these, often referred to as **geometric attenuation,** arises from the spreading of the radiated sound energy over a sphere of increasing area as the wave front propagates away from the source. As the total area of the spherical wave front increases in proportion to the square of the distance, r, from the source, the sound intensity and resultant sound pressure level decrease at a rate inversely proportional to r^2. Equation 2 provides the relationship between the sound power level of the source, L_w, and sound pressure level, $L_p(r)$, at a distance of 'r' meters from that source. The effect of this relationship in terms of the attenuation of sound pressure level with increasing distance is provided by equation 3. This equation shows that the sound pressure level, L_p, decreases by a constant 6 dB per doubling of distance away from the source. The attenuation arising from geometrical spreading is independent of the frequency of the propagating sound wave.

$$L_p(r) = L_w + 10\log[1/(4\pi r^2)] \qquad [2]$$

$$L_p(r_2) = L_p(r_1) + 20\log [r_1/r_2] \quad [3]$$

The Effect of Atmospheric Absorption (A2)

The second attenuating effect that occurs as sound propagates through a homogeneous atmosphere arises from **atmospheric absorption**. Sound waves comprise the regular and ordered oscillation of air molecules. The higher the frequency of the sound, the greater is the rate of oscillation of the molecules about their equilibrium position. This vibration of the air molecules results in two distinct dissipative mechanisms by which energy can be lost from the propagating sound wave. These mechanisms comprise frictional losses (sometimes termed 'classical attenuation', these include losses through viscous action and through heat conduction) and 'molecular relaxation' (these involve the interaction of water vapour with the resonance of oxygen and nitrogen molecules). Both of these atmospheric absorption processes are complex but, using the equations of fluid dynamics and statistical mechanics respectively, the overall effects on the attenuation of sound pressure with distance are calculable. As such, standardised techniques have been developed by which the attenuation of sound due to atmospheric absorption effects can be calculated. The most commonly adopted standard for this purpose is ISO9613-1 [15] which presents equations for the foregoing calculations together with summary tables of typical atmospheric absorption attenuation rates with distance. What is important to realise here is that the governing equations relating the losses to propagating waves are frequency dependent. This dependency largely arises from the increasing velocity gradients experienced by the oscillation of air molecules supporting the propagation of sound waves of increasing frequency. The controlling processes are also highly dependent on the temperature and humidity of the air, where again the magnitude of the influence of these parameters is frequency dependent, with higher frequency sound experiences greater atmospheric attenuation than lower frequency sound. Thus the relative contribution of higher frequency sounds to the overall A-weighted sound pressure levels at larger propagation distances becomes increasingly less significant.

The Effect of a Bounding Ground Plane or Commonly Ground Effect (A3)

The introduction of a ground plane into the situation considered above adds the possibility not only for a sound to propagate directly from the source to the receiver, but additionally for a secondary propagation path resulting from a reflection off the ground plane. This secondary propagation path can result in the occurrence of interference effects at the receiver location between the direct and reflected waves. Depending on the relative amplitudes and phases of the direct and reflected waves this interference effect may be either constructive or destructive: in other words the interference may result in either enhancement or attenuation of the received sound pressure level. The reader is referred to the texts by Salomons [1] and Attenborough et al [2] for a more detailed discussion of this complex effect, thus restricting the present discussion to the general principles involved and the potential magnitude of effects on received sound pressure levels. Figure 2 shows the geometry of the simple situation being considered, namely that of a spherical source radiating into a homogeneous atmosphere above a perfectly flat, homogeneous and locally reacting ground surface to a single receiver location which is also located above the same ground surface.

When considering the situation shown in Figure 2, the first point to recognise is that both the direct and reflected paths are subject to the geometric and atmospheric attenuating effects considered in the preceding section. The only difference between the direct and reflected waves at the receiver location is that the reflected wave has travelled a slightly longer path and, in the process, has been reflected from a surface. The precise relationship between the direct and reflected waves at the receiver critically depends on the following three factors:

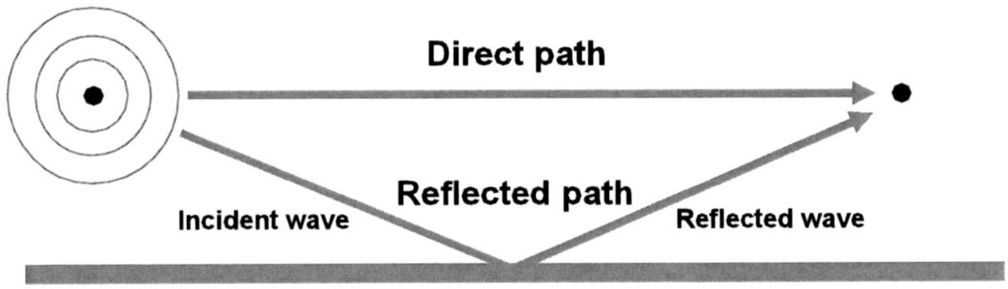

Figure 2. Simple geometry used for describing the effects of the ground on the received sound pressure levels arising from the interference between direct and reflected propagation paths from a point source to a single receiver, both being located in a homogeneous atmosphere above a ground plane.

- the difference between the direct and reflected path lengths (which is a function of the source and receiver separation distance and the heights of the source and receiver above the ground);
- the wavelength of the sound being considered (which is inversely proportional to its frequency and proportional to the speed of sound in the atmosphere);
- the effect that the ground reflection has on the amplitude and phase of the reflected sound wave relative to the incident sound wave.

It is useful to consider the extremes of what may occur to gain a basic picture as to the possible effects that the presence of the ground may have on the resultant sound pressure field at the receiver. This will be done in the first instance by considering the specific case of a point source which emits a pure tone at a frequency of 570Hz into a homogeneous atmosphere in which the speed of sound is $340ms^{-1}$. Thus the wavelength of the sound is 0.6m.

If the ground is perfectly acoustically absorptive then all the energy in the incident wave is absorbed by the ground and no reflected wave exists. Therefore there can be no interaction between the direct and reflected waves and the resultant sound pressure level at the receiver is that which would exist if no ground were present.

At the other extreme, if the ground is perfectly acoustically reflective then all the energy in the incident wave is reflected from the ground[1].

The consequence of the incident wave being perfectly reflected from the ground is that, at the receiver location, the resultant sound pressure is a combination of that arriving via the direct path and that arriving via the reflected path. The reflected sound pressure is reduced in amplitude relative to the direct pressure wave (by the ratio of the direct path length to the total reflected path length) and it also suffers a time delay (equal to the path length difference between the reflected and direct paths divided by the speed of sound in the atmosphere). It is the relationship between this path length difference and the wavelength of the sound in question that determines how the direct and reflected waves combine.

For the case of the 570Hz pure tone being considered in the present example, the wavelength is 0.6m. Thus if the path length difference is 0.0 m (which can only occur when the receiver lies in the ground plane itself), 0.6 m, 1.2 m or any other integer multiple of the sound's wavelength of 0.6m, then the pressure peaks or troughs of the direct and reflected waves will arrive simultaneously at the receiver location and the waves will interfere constructively. Ignoring any reduction in amplitude of the reflected pressure wave due to the difference in direct and reflected path lengths, for the case of the pure tone being considered this constructive interference will result in a pressure doubling, which equates to an increase in sound pressure level of 6dB. It is also worth noting here that the enhancement of incident sound pressure levels may, under some special circumstances, increase above 6dB. Where this occurs it is due to the presence of so-called 'surface waves' which are addressed separately at the end of this section.

[1] note that whilst, for ease of visualisation, the situation is represented here by a single ray path, in reality a point source creates a spherical wave front and therefore the reflection must, for all practical purposes, be determined from equations describing the reflections of a spherical-wave from a surface layer

In contrast to the above, if the path length difference is 0.3 m (i.e. half the wavelength of the sound) or any other odd integer multiple of the half wavelength, for example 0.9 m or 1.5 m,, then the arrival of a pressure peak of the direct wave will be matched with the arrival of a pressure trough of the reflected wave, and vice versa. Again ignoring any differences between the amplitudes of the direct and reflected waves, the two waves will destructively interfere. This destructive interference will theoretically result in zero sound pressure, which equates to an infinite reduction in sound pressure level. It should be noted, however, that even in this highly idealised example the difference in amplitudes between the direct and reflected waves, no matter how small, would not allow this perfect cancellation to occur and the reduction in sound pressure level would be limited to a finite level of typically around 30dB to 40dB.

For any other path length differences, depending on the actual difference involved compared with the wavelength of the sound of interest, some intermediate degree of partial constructive or destructive interference will occur and the increase or reduction in received sound pressure level will lie somewhere between the extremes just discussed. Note that the important parameter here is the difference between the path length difference (which is a function of source height, receiver height, and source to receiver separation distance) and the wavelength of the sound being considered.

Figure 3 illustrates the foregoing effect by showing the calculated ground effect attenuation for propagation over an acoustically perfectly reflective ground across the frequency range from 63Hz to 4000Hz. The figure shows the results for the case of a source and receiver separated by a distance of 500m, with the receiver being positioned at 1.5m height and source 50m height above the ground plane. The first attenuation dip occurs at the frequency of 575 Hz (see also Table 1) with a theoretical attenuation of around 40 dB. What is of interest here is the reduction in attenuation as the frequency of the sound moves away from the 575 Hz at which optimal destructive interference occurs. Inspection of Figure 3 on this basis reveals that the dominant ground effect dip centred around 575 Hz is approximately one octave band wide, with a theoretical maximum reduction at 575 Hz of around 40 dB. It is important to note, however, that this very large attenuation relates to a narrow band analysis such as may be undertaken if the sound of interest is dominated by a pure tone. Based on a third octave band analysis the maximum reduction of the sound pressure level in any one third octave band is less than 15 dB. These quoted ground effect attenuations compare to ground effect enhancement effects of up to 6dB due to constructive interference at frequencies between those at which destructive interference occurs.

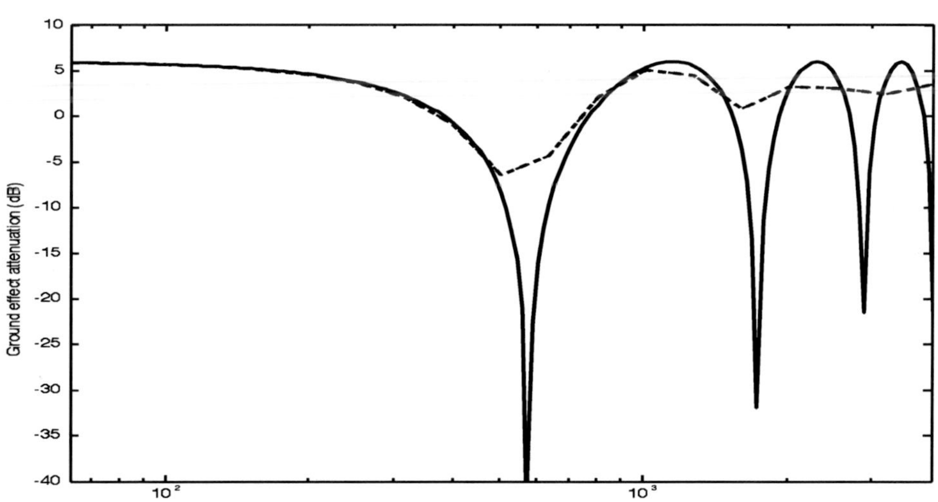

Figure 3. Calculated attenuating effects of an acoustically perfectly reflecting ground plane as a function of frequency from 63Hz to 4000Hz for the following conditions: source 50m height and receiver 1.5m height source to receiver separation distance 500m, ground plane perfectly acoustically reflecting. Results are shown both for discrete, narrow band frequencies and also for third octave band frequencies.

It remains to place into context the physical situations whereby path length differences large enough to allow significant attenuation of received sound levels may occur. Table 1 presents a list of path length differences for a number of different source/receiver heights and source/receiver separation distances along with the corresponding sound frequencies whose half wavelength matches the calculated path length differences. All calculations have been based on a sound speed of $340ms^{-1}$. The results of Table 1 indicate that the typical range of frequencies at which destructive interference may occur are, as expected, very dependent on the exact source/receiver configuration.

Table 1. Sample path length differences between direct and reflected waves for various source/receiver height and source/receiver separation distances. The last column also shows the lowest frequency of the sound having the same half wavelength of the path length difference of each source/receiver configuration based on reflection from an acoustically perfectly reflective surface. Sound speed c = 340 m/s

Source height, m	**Receiver height, m**	**Distance between source and receiver, m**	**Difference between direct and reflected path lengths, m**	**Frequency of sound with half wavelength equal to the path length difference, Hz**
50	1.5	500	0.30	570
50	1.5	1500	0.10	1700
50	1.5	3000	0.05	3400
80	1.5	500	0.47	360
80	1.5	1500	0.16	1060
80	1.5	3000	0.08	2120
80	4.0	500	1.26	135
80	4.0	1500	0.43	400
80	4.0	3000	0.21	797

The preceding examples have shown the theoretical extremes of what may occur due to the presence of a perfectly reflecting ground compared to the situation that exists in an unbounded atmosphere: namely a potential increase in received sound pressure level of up to 6dB and a potential decrease in sound pressure level of a much greater magnitude for perfectly reflective ground and zero change for perfectly absorbing ground. The examples have also provided a basis for understanding the basic physical mechanism behind the 'ground effect'; namely that it arises from interference effects between direct and ground reflected waves. The question that must now be addressed is what actually happens between these extremes in practical situations.

The acoustic properties for virtually all practically encountered ground surfaces lie somewhere between the two extremes of being a perfect absorber or a perfect reflector of sound[2]. What this means in practice is that, when an incident sound wave strikes the ground it is able to penetrate the porous surface. In penetrating this porous surface, and in overcoming the frictional energy losses in so doing, a proportion of the sound wave's energy is converted into heat energy. Thus the amplitude of the reflected wave is reduced relative to the amplitude of the incident wave by an amount that is determined by the energy transferred through frictional losses. Furthermore, the reflected sound wave can also experience a time delay, i.e. a phase shift, relative to that which would be expected from a perfectly hard reflective surface. This phase shift is due to the time taken between the sound wave

[2] One particular exception to this is water which can act as a nearly perfect reflector of airborne sound. For this reason the calculation of sound propagation over water using models that have been derived empirically based on measurements made over land can lead to significant errors [16] [29]

entering and exiting the porous surface. In short, the governing factor is the 'ease' with which the sound wave can enter and leave the porous surface, and this is defined by the so called 'flow resistivity' of the surface. Flow resistivity represents the ratio of the applied pressure gradient to the induced volume flow rate per unit thickness of material. Surfaces having a high flow resistivity do not easily admit sound waves due to their low surface porosity, for example water and concrete, whilst surfaces having a low flow resistivity more easily admit the incident sound wave due to their higher surface porosity, such as freshly fallen snow [2].

The manner in which the acoustical properties of the ground may affect received sound levels when considering environmental sound fields has received considerable attention over the past few decades. This is not only from a theoretical point of view but also through experimental validation of the models that have been developed, some of which involve many parameters to describe the acoustic properties of the ground. It is important to realise that not only may the ground's acoustic properties change as a function of frequency, but they may also vary as a function of angle of incidence of the sound. Furthermore, some investigations not only consider the local acoustical characteristics of the ground surface itself, but also whether or not the ground may exhibit an 'extended' reaction to an incident sound wave, i.e. whether a sound wave incident at one particular location can have an effect at another location due to the extended response of sound wave/surface interaction. These and other developments in the understanding of ground effects are covered in considerable detail in the recent publication by Attenborough et al [2] as key researchers in this area. The interested reader is strongly urged to refer to that publication for a more detailed discussion of this very detailed subject which is necessarily only dealt with on a summary basis here.

It needs, therefore, to now be considered how the foregoing factors associated with practically encountered ground surfaces may affect the situations considered earlier, where the principles of the ground effect were discussed by reference to the extreme example of an acoustically perfectly reflecting ground surface.

The results listed in Table 1 confirm the expected trend that the greater the path length difference between the direct and reflected propagation paths, the lower the minimum frequency at which destructive interference may occur. It is important to note here that an increase in propagation path length difference equates to an increase in the time delay between the arrival of the direct and reflected paths. One of the points already made about porous ground surfaces is that they introduce an additional time delay in the reflected propagation path due to the phase lag between the incident wave entering the porous surface of the ground and the reflected wave leaving the porous surface layer. This additional delay therefore makes the effective path length difference between the direct and reflected paths even greater than would be the case for the same geometrical configuration of source and receiver over a perfectly reflecting ground surface. For this reason it may reasonably be expected that the minimum frequency at which destructive interference occurs will be reduced for propagation over a porous ground compared to that predicted over a perfectly reflecting ground for the same source/receiver configuration. This reduction in frequency will be directly related to the effective increase in path length difference between the direct and reflected paths, taking into account the additional time delay introduced at the surface of the porous ground itself. This additional delay is related to the acoustic impedance of the ground surface relative to the acoustic impedance of the air, and in particular to the flow resistivity of the ground.

Several models have been developed for the derivation of ground surface acoustic impedance characteristics. Again it is not proposed to go into any detail on these models here. Instead the interested reader is again directed to the recent texts by Attenborough for more detailed information on this matter. However, for reference purposes, Figure 4 shows the calculated ground effect attenuations for the same physical configuration of source and receiver as detailed in Figure 3, but now with a non-acoustically 'hard' ground surface having a flow resistivity corresponding to that of typical grassland, or 200 kPA.s m^{-2}.

The results of Figure 4 confirm that the additional time delay between direct and reflected waves introduced by the porous surface results in an increase in the effective path length differences between the direct and reflected waves and thus a decrease in the frequency of the main ground effect dips. However, not only have the frequencies of these dips been lowered, but their maximum depths have also been reduced. This is due to the fact that the reflected wave is of significantly lower amplitude when compared to the direct wave as a result of losses incurred on reflection from the ground surface. Thus the destructive interference between the direct and reflected waves is less complete than for the case where the reflected wave was reflected from an acoustically perfectly reflective surface, even for the case when the phase match between the two waves is perfect. Conversely, the potentially significant enhancements in received sound levels due to the ground effect shown for the acoustically perfectly reflective surface are also reduced in magnitude as the frequency increases.

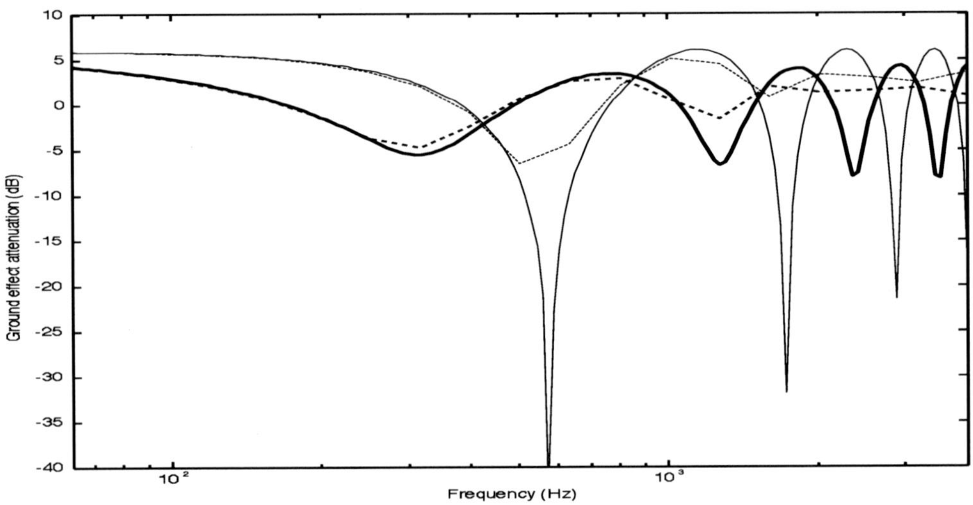

Figure 4. Calculated attenuating effects of an acoustically partially absorbing ground plane as a function of frequency from 63Hz to 4000Hz for the following conditions: source and receiver height both 2m, source to receiver separation distance 32m, ground plane flow resistivity 200 kPa.s.m^{-2}. Results are shown both for discrete, narrow band frequencies and also for third octave band frequencies. The results obtained from Figure 3 based on an acoustically perfectly reflecting ground surface are also shown as the thinner lines for comparison.

One final effect of spherical propagation of sound over a porous ground surface is the possible existence of a 'surface wave' in addition to the direct and reflected waves thus far considered. This facet of sound propagation is dealt with in some detail by Attenborough who concludes that the surface wave can only propagate effectively if the ground possesses certain special impedance characteristics.

Physically, the surface wave comprises the motion of air particles in an elliptical pattern close to the surface region. The wave is separate from the main body of the propagating sound wave and moves at a different (slower) speed than it. It can only exist if the acoustic impedance characteristics of the ground surface are such that its reactance is considerably larger than its resistance. This condition means that the induced sound field will suffer relatively little dissipative energy loss as a result of its interaction with the surface. The satisfaction of this condition relies on a combination of the ground surface's material properties, and the frequency and angle of incidence of the sound. It is the presence of this special impedance condition that allows the elliptical particle motion to be supported, owing to the combination of particle motion parallel to the surface with particle motion in and out of the surface pores in a direction that is normal to the surface. Given the requirements for this interaction condition to exist to support the surface wave, it can only exist close to the surface, where the term 'close' is defined relative to the wavelength of the sound being considered.

As pointed out by Attenborough, the resistive component of most practically encountered ground surfaces is such that any surface waves decay rapidly as they propagate along the surface. However, for certain special conditions surface waves have been positively identified and measured in practical outdoor experiments. One example of this was in the experiments reported by Albert [5] which involved the use of a pistol shot as the source to investigate the propagation of impulses over newly fallen snow on frozen ground. This particular scenario, of a thin porous surface layer over an effectively hard backing, is precisely that which is required to support particle motion in and out of the pores normal to the surface whilst also offering the low resistive losses required to allow the surface wave to propagate over larger distances with a low rate of attenuation. Due to the slower propagation of the surface wave relative to the direct and reflected propagation paths, Albert was able to positively identify sound energy arriving at the receiver location via the surface wave separately from the energy arriving via the direct and reflected paths.

The consequence of the above is that the surface wave is a real effect and its presence can, under certain special circumstances and at low frequencies, result in an enhancement of sound pressure level measured close to the ground surface. This means that the theoretical maximum limit of +6dB enhancement of sound pressure level above a ground surface resulting from the constructive interference of direct and reflected waves alone may, under these special set of circumstances, be exceeded.

The Effect of Wind and Temperature on Sound Speed (A4)

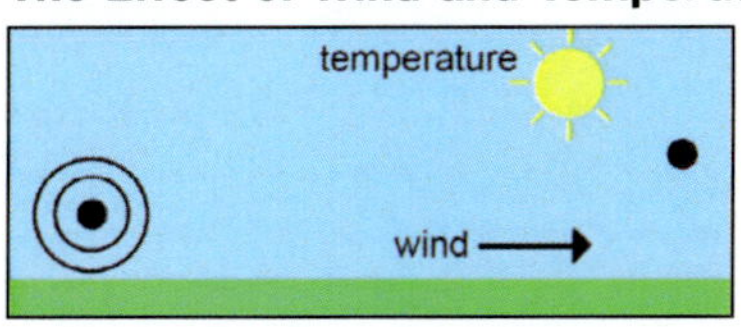

The discussions of the preceding two sections have accounted for the attenuating effects of the atmosphere in terms of atmospheric absorption, but they have somewhat simplistically assumed that the atmosphere is acoustically homogeneous. In virtually all practically encountered situations this isn't the case. The wind, for instance, is slowed down by friction as it passes over the ground surface thereby resulting in lower wind speeds at lower heights above ground level. Also, convection within the atmosphere and the radiative heating/cooling effects of the mass of the ground often result in quite different air temperatures close to the earth's surface than at heights even a few meters or tens of meters above the surface. It is therefore true to say that, during most weather conditions, both the temperature and wind vary with height above the ground.

As both wind and temperature affect the speed of sound in air, these variations result in an acoustically non-homogenous atmosphere with sound speeds either decreasing or increasing with height above the earth's surface. Also, not only is the atmosphere non-homogeneous with regards height, but the presence of a wind direction also affects the directional propagation of sound. This section discusses these in-homogeneities and their effects on the propagation of sound.

In the presence of a vertical gradient of sound speed, sound waves are refracted in the direction from higher sound speeds to lower sound speeds. Depending on the particular circumstances, there are two possible curving trends, downwards or upwards.

Downward Refraction

The downward propagation of acoustical paths usually occurs under a temperature inversion (stable atmospheric conditions) or under downwind propagation. The resulting propagation of sound under these two atmospheric phenomena is essentially very similar, but could differ if the shapes of the wind and temperature vertical profiles are different. Any sound speed function which increases with increasing height will generate a downward sound propagation, as shown by Figure 5.

However, depending on the relationship between sound speed and height, the downward propagation will have specific properties. A linear sound speed variation with height will make sound rays travel from source to receiver along circular arcs with centres of curvature lying on a horizontal

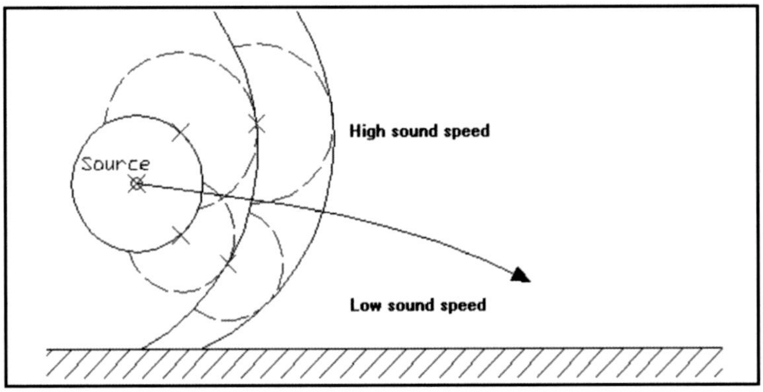

Figure 5. Sound curvature under temperature inversion and/or downwind conditions

line at a distance $1/\gamma_T$ from the surface (γ_T being the rate of increase in sound speed with height). In contrast to the linear case, logarithmic sound speed profiles do not generate a homogeneous sound path throughout the atmospheric boundary layer. Sound rays are almost circular arcs at high altitudes, where the logarithmic sound speed function is nearly linear, but at heights close to the ground it becomes much more difficult to assign an analytic description of their behaviour as the logarithmic dependency demonstrates a less linear behaviour. Other, less common, profiles may also be considered such as power or exponential relations.

The main effect of downward sound propagation is the possible generation of multiple ground reflected rays from source to receiver. Whilst it is not the purpose of the present sections to establish preferred modelling techniques for representing the features being discussed, it is nevertheless useful to use one of the techniques, ray tracing, to demonstrate what the typical magnitude of effects on received sound levels may be as a result of the downward refraction of sound waves as they propagate through a non-homogeneous atmosphere.

Using ray tracing it is possible to evaluate the trajectory through the atmosphere of the sound waves as they propagate from source to receiver. It is also possible to account for the effects of reflection of these waves from a ground plane of known, finite impedance. Finally, it is possible to evaluate the cumulative sound intensity of waves at the receiver. The method works for the case considered here of a linear sound speed profile with positive slope, thus relating to downwind or temperature inversion conditions. Under these conditions, the ray paths are circular concave arcs with centres of curvature lying on a horizontal line at a distance $1/\gamma_T$ below the surface. Making use of this geometrical property it is possible to establish all possible ray paths from source to receiver. For the general case of finite height for source and receiver, h_s and h_r, there are a total of four reflected ray paths for each number of reflections per ray, n, greater than one. Figure 6 shows the case of n = 1, where the ray group consists of 3 ray paths instead of 4. Figure 7 shows a set of ray paths for n = 2, where now the four rays per group are evident.

More detailed analysis indicates that if the height reached by a predicted ray, H, is smaller than the minimum height of either the source or receiver, then the ray cannot exist. As the zenith height of rays is a function of ‘n’, this fundamental requirement therefore defines an upper limit to the number of possible reflections that may exist under any given set of conditions. The relevant controlling parameters are the source and receiver heights and their separation distance, together with the vertical sound speed gradient.

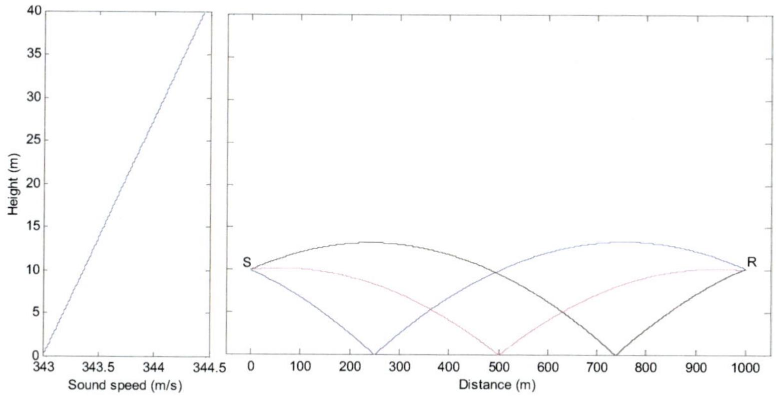

Figure 6. The grouping of individual rays according to ray theory. In particular, 3 ray paths form the group of rays that reaches the receiver when only one reflection from the ground (n=1) occurs. With n≥2 each group is formed by 4 rays instead of 3.

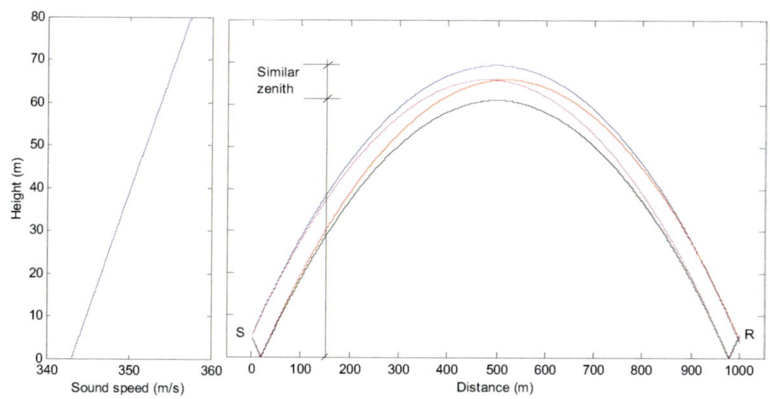

Figure 7. The grouping of individual rays according to ray theory. In particular these 4 ray paths form the group that reaches the maximum height at zenith.

Figures 6 and 7 illustrate the number of possible ray paths where the source and receiver are at similar heights above the ground. For the particular case where the source and receiver are 80m and 1.5m above the ground respectively it can be shown that a number of curved ray paths or sound paths are possible in strong downwind conditions. Where the source is elevated and the distance between source and receiver is 1000m, apart from the singular ground reflection near the receiver two ray paths may undergo 1 ground reflection, four ray paths take 2 reflections and two further ray paths undergo 3 ground reflections between source and receiver. As the receiver distance from the source decreases the total number of possible ray paths also decreases. For example at 500m there may be two ray paths associated with two ground reflections to the receiver plus the single ground reflection ray but at 100m only the simple "image" ground reflection ray exists.

Once the number of possible ray paths and reflections are known, it is possible to calculate the intensity and excess of sound pressure level over neutral conditions at the receiver through the summation of the contributions of each ray. This approach was first developed by Embleton [4]. For the case of a reflecting hard ground surface and an infinite sound ray path situation ($h_s = h_r = 0$) it can be determined that the maximum correction to be added in an inversion/downwind situation over a neutral one is approximately + 2.2 dB. A lower increase than 2.2 dB is obtained with soft ground by assuming a reflection coefficient less than unity for the intermediate reflections at the ground surface.

There are clearly limitations to the foregoing conclusions, such as the simplifying assumption of a linear sound speed profile and the assumption that this same profile exists along the whole propagation path from source to receiver. However, the analysis places into context the typical upper range of the increase in overall sound pressure level that may be expected to occur as a result of propagation under downwind or temperature inversion conditions when compared to propagation between the same source and receiver under neutral conditions.

It is a frequent subjective observation that downwind propagation appears to increase received sound levels significantly, whereas the foregoing analysis would suggest otherwise. In addition, reported measurements [18] [33] have confirmed that the effect of downwind propagation results in only a marginal increase in overall noise levels of typically less than 3 dB when compared to propagation under acoustically neutral conditions. The question must therefore be raised as to the cause of this apparent discrepancy between subjective perception and measured/predicted scenarios. The details contained in the following section will show that received noise levels under upwind propagation conditions can be 10 dB or more lower than the received noise levels under neutral conditions. It is thus highly probable that, when claims of very significant increases in received noise levels under downwind conditions are made, the comparison is actually being made with the greatly reduced received noise levels under upwind conditions rather than those experienced under neutral propagation conditions.

One additional point when considering the foregoing conclusions is that they relate only to broadband overall noise levels. For pure tones or much narrower bandwidth noise the underlying principles of constructive and destructive interference are just the same as already discussed in some detail in relation to interference between direct and ground reflected waves. Therefore, whilst the increase in the number of sound ray paths between source and receiver generally results in the moderate amplification of the received overall noise levels, the varying lengths of the reflected rays and their several interactions with the ground can also cause more significant constructive and destructive interference effects at some frequencies.

Upward Refraction

When the sound speed decreases with height, the sound rays are bent upwards away from the ground, as depicted by Figure 8. For typical sound speed profiles there is a limiting ray leaving the source which grazes the ground. Above this limiting ray the sound field is composed of direct and ground-reflected waves, whereas below the limiting ray there is an acoustical shadow in which sound waves do not exist, at least according to ray theory, as shown in Figure 9. In practice, however, some sound energy does penetrate these shadow zones.

A number of different studies have considered in greater detail the process of sound penetration into shadow areas. Salomons [1] considers that sound waves are mainly scattered into shadow regions by the refraction of sound waves caused by small random changes of the propagation direction as a result of local air turbulence, as discussed in greater detail in the next section., A diffractive penetration mechanism also causes some transfer of sound energy into the zones. This diffractive mechanism is described as being analogous to the diffraction of sound waves at the top of an acoustic barrier, beyond which a shadow zone where no sound energy penetrates should theoretically exist. However, such theoretical conclusions are based on frequency independent ray theory. Modelling based on wave theory supports the presence of a frequency dependent diffraction of sound energy into the shadow zones, with low frequencies being diffracted to a greater extent than higher frequencies, just as in the analogous case of diffraction over barriers. Due to a combination of the foregoing effects, practical attenuations in shadow zones are limited to no more than around 20 dB.

It is difficult to accurately quantify the potential magnitude of the various effects referred to above from measurements. This is due to the difficulties associated with separating out the different effects and also in quantifying the controlling parameter (sound speed) along the entire sound propagation path.

An alternative approach is to model the effects theoretically. In practice it is known that any measurement of environmental sound over time will yield a spread of sound levels distributed in some manner about a central tendency. To be useful in establishing sensitivity to meteorological parameters, any theoretical model must be capable of adequately deriving both this central tendency and the scatter of sound levels about this central value. Using such a model enables a parametric sensitivity analysis to be performed to quantify the effects of changes in each meteorological parameter on received sound levels. Measurements of the statistical distributions of controlling meteorological parameters can then be input to the model to estimate the potential variability in received sound levels under practically encountered conditions.

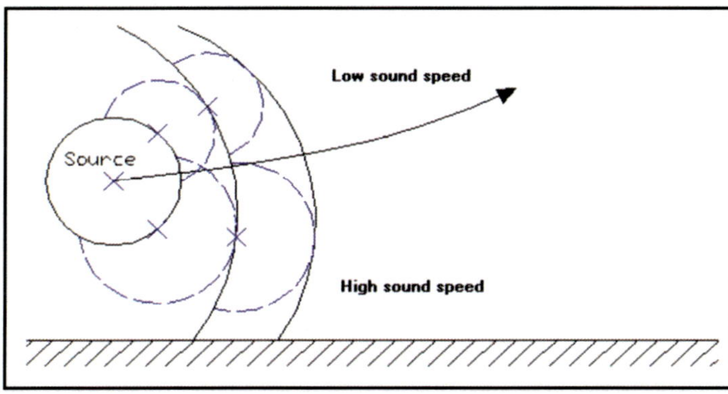

Figure 8. Sound curvature under temperature lapse and/or upwind conditions

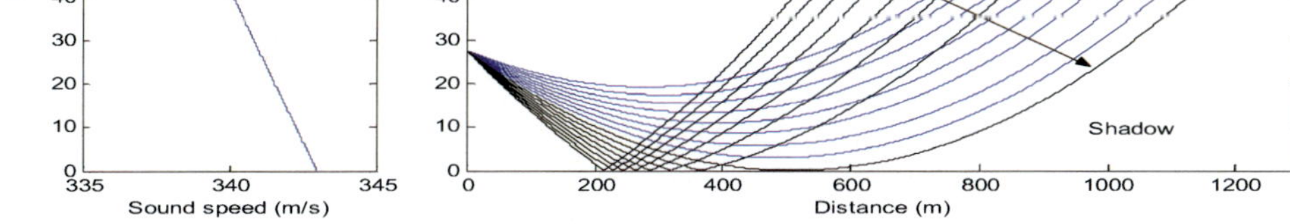

Figure 9. Sound propagation ray paths under temperature lapse and/or upwind conditions

In conclusion to this section, refraction in the atmosphere caused by vertical sound speed gradients (be these caused by vertical wind speed or temperature gradients, or some combination of the two) may result in increases in overall sound pressure levels of typically 2 dB for the case of positive vertical sound speed gradients when compared with acoustically neutral propagation conditions, but decreases in overall sound pressure levels of 10 dB or more, but typically no greater than 20 dB in all practical circumstances, in the case of negative vertical sound speed gradients.

The Effect of Turbulence

Whilst the discussion of the preceding section has accounted for the effects of sound speed gradients, it has assumed that these gradients are constant, both in time and space. In reality air turbulence introduces perturbations in local sound speed, and therefore the assumed sound speed profiles may only be considered as spatial averages over typical periods of 10 minutes or more. Turbulence arises from atmospheric instability of which there are two responsible types: shear and buoyancy [2]. Shear instabilities arise primarily from the disturbance of the air by the wind and are therefore prevalent in higher wind conditions. Buoyancy effects arise due to the presence of larger temperature differences between the ground and the air such that convective air circulation occurs. Thus convective turbulence tends to be most prevalent when the ground is much warmer than the air immediately above it such as on a hot sunny day.

The effect of turbulence on the propagation of sound is that, as a wave front passes through a turbulent region, it sees a local change in the sound speed. As already discussed in the previous section, any change of sound speed will result in local refraction, i.e. a local change in direction, of the wave. By its very nature, however, turbulent effects are quite random. Therefore, unlike the systematic upwards or downwards refraction that occurs under conditions of a steady change in temperature with height, turbulence induces random small scale changes in the direction of the propagating wave. This effect is generally termed 'wave scattering' as it scatters the wave energy about its mean direction propagation.

The extent to which any wave is scattered by turbulence depends very much on the relationship between the wavelength of the wave and the length scale of the turbulence, taken together with the turbulence intensity. As a guide, the typical length scales involved range from around 20 m for frequencies of 100 Hz to 0.2 m for frequencies of 10 kHz.

One of the features of turbulent scattering is that it is a two way process: just as energy is scattered out of the direction of propagation, so too is energy scattered back. Thus one of the effects of turbulence is that it results in short term fluctuations about the mean received sound pressure level. It is sometimes assumed that the further away the receptor point is from the source, the greater the amplitude of these short term fluctuations will become. However, the full effects of turbulent scattering of sound energy are realised relatively close to the source and then do not increase above this with distance.

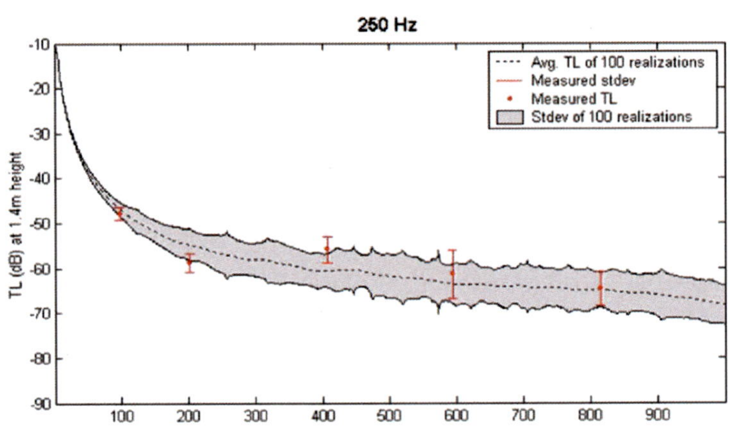

Figure 10. Short term variations in transmission loss (TL) due to the effects of turbulent scattering plotted as a function of distance from a pure tone sound source. Both measured and predicted results are shown for a frequency of 250Hz [34]

Figure 10 demonstrates this feature by showing a comparison of measured and calculated variations in one hundred samples, each one second long, of the received sound pressure level as a function of distance up to around 1 km away from the source for a discrete frequency of 250 Hz. The results of both measurements and predictions, which are based on a Parabolic Equation model incorporating the effects of turbulence through a Monte Carlo simulation, show two significant features. First, once the effects of turbulent scattering have developed with distance, which in this example is at about 200 m, the magnitude of the effect thereafter remains relatively constant. Second, the standard deviation of the received sound pressure level is typically no greater than around 10 dB, even at 1 km. For overall levels of broadband sound these variations will be significantly less, with the standard deviation of short term variations in received sound levels being more likely to be around 5dB.

Turbulence also impacts on other facets of sound propagation. All the effects thus far considered have relied on idealised descriptions that make assumptions concerning the geometry and temporal stability of the situation. The discussion of the ground effect, for instance, has assumed perfectly spherical wave fronts when considering the combination of direct and reflected waves. This has resulted in conclusions of very significant ground dip excess attenuations of up to around 40dB for pure tone sources. The introduction of turbulence to the analysis means that, even though the physical geometry of the source and receiver may provide the effective path length difference between direct and reflected waves required to maximise the effects of destructive interference at the receiver, turbulent effects may change local sound speeds and also cause local angles of incidence on the ground surface to vary, thereby causing short term variability in the effective path length. The resultant effect of this variability is that the precise conditions required for the exact phase matching of sound travelling via multiple propagation paths will never be maintained. Thus the maximum excess amplifications and attenuations resulting from constructive and destructive interference effects will, in practice, generally be limited to lower levels than those that theoretically exist in the absence of any turbulence. However, whilst turbulence will generally limit such effects to below their theoretical maxima, the special case of tonal noise propagation needs separate consideration. Constructive or destructive interference of tonal noise is highly sensitive to the precise relationship between the wavelength of the tone and the difference in path lengths between multiple propagation paths. The presence of turbulence may cause the path length differences to vary with time. Due to the sensitivity of tonal noise to the path length difference, the resultant effect is that tonal noise may vary considerably in level as turbulence causes the condition for exact constructive interference to occur and then to not occur, possibly within periods of seconds of each other. Thus tonal noise and, albeit to a lesser degree, band limited noise from wind turbines is always likely to be more variable than broad band noise over shorter time periods.

The short term variability in received sound pressure levels due to local changes in propagation paths is one of the effects of turbulent scattering. It is also possibly the effect that can be most easily appreciated from a subjective point of view. One manifestation of the effects of turbulent scattering is that of standing some distance from a source that varies in audibility over periods of a few seconds, with certain frequencies bands of the sound appearing to be enhanced more than others. The effect is particularly evident with public address systems which, when heard at large distances from the source, can vary between being clearly audible with intelligible speech to being barely audible, with these changes occurring over periods of just a few seconds. In such cases the effects of turbulence close to the ground can have particularly significant effect due to turbulence modifying the phase shift associated with ground reflected waves. The resultant short term variations in the effective path length differences between direct and reflected waves causes the dominant frequency ranges for constructive and destructive interference to vary over short periods of time, sometimes with subjectively quite noticeable effects.

Another area where turbulence plays a major role in modifying the conclusions presented in the earlier sections is in the propagation of sound under upwind and/or temperature lapse conditions. This scenario has already been shown in Figures 8 and 9. Under these conditions sound waves are refracted

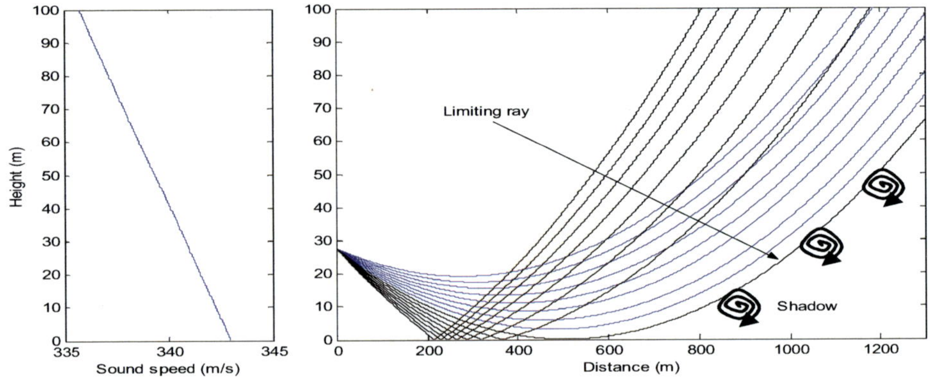

Figure 11. Sound curvature under temperature lapse and/or upwind conditions, with the additional effects of turbulence shown which scatter sound energy into the 'shadow' zone.

upwards such that, in theory at least, beyond some limiting ray there is a 'shadow' region where no sound energy from the source can enter, thereby resulting in an infinite excess attenuation due to the effects of the refracting atmosphere. What happens in practice is that turbulence causes scattering of sound energy into the shadow zone. Whilst this scattering may only result in a small fraction of the sound energy leaking in to the shadow zone, its effects in limiting the excess attenuation to more realistic levels is very significant. The effect may be likened to the effect of opening a sound insulating window or door by a tiny amount: the reduction in sound insulation performance is quite dramatic. The magnitude of the effect of scattering sound energy into shadow zones is both frequency and situation specific. However, in practical terms it will typically limit the excess attenuation of overall sound levels due to an upward refracting atmosphere to no more than around 20dB. The effect is shown in Figure 11.

In conclusion to this section, the effects of turbulence potentially impact on all areas of sound propagation. All the foregoing analyses have assumed that the sound speed profile remains constant between the source and receiver, both in time and space, also ignoring any localised changes in sound speed. However, it is highly improbable that atmospheric conditions would remain constant during the course of the propagation of sound waves over the practical distances considered in wind turbine noise assessments, there will always be some perturbations. Thus turbulent effects will always limit the extreme excess attenuations predicted to arise from idealised models (be these due to ground effects, barrier effects, refraction effects, or whatever other effects may be considered) if those models do not take any account of turbulence and other local variations in sound speed. Of course the consequence of this conclusion is that, in order to model any scenario accurately, the atmospheric factors affecting local sound speed variations must be known precisely at every moment and at every point along the propagation paths. Clearly this is an onerous requirement that is outside the capabilities of any existing measurement schemes. It is therefore inevitable that any sound propagation modelling exercise must include some degree of averaging in its approach, certainly temporally and possibly also spatially, coupled with a statistical estimate as to the degree of variation that may be expected to occur around the calculated central tendency.

The Effect of Barriers and Terrain (A5)

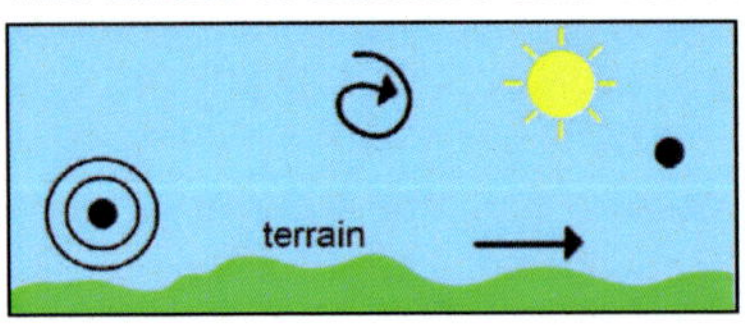

A barrier is any solid object located between a source and receiver. The presence of the barrier interrupts the sound propagation path between source and receiver such that the received sound pressure level at the receiver is reduced compared to the level that would otherwise have resulted from the same source.

Barriers may be introduced purposefully for the attenuation of sound, or they may occur as natural features such as terrain. The purpose of any introduced barrier is generally to interrupt the direct sound propagation path. In order to achieve this the barrier will normally be designed such that it wholly interrupts the direct line of sight between the source and receiver.

Methdologies for calculating barrier insertion losses are well developed and available in numerous published texts. For this reason they will not be discussed in any greater detail here. Instead, this section will concentrate on the interaction of barrier effects with the other sound propagation effects already considered, as it is often the case that calculated excess attenuations due to barriers in a neutral atmosphere can be significantly reduced under conditions commonly encountered in outdoor sound propagation. It is also the case that the introduction of an acoustic barrier into a propagation path may have an influence on other excess attenuations. Most notably, the removal of a ground reflected path from the situation can result in the total loss of any ground effect.

The excess attenuation provided by a simple, infinitely long, acoustic barrier under neutral propagation conditions, as shown schematically in Figure 12, is a function of the following variables:

- the distance from the source to the barrier
- the distance from the barrier to the receiver
- the heights of the source, receiver and barrier
- (all the above combine to make up the path length difference between the direct path in the absence of the barrier and the shortest path over the top of the barrier)
- the frequency (wavelength) of the propagating sound

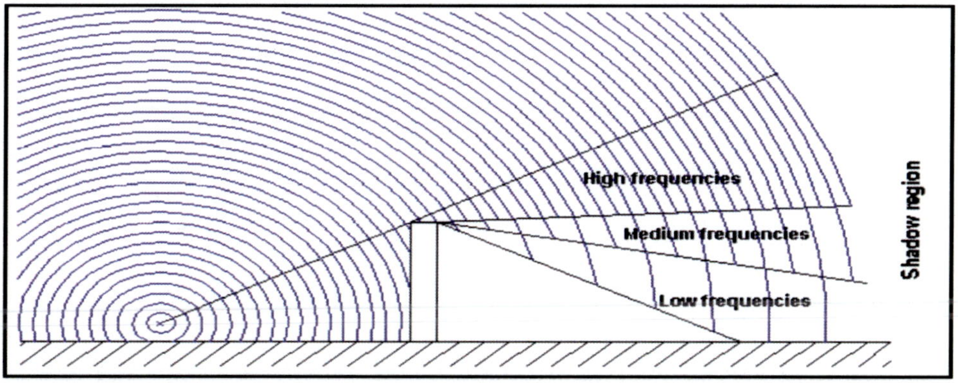

Figure 12. Schematic diagram of a simple, infinitely long acoustic barrier showing the frequency dependent effects of the diffraction of sound energy into the shadow region

On the assumption, based on ray theory, that sound in a homogenous atmosphere travels in straight lines, it is theoretically the case that no sound energy should be able to transmit to any point on the far side of the barrier that has no direct line of sight with the source. However, through the process of diffraction based on a wave based approach, a series of virtual sources are effectively created along the top of the barrier. These virtual sources then allow sound energy to be radiated into what would otherwise be the shadow region. This diffraction effect is, however, frequency dependent with low frequency sounds being re-radiated deeper into the shadow region than high frequency sounds, as indicated in Figure 12. It is for this reason that excess attenuations calculated for barriers in neutral atmospheres increase with increasing frequency. Typical overall excess attenuations for an effective, purpose built acoustic barriers installed relatively close to, say, a busy road can be anything up to around 20 dB(A). The actual attenuation achieved will depend on the exact configuration of the barrier relative to the source and receiver geometries and also on the source spectrum. However, in achieving an overall reduction in sound level of 20 dB(A), the excess attenuation at 63 Hz will be only around 5 dB whereas the attenuation at 8 kHz will be around 30 dB.

When considering the installation of a barrier as a possible outdoor noise control solution there are two important considerations to account for when estimating by calculation the performance of that barrier under practical circumstances:

- the impact of non-neutral sound propagation conditions and the refraction of sound waves.
- the impact of the barrier on the excess attenuation already provided by the ground effect;

Considering the issue of the non-neutral sound propagation first, it has been shown earlier that, under temperature inversion or downwind conditions, the positive vertical sound speed gradient causes the downwards refraction of sound waves. The resultant effects are shown in Figures 6 and 7 which indicate that sound waves no longer travel in a straight line between source and receiver but instead via curved paths, either directly or via reflections off the ground. The curvature of these paths is dependent on the prevailing atmospheric conditions, but it is easy to see how a barrier that provides significant excess attenuation under neutral propagation conditions may provide virtually no excess attenuation when downwards refraction causes the curved sound propagation path to pass over the top of it. Whether or not this reduction in excess attenuation will occur in practice in any given situation depends largely on the relative separation distances between the source, receiver and barrier. Where all three are located relatively close together (i.e. within a few metres of each other) then there will be insufficient distance for the curvature of the sound propagation path to refract downwards into the shadow region. However, where separation distances extend to tens of metres or more, it is possible that the reductions in excess attenuation experienced under conditions of temperature inversion, and particularly under downwind conditions when the curvature of the sound waves can be tighter, may be very significant, to the extent of possibly being totally negated.

Now considering the issue of the ground effect, the earlier examples have indicated that excess attenuation arising from this effect results from constructive and destructive interference between the direct and reflected propagation paths (in other words the 'attenuation' can be both positive and negative, depending on the geometrical configuration of the source, ground, receiver layout and the frequency of the propagating sound). If the presence of a barrier removes this ground reflected path, but a direct path still exists through downward refraction, then it is possible that in those frequency regions where a ground effect dip would otherwise have existed, the barrier may result in the noise level actually increasing above that which would exist in the absence of the barrier. It is for this reason that schemes for predicting outdoor sound propagation that rely on the additive effects of each attenuation mechanism calculated separately can significantly overestimate the actual performance of barriers.

Due to the elevated level of the source of noise when considering wind turbines, the application of dedicated barriers as a means of noise control is extremely limited. However, barrier effects can also arise from topographical features that interrupt the direct line of sight between a receiver location and a wind turbine. As a consequence this potential attenuating effect is now considered.

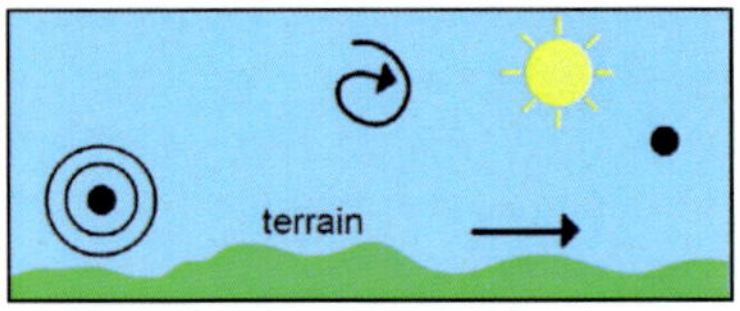

The manner in which the presence of a ground plane can significantly modify received sound levels has already been discussed. This modification results primarily from the interaction of sound waves that arrive at the receptor location directly, and those sound waves which arrive via one or more reflections off the ground plane. It was shown that the local acoustical properties of the ground where a reflection takes place can have a significant impact on the magnitude of any such interference effects. In developing that discussion it was assumed that the ground was perfectly flat, such that the angle of reflection of any incident sound wave would be the same as the angle of incidence, both angles being expressed relative to the general ground plane.

However, in most practical circumstances, the ground lying along any propagation path between a source and receiver will not be perfectly flat, but will instead exhibit at least some degree of variation in height from point to point.

Variations in ground height may range from local undulations of a few millimetres, on what may otherwise appear to be generally flat land, to differences of tens or even hundreds of meters in naturally hilly or mountainous areas. Depending on the scale of the variations in ground height, different effects may be seen in the received sound levels. The situation where significant variations in ground height exist is considered first.

The greatest effects occur when the ground rises up as a hill or other ground feature to such a degree that it actually blocks direct line of sight between the source and receiver. This situation is illustrated in the upper drawing of Figure 13. In such cases the hill can be considered as an acoustic barrier beyond which an acoustic shadow region exists.

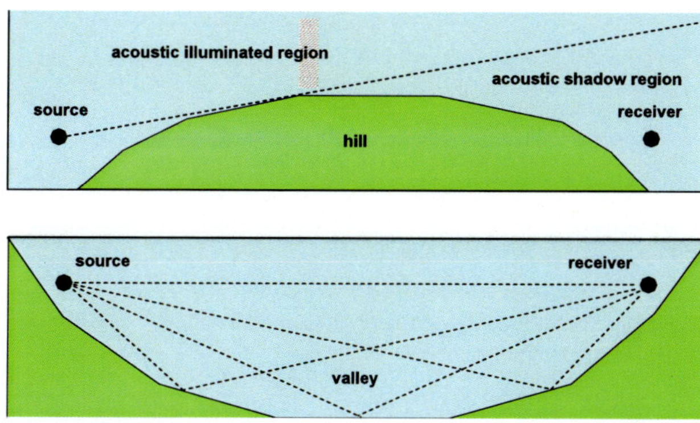

Figure 13. Schematic diagrams of the acoustic shadow region caused by the presence of a hill between source and receiver (upper drawing) and the possibility of multiple reflection paths for sound propagation across a valley (lower drawing)

The effects of barriers have already been discussed in the preceding section, where it was demonstrated that the magnitude of any reduction in sound levels in the shadow region will be dependent on a number of factors: the distances of the source and receiver from the obstruction, the extent to which the obstruction interrupts the direct line of sight between the source and receiver and the frequency of the sound being considered.

At the other extreme to a hill, it may be that the source and receiver are located on opposite sides of a valley, in which case the ground falls away significantly between the two. This situation is illustrated in the lower drawing of Figure 13. The manner in which reflections occur off this concave ground surface path means that an enhanced number of reflected paths may exist between the source and receiver when compared with a flat ground situation, thus potentially enhancing the overall received sound level.

In order to gain a better understanding of the magnitude of attenuation or amplification effects that may be experienced under practically encountered conditions in the presence of hills or valleys, it is useful to consider the effects just described in another manner, namely through the use of conformal mapping [1]. Conformal mapping allows the convex or concave ground plane to be modelled as a flat surface and the distance related curvature effects to be re-introduced (or 'mapped') into the sound wave paths. The great value of this technique is that it allows these topographical effects to be better understood in the context of the other propagation effects considered here.

Figures 14 show schematically the results of conformal mapping to a flat ground plane the situations that exist for propagation over a hill and a valley respectively. Thus it may be seen that propagation over a hill is the same as propagation over flat ground in an atmosphere that refracts sound waves upwards, such as may occur for upwind propagation or when temperature lapse conditions exist. Propagation over a valley is the same as propagation over flat ground in an atmosphere that refracts sound downwards, such as may occur for downwind propagation or when temperature inversion conditions exist.

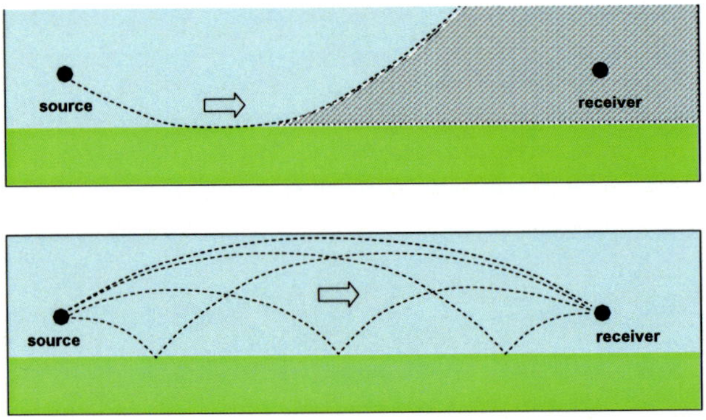

Figure 14. Schematic diagrams showing the results of conformal mapping the scenarios shown in Figure 13 to the equivalent flat ground cases. The upper drawing shows the effective upwards sound ray curvature that results from conformal mapping sound propagation over a hill and the lower drawing shows the effective downwards sound ray curvature that results from conformal mapping sound propagation across a valley

The potential magnitudes of the effects shown in Figure 14 have already been discussed in the context of sound speed gradients. That discussion concluded that excess attenuations of 10 dB to 20 dB in overall sound level may be experienced in the acoustic shadow zone created by upward refracting sound waves, whilst enhancements in received overall sound levels due to downward refracting sound waves were limited to typically 2 dB. It should again be noted, however, that the foregoing conclusions relate to overall sound levels. For specific frequency ranges, these frequency ranges being dependent on the physical dimensions of the situation being considered, greater attenuations or enhancements may be encountered in received sound levels due to specific interference effects.

There also exists one other key consideration when assigning attenuation losses to shadow regions beyond topographical features such as hills. The introduction of a downward refracting atmosphere into the upper drawing in Figure 13, as would result from the presence of a positive sound speed gradient caused either by a temperature inversion or by downwind propagation, will cause sound to be refracted over the hill, thus entering into the acoustic shadow region beyond. The refracted sound energy would thus reduce the effective attenuation properties of the topographic feature. The actual reduction in attenuation will depend on the relative magnitudes of the two curvature effects (i.e. the effective upwards ray curvature resulting from the topographical effects and the effective downwards ray curvature resulting from the meteorological effects) and the separation distances between either the source or receiver and the topographic screening feature. As a general rule, practical measurements indicate that where direct line of sight is interrupted by gently undulating hills (as opposed to shear sided mountains or cliffs which offer a much more pronounced acoustic cut-off to the shadow region) then meteorological effects can all but negate any excess attenuation effect of the hills [33].

The final consideration relating to non-flat ground is that which exists when the ground is essentially flat on a larger length scale, but on a smaller length scale it varies. This situation effectively scatters sound waves that are incident on the ground, because the angle of reflection of each sound wave relative to the general ground plane depends on both the angle of incidence of that wave and the local angle of the undulating ground where the wave strikes the ground. Thus the previously eluded to effects of constructive or destructive interference between direct and reflected waves are reduced in their potential magnitude.

REVIEW OF OUTDOOR SOUND PROPAGATION MODELS

The complexity of atmospheric conditions makes it impractical to accurately define all relevant environmental parameters that may influence local sound speed along all possible sound propagation paths between a source and receiver. Any practical model therefore requires that at least some assumptions and simplifications be adopted. This has led to the existence of a variety of sound propagation models, of which some are inter-related, others are hybrids and others unique, but all are limited in capability in some respects. Generally, all these models can be classified by the three categories listed below. The models are sorted here by increasing complexity and potentially increased accuracy. However, it should be appreciated that the more complex the model, the more demanding the requirement becomes for the accurate specification of an increasing number of input parameters. The promise of greater accuracy of more complex models can only be realised if the values of all input parameters adequately represent the situation being modelled.

Practical engineering methods: The technique adopted by these models involves the calculation of noise levels by adding the separate contributions that each sound attenuation factor has on noise propagation. The common factor in all these models is that they are mainly based on empirical results. In general, they are simple and easy-to-use, but at the same time their considerable amount of assumptions and simplifications make them less accurate than other models. However, in line with discussion contained in the introduction to this chapter, whether or not this apparent lack of accuracy makes them less suitable for practical application depends on the use to which their output results are being put.

Approximate semi-analytical methods: These methods retain the same practical structure as engineering methods, but are based on simplified analytical solutions of the wave equation rather than empirical results. While the practical engineering methods only take into account averaged meteorological effects, these methods allow a better tracking of the influence of specific meteorological conditions on noise levels, such as upwind or downwind situations. Simple ray tracing models are the most popular methods within this category.

Numerical methods: This group includes a range of methods such as the Fast Field Program (FFP), the Parabolic Equation (PE) and other models based on the direct solution of the wave equation. In general, all these methods allow the calculation of sound propagation over non-complex level terrain with any user-specified atmospheric conditions. They are extremely useful for analysing sound propagation under specific meteorological conditions. The problem is that they yield results for only those specific conditions and give little indication of statistical mean values of sound levels. Also, the user must provide substantially more information which can be difficult to generate, such as complete profiles of wind and temperature.

Early work comparing the output of engineering method predictions against measurements for the purpose of developing a wind farm noise propagation prediction code were reported by Bass et al in 1998 [33]. However, comparisons of engineering methods against measurements for wind turbine noise have seen a significantly increased profile in the public domain lately. Publications to which the reader may wish to refer include reports on the use of Nord2000 by Plovsing and Sondergaard [18] [40] [41], Forssen at al. [27] and HARMONOISE by Jonnson [26]. Bolin et al [29] [35] have also published the results of further studies, and Bullmore et al [37] have similarly presented the results of more recent comparisons between predictions and measurements in the UK.

OVERVIEW OF ENGINEERING METHODS

There are numerous practical engineering (empirical) methods which have been developed over the years which are applicable for sound propagation modelling. Commercial software packages which implement standard and other well established non-standard engineering methods are common place. Available and commonly used methods include:

- ISO 9613-2: 1996
- CONCAWE 1976 (UK)
- BS 5228 2009 (UK)
- NORD 2000 (Joint Nordic)
- HARMONOISE 2002 (EU)

The following overview of methods covers ISO 9613-2 [16], HARMONOISE [11] and NORD2000 [18] with some brief remarks on other hybrid methods

ISO 9613-2

ISO 9613 has been established for approximately 15 years as the primary International Standard for the calculation of environmental sound propagation. The standard is divided into two parts: Part 1 provides a method for calculating the attenuation of sound due to atmospheric effects as it propagates through the atmosphere, while Part 2 provides a general method of calculation for environmental sound propagation. The introduction to Part 2 describes the calculation method as being general in that it may be applied to a wide variety of noise sources and it covers most of the major mechanisms of attenuation.

The ISO 9613-2 method was based on a combination of the modelling of certain effects, where these could be directly modelled, and then using measurement data to refine the model empirically. The procedure provides a method by which, under conditions of a positive sound speed gradient (i.e. downwind or temperature inversion), the equivalent continuous octave band sound pressure level L_T(DW) is calculated at a receiver location in accordance with equation 4, where L_W is the octave band sound power output of the source, D_C is a source directivity correction and A is the total octave band attenuation that occurs between source and receiver. Downwind propagation conditions are quantified as meaning within a wind direction of +/- 45 degrees of the direction connecting the source with the receiver with the wind blowing at a speed of between approximately 1 m/s and 5 m/s (measured at a height of 3 m to 11 m above the ground) from the source to the receiver. The methodology is also stated to hold for propagation under a well-developed moderate ground based temperature inversion. This is expected as the effect being modelled is actually that caused by a positive sound speed gradient, and both downwind conditions and temperature inversion conditions result in an increase in sound speed with increasing height above the ground.

$$L_T\,(DW) = L_w + D_c - A \qquad [4]$$

The general form of equation 4 is similar to that already presented in this chapter as equation 1. The total octave band attenuation term, A, in equation 4 is made up of five components, as indicated in equation 5:

$$A = A_{div} + A_{atm} + A_{gr} + A_{bar} + A_{misc} \qquad [5]$$

where (relating these factors to the previously identified factors A1 to A5):

A_{div} is the attenuation due to geometric divergence = **A1**;
A_{atm} is the attenuation due to atmospheric absorption = **A2**;
A_{gr} is the attenuation due to the ground effect = **A3**;
A_{bar} is the attenuation due to a barrier or topographic features = **A5**;
A_{misc} is the attenuation due to propagation through foliage, industrial sites, and housing.

The attenuation factor **A4** for meteorological effects is accounted for by default on the basis that the prediction methodology is relevant only to downwind or meteorologically equivalent propagation conditions.

The Standard contains methods for calculating each of the attenuation terms which, with the exception of A_{div}, are all frequency dependent.

A1 Geometric divergence For a point source of a spherical wave the attenuation is not dependent on frequency and is calculated by the formula

$$A_{div} = 20\log_{10}(d/d_0) + 11 \qquad [6]$$

where d is the distance from the source to the receiver and d_0 = 1m is a reference distance.

A2 Atmospheric absorption attenuation is calculated in accordance with Part 1 of the Standard although a sub-set of 'standard' attenuation factors are also incorporated into Part 2.

A3 Ground effect attenuation is based on a series of empirically derived factors that have been established from the results of measurements. The procedure involves dividing the propagation path into three distinct sections, these being a source region, a receiver region and a middle region and then calculating octave band ground attenuations for each region before summing the effects to arrive at the total ground attenuation in each octave band. The horizontal extents of the source and receiver regions are thirty times the source and receiver heights respectively, and the ground effect calculation also involves the mean propagation height between source and receiver. The calculation of ground effects also requires the specification of a ground type, G, between source and receiver. A ground type of G = 0 indicates acoustically hard ground, such as paving, water, ice, concrete and tamped ground. A ground type of G = 1 indicates acoustically soft ground which is stated to include ground covered by grass, trees or other vegetation or arable farming land. The actual value of G used is calculated separately for each of the source, receiver and middle regions on the basis of the relative proportion of hard and soft ground types that exist in each region.

A5 Barrier effect attenuation is calculated on the basis of the dimensions of the barrier and the effective path length difference relative to the typical wavelength in each octave band that the barrier introduces compared to the path length that would have existed for the sound propagating direct from the source to the receiver in the absence of the barrier. Barriers can include topographic features although the practical implementation of this calculation for terrain is not clear from the standard which is more concerned with 'standard' barriers.

Once the calculation of each of the octave band attenuation factors has been completed, the overall long term downwind sound pressure level is calculated through the combined use of equations 4 and 5. The Standard additionally allows the application of a meteorological correction factor, C_{met}, based on the average long term meteorological conditions experienced at the site. The Standard is not entirely prescriptive as to how this correction factor should be calculated, although it is stated that positive correction values in excess of 2 dB should be considered exceptional.

In terms of the stated limitations of use of the Standard it is intended to provide an estimate of noise immission levels for two cases only: during propagation under favourable (effective worst case) conditions or as longer term average based on a likely range of meteorological conditions. The estimation of noise levels under any other specific set of conditions, such as upwind propagation or temperature lapse, is precluded. Estimates of accuracy of the method based on the aforementioned limitations of use are presented in Table 5 of the Standard. These error bounds are stated in terms of overall A-weighted L_T(DW) predictions only, where an upper bound of +/- 3dB is presented for a maximum source height of 30 m and a maximum propagation distance of up to 1,000 m, although the possible use of the method beyond these dimensions is not precluded.

In respect of its application to wind turbine sound, studies have shown [33] [34] [36] [37] that, despite its relative simplicity, the ISO 9613-2 method is capable of predicting noise from wind turbines with a good degree of reliability in that the predicted levels represent well the upper level of noise immission levels measured at typical receiver locations under downwind propagation conditions. However, the same studies identified that such reliability could only be ensured through a consistent choice of the values of some of the parameters that are left open for users of the Standard to select, and also provided that the ground and barrier attenuations were limited in their magnitude. These factors have subsequently been summarised in a UK Institute of Acoustics Bulletin Article [38] and comprise:

- the adoption of a G = 0.5 ground factor coupled with source sound power data includes a warranty margin or the test data uncertainty or, alternatively, source sound power data based on the reported test levels coupled with the use of a G = 0. ground factor;
- the adoption of a receiver height of 4.0 m (approximating first floor window height) which has the effect of reducing the potential over-sensitivity of the calculation to the receiver region ground factor for lower receiver heights;
- the limitation of topographic screening effects to no more than 2 dB, and then only if there is no direct line of sight between the highest point on the turbine rotor and the receiver location, although higher barrier attenuation values may be appropriate in the case of significant screening from an landform barrier located very close to the receiver.

The results of the Joule study [33] also suggested the addition of 3 dB to the calculated overall A-weighted noise level for propagation across a valley, where the ground falls away significantly between the turbine and the receiver location.

A key limitation of the ISO 9613-2 method is that it only predicts noise immission levels under effective 'worst case' meteorological conditions, as illustrated later in the sections on parabolic equations and model validation. This is generally useful for cases where the demonstration of compliance with noise limits is required, but may be overly restrictive when it needs to be determined what the typical noise level may be in practice when the receiver is located crosswind or upwind of the wind turbines. Such a requirement may occur, for example, if the cumulative effects of two wind farms, one lying either side of a receiver, needs to be assessed. In such instances the receiver clearly can not lie simultaneously downwind of both wind farms, so to assess cumulative impact on the basis of the output of an ISO 9613-2 noise model would be overly pessimistic. In such cases a simple directivity correction is sometimes assumed, where this correction is applied to the overall A-weighted level calculated using the ISO 9613-2 model. Actual assumptions for directivity vary, but typical reductions of 2 dB in crosswind conditions increasing to 10 dB in upwind conditions are frequently assumed. Another example would be the requirement in some countries to assess wind farm noise not in terms of worst case levels but in terms of the annual average L_{DEN} level. This would require knowledge of the annual distribution of meteorological parameters and the appropriate decibel weighting that would need to be applied to the predicted worst case levels to account for the estimated effects of these varying meteorological conditions. It is when faced with such a requirement that more complex noise propagation models able to account for varying meteorological conditions, as discussed subsequently in this chapter, may prove to be a more appropriate choice.

HARMONOISE and Nord2000

Harmonoise and Nord2000 are ray-based rather than wave-based methods and calculate the sound pressure levels in third-octave frequency bands. They are among the most prominent and accurate engineering models for outdoor sound propagation. The aim of both models is to predict sound pressure levels quickly and accurately for a variety of situations.

The Nord2000 model is the result of cooperation between the Nordic countries starting in the mid 90's. It has been validated based on the results of measurements and by comparison with results from calculations using accurate numerical models. The model is widely used in the Nordic countries. It was established for traffic noise and allows for the calculation of short-term sound levels like short-term equivalent sound pressure levels or maximum levels relating to site-specific weather conditions. The model predicts sound propagation from a point source in one-third octave bands from 25 Hz to 10 kHz at a receiver. The main objective with the methodology was to apply theoretical bases in frequency band calculations instead of empirical formulae.

The Harmonoise model is more recent and is the result of cooperation between a number of European countries. A part of the project was to develop an accurate reference method based on numerical methods such as BEM and PE. The engineering model has been extensively validated with this reference method in addition to validations with measurement results. Since the Harmonoise model has not been around as long as the Nord2000 model it is not presently used as widely as the latter.

The methodologies behind Harmonoise and Nord2000 are two very similar engineering models for predicting environmental noise levels. Both models predict the sound pressure level at the receiver position in one-third octave bands from 25 Hz to 10 kHz from the sound power level of the source. The effects of various factors that influence the sound propagation are calculated separately (in decibels) and subtracted from the sound power level. These factors are: spherical geometric divergence, air absorption, reflections from the ground and diffraction from barriers, energy losses during side reflections, and effects of scattering zones. Although the aspects of propagation are the same in the two models, the approaches used to evaluate them differ. The effects of spherical divergence, air absorption and energy losses during reflections are handled in the same way by the two models.

To solve the problem of approximating the effects of the terrain surface the two models use, in principle, a similar approach. The ground models are basically two-dimensional where the terrain shape used is determined by the vertical cross section along the propagation path from the source to the receiver which is approximated by a number of straight line segments. Details of the two methodologies are provided below:

A1 Geometric divergence For a point source of a spherical wave the attenuation is not dependent on frequency and is calculated by the formula

$$A_1 = 10\log_{10}\left((d/d_0)^2\right) \qquad [7]$$

Where d is the distance from the source to the receiver and d_0 = 1m is a reference distance.

A2 Air absorption The effect of air absorption is evaluated in the same way as in many models. It is calculated on the basis of ISO 9613-1:1993 which predicts pure-tone attenuation from 50 Hz to 10 kHz. Pure-tone attenuation is, however, not always applicable in the prediction of broadband attenuation since the band attenuation depends on the width and shape of the bandpass filters and the frequency-dependent characteristics of atmospheric attenuation. Both models therefore employ a method to account for octave bands.

A3 Ground effect The *ground effect* is determined by geometrical ray theory at a flat homogenous ground surface. In cases with non-flat and non-homogenous terrain, heuristic semi-analytical modifications have been added to convey a general model. Nevertheless the model gives more accurate results for a simple terrain than a more complex location. Classifications of ground types can be found in many acoustic texts, whereby the surface acoustic impedance of each ground segment is generally characterised by the ground flow resistivity of that segment using the Delany and Bazely one-parameter impedance model [17]:

$$Z = 1 + 9.08\left(\frac{f}{\sigma}\right)^{-0.75} + i11.9\left(\frac{f}{\sigma}\right)^{-0.73} \qquad [8]$$

where f is the one-third octave band centre frequency in Hz and σ is the effective flow resistivity in $\mathrm{kNsm^{-4}}$.

In addition, Nord2000 defines a roughness parameter for each ground segment. This parameter quantifies terrain unevenness characterised by random height variations. Due to a lack of experimental data this parameter has yet to be validated and its accuracy is therefore still uncertain.

To calculate the effect of the terrain properties on the reflected sound both models use the concept of Fresnel-zones. It is an approximate solution used to predict sound propagation over flat terrain with varying surface properties. The sound field at the receivers is assumed to be determined by the surface conditions in a region around the reflection point, i.e. the *Fresnel-zone.*

The *Fresnel ellipsoid* is used in Nord2000 and HARMONOISE to quantify the effective size of the sound field, as indicated in Figure 15. Only the part of the sound field enclosed by the Fresnel ellipsoid will generate sound at the receiver, generally. The area that occurs in the intersection of the plane and the ellipsoid is called the Fresnel-zone. When calculating the ground effect, the surface properties at the Fresnel-zone determine the sound field at the receiver.

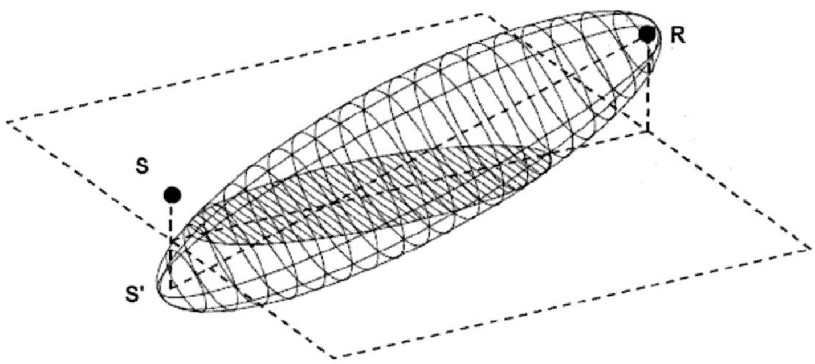

Figure 15: Fresnel ellipsoid and Fresnel-zone (R=receiver, S=source) from the Nord2000 method [18]

If two or more surfaces with different properties exist within the Fresnel-zone their importance in the overall effect of the ground is determined by their Fresnel-zone weight. The Fresnel-zone weight of a surface is defined as the size of that area relative to the size of the entire Fresnel-zone. In a two-dimensional propagation model, as used by both Nord2000 and Harmonoise, the Fresnel-zone are essentially one-dimensional, that is line segments, and the weighting coefficients become a ratio of the line segment lengths.

A4 Wind and Temperature effects The effect of atmospheric refraction is handled in fundamentally different ways by the two models. Nord2000 replaces the straight rays in the model for a homogeneous atmosphere by curved rays, thereby simulating the actual phenomenon of refraction. Harmonoise uses straight rays but curves the ground using conformal mapping to simulate the same effect. The curved ground analogy used by Harmonoise is illustrated in Figure 16. The curvature is, furthermore, calculated differently. Harmonoise calculates the radius of curvature based on the maximum height of the curve, while Nord2000 uses a rather more complicated method of calculating the length of the curve and the angle it makes relative to the ground. The importance of meteorological conditions is strong with wind turbine noise, hence it is expected that the two methods would provide different results.

However both methods employ similar models for meteorological conditions where the model parameter sound speed varies with height. This includes a linear sound speed variation with the height being assumed, but with a logarithmic sound speed profile often being more accurate the principle has been elaborated. Harmonoise determines the radius of curvature from the maximum height of the curve, whereas Nord2000 uses a more complicated method of calculating the length of the curve and

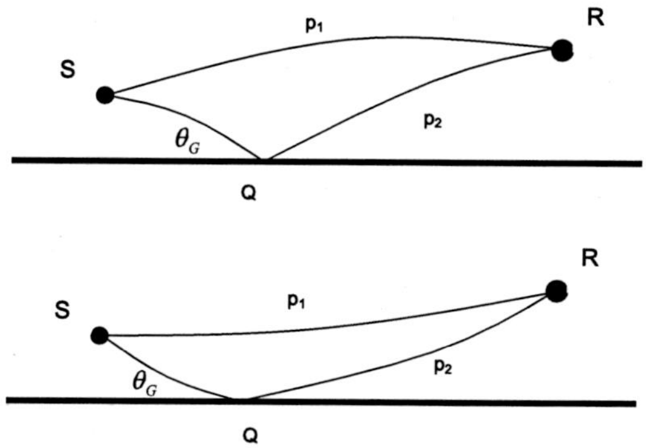

Figure 16. Curved ray paths. Top: Downward refraction. Bottom: Upward refraction

its angle relative to the ground. The principle determines the equivalent linear sound speed profile for those cases. Nord2000 is generally only valid for modest refraction, i.e. weather conditions that are not conducive to domination of multiple ground reflections and shadow zones. The effective sound speed c(z) is approximated by

$$c(z) = B \cdot \ln\left(\frac{z}{z_0}+1\right) + Az + C \qquad [9]$$

where z is the height above ground, z_0 is the roughness length, B is the logarithmic coefficient, A is the linear coefficient and C is the speed of sound at z=0m

Finally both models take account of the effects of meteorological conditions such as the vertical temperature and wind profile, diffraction and atmospheric instability, but the approaches used to simulate refraction and turbulence are very different.

A5 Terrain and Barrier Effect

The loss caused by a screening object is calculated by Harmonoise based on the Fresnel number N_f given by

$$N_f = sign\left(h_{eff}\right)2\left(R_s + R_R - R\right)/\lambda \qquad [10]$$

where h_{eff} is the effective height of the barrier, R_S is the path length between the source and the top edge of the barrier, R_R is the path length between the top edge of the barrier and the receiver and λ is the acoustic wavelength. The attenuation due to each edge is accounted for by a series of empirical formulae, for example, where the Fresnel number is small but positive

$$A_b = 6 + 12\sqrt{N_f} \; where \; 0 \le N_f < 0.25 \qquad [11]$$

Nord2000 handles diffractions from hills and screens quite differently. Diffraction effects are based on the wedge diffraction solution, commonly called the Hadden and Pierce model [31] as modified to include an absorbing wedge by assuming it has finite impedance.

For both Nord2000 and Harmonoise the *terrain* is defined by a number of straight-line segments, as shown by the examples in Figure 17. There is no limitation in the permissible quantity of segments, but for a general calculation a maximum of 10-15 is recommended. For each segment acoustic impedance, roughness and porosity may be separately defined.

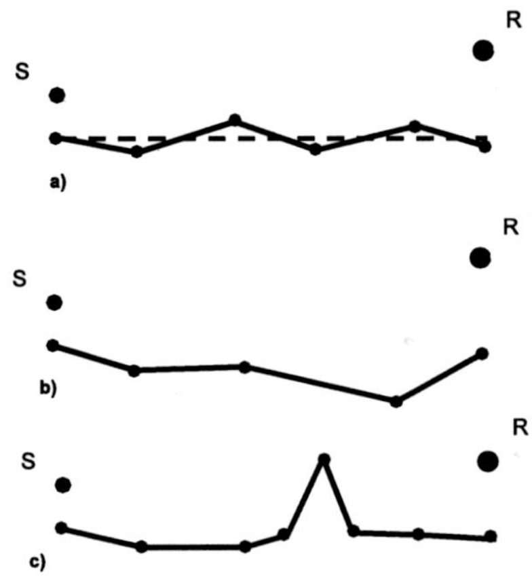

Figure 17. Segmented terrain a) flat terrain, b) valley terrain, c) hill-shaped terrain

To obtain the total terrain effect the models combine the results obtained from each of the previously described terrain models. The total terrain effect between the source and receiver used in Harmonoise indicates that the ground effect is limited to a maximum attenuation of 25 dB. This is to prevent unrealistically large ground attenuation caused by interference dips in the predicted ground effect spectrum. A similar precautionary approach is taken by Nord2000 by modifying the valley model. Nord2000 uses a slightly different approach to combine the results from the terrain models. In essence a weighted average of the component parts are summed.

WAVE MODEL – The Sound Field Above Absorbing Ground

The behaviour of sound propagation over ground of finite impedance is derived through analogy to electromagnetic wave theory. The geometry for this situation is illustrated in Figure 3. When airborne sound is incident on a locally reacting fluid, the transmitted wave is refracted at right angles into the surface. The reflected portion of the wave leaves the surface at the angle of incidence, with its amplitude and phase modified by the impedance of the surface. The reflected wave propagates to the receiver, along with the direct wave from the source. Depending on their relative phases and amplitudes, they may constructively add or destructively interfere. As already discussed in this chapter, in the limiting case of the receiver near the ground and perfect reflection and no atmospheric turbulence (i.e. coherent addition) the sound level at the receiver will be increased by 6 dB, which equates to an excess attenuation of minus 6dB. Effectively the receiver sees two sources, the actual source and a reflected or "image" source, and the sound pressure is therefore doubled.

Using the electromagnetic wave theory, the governing equation for the pressure amplitude p at the receiver, assuming a uniform medium (no refraction), over level ground is given by the formula:

$$p(x,y,z) = \frac{e^{ikR_1}}{4\pi R_1} + R_p \frac{e^{ikR_2}}{4\pi R_2} + \left[\left(1 - R_p\right)F(w)\right]\frac{e^{ikR_2}}{4\pi R_2} \qquad [12]$$

The first term is the direct wave. The second term describes a reflected wave after its amplitude and phase have been modified by the plane wave reflection coefficient. The third term accounts for the difference between the reflection of a plane wave and that of the actual case of a spherical wave. In the terminology of electromagnetic wave propagation, this last term is called the ground wave. The added component compensates for the plane-wave reflection coefficient term to match the boundary conditions of the spherical wavefront to the boundary surface. The quantity *F(w)* is called the boundary loss factor and is actually a first term of a more accurate asymptotic series solution. Algorithms defined by Chien & Soroka [42] for computation of this expression are available. For certain types of ground, it has been shown that a surface wave may develop in addition to the ground wave, as discussed earlier in this chapter. This expression appears as an augmented term in the boundary loss factor $F(w)$ as a surface wave and exists in a region above the ground for a certain angle of wave reflection incidence. This angle depends on the resistive and reactive parts of the impedance condition that describes the ground type. For grazing incidence this becomes a simple well-known condition but is rather more complicated for sources equivalent to wind turbine heights.

This equation provides an effective method to predict the excess attenuation due to ground absorption if the surface impedance is known as a function of frequency. Measurements of the impedance of various surfaces have been made and curve-fitting techniques applied to the experimental results. While more complex models have been proposed, it has been found that it is sufficient to use a single parameter, the flow resistivity of the surface, to describe the absorptive properties of a surface. The flow resistivity model is only valid for surfaces of constant porosity. The value of the effective flow resistivity may be taken from standard tables, for example: $\sigma = 63$ kNsm^{-4} for uncompacted loose ground, $\sigma = 250$ kNsm^{-4} for normal uncompacted ground, $\sigma = 630$ kNsm^{-4} for a compacted field and gravel and $\sigma = 2000$ kNsm^{-4} for compacted dense ground.

The wave approach is an exact frequency dependent model which takes into account any surface waves that may exist for certain ground conditions. This is advantageous for most sound propagation scenarios but where there is an elevated source such as a wind turbine any attenuation from the ground surface wave is not significant if a model only considers one reflected wave apparent near the receiver location. For octave band results this model requires solutions at several frequencies before combining into octave or third octave bands.

Meteorological conditions are not taken into account in this model.

SAMPLE NOISE IMMISSION LEVEL PREDICTIONS FROM WIND TURBINES

To show the effect of using various current prediction methods, comparative results are provided for sound propagation predictions from an elevated source in favourable wind conditions. For the various methods compared, a wind turbine having a hub height of 80 m with a sound power output as set out in Table 2 has been assumed. Receptor heights of both 1.5 m and 4.0 m have been considered located at distances of 0.5 km and 1.5 km from the source across flat ground. The ground attenuation for the ISO 9613-2 model (A_{gr}) has been assigned the values of G = 0 and G = 0.5 to represent hard and 'mixed' ground. For the Harmonise and Wave models a representative flow resistivity value of $\sigma = 250$ kNsm^{-4} has been assumed corresponding to absorbing ground. Where applicable, a 10 m height wind speed of 4 m/s has been selected to represent favourable downwind conditions . A ray-path calculation indicates there to be no extra ray-path reflections under the assumed conditions.

Table 2. Octave band sound power levels for a typical large-scale wind turbine

Frequency, Hz	**63**	**125**	**250**	**500**	**1000**	**2000**	**4000**	**8000**	**dB(A)**
Octave band sound power level, dB	118	113	109	106	103	99	92	82	**108**

The results of the calculations relating to the 500m propagation distance over hard ground are shown in Table 3.

The results of Table 3 illustrate the varying differences for this most basic situation comprising flat rigid ground and neutral wind. The overall A-weighted calculated levels are around 44 dB(A) at the receiver location 500 m from the source. However, on closer inspection of the octave band results, ISO 9613-2 underestimates the levels at lower frequencies but compensates at mid-frequencies. According to Table 1, the source and receiver locations suggest that interference effects between direct and reflected waves occur at around 360 Hz. This effect accounts for the attenuation in levels in the 250 Hz and 500 Hz bands for both the Wave and Harmonoise methods.

Table 3. Octave band linear and overall A-weighted sound pressure level predictions at 500 m separation distance in downwind conditions over hard ground at a receiver height of 1.5 m above ground level

Frequency, Hz	**63**	**125**	**250**	**500**	**1000**	**2000**	**4000**	**8000**	**dB(A)**
ISO 9613-2	56.1	50.8	46.0	42.9	38.8	32.3	16.6	-	**45**
Wave Solution	58.4	52.5	44.6	40.3	38.7	31.6	20.8	-	**44**
Harmonoise	58.6	52.8	45.0	40.0	39.7	33.5	17.6	4.9	**44**

The preceding example accounts for acoustically hard ground between source and receiver. Introducing absorbing ground having a flow resistivity corresponding to that of typical grassland, or $\sigma = 250$ kNsm^{-4}, whilst maintaining the same receiver height of 1.5m and separation distance of 500 m, the results of Table 4 are calculated. According to ISO 9613-2, any ground surface of low porosity is to be considered as acoustically hard (G = 0) and any vegetation covered ground, including tree coverage is to be considered as acoustically soft (G = 1). This approach in general simplifies a considerable range of properties and effects down to a simple choice of one of two extreme values. When predicting outdoor sound propagation over a long-range, including mixed-ground and natural changes of topography, a simple model is generally adequate for A-weighted indices. Taking a value of G = 0.5 for mixed ground seems to be a good compromise choice when studying propagation over grassland or general arable land based on experience gained with the ISO 9613-2 procedure. This is therefore the value of G adopted in the present exercise.

Table 4. Octave band linear and overall A-weighted sound pressure level predictions at 500 m separation distance in downwind conditions over grassland at a receiver height of 1.5 m above ground level

Frequency, Hz	**63**	**125**	**250**	**500**	**1000**	**2000**	**4000**	**8000**	**dB(A)**
ISO 9613-2	56.1	47.8	40.9	38.9	37.0	30.8	15.1	-	**42**
Wave Solution	59.7	52.0	43.3	40.4	37.1	30.0	18.9	-	**43**
Harmonoise	59.9	52.2	43.8	40.5	38.0	31.3	21.1	4.8	**44**

The calculated overall A-weighted levels are around 43 dB(A) at the receiver location 500 m from the source over absorbing ground. The overall level has not decreased significantly from the hard ground case in Table 3 except for the ISO 9613-2 case where a reduction of 3 dB is seen to occur.

Table 5 shows results for the basic situation already considered in Table 3 of flat rigid ground and neutral wind but here the receiver is located at a distance of 1500 m from the source. In this case interference effects account for attenuation in the 1000 Hz band, as evidenced by the significant differences seen between the ISO results and the Wave and Harmonoise results.

Table 5. Octave band linear and overall A-weighted sound pressure level predictions at 1500 m separation distance in downwind conditions over hard ground at a receiver height of 1.5 m above ground level

Frequency, Hz	**63**	**125**	**250**	**500**	**1000**	**2000**	**4000**	**8000**	**dB(A)**
ISO 9613-2	46.5	41.0	35.4	31.1	25.3	14.0	-	-	**33**
Wave Solution	49.2	43.8	39.0	32.3	19.1	16.2	3.3	-	**34**
Harmonoise	49.4	44.1	39.6	33.1	19.0	19.1	3.4	4.9	**35**

Table 6 shows the results of including absorbing ground in the scenario otherwise considered in Table 5 for propagation over a distance of 1500 m.

Table 6. Octave band linear and overall A-weighted sound pressure level predictions at 1500 m separation distance in downwind conditions over grassland at a receiver height of 1.5 m above ground level

Frequency, Hz	**63**	**125**	**250**	**500**	**1000**	**2000**	**4000**	**8000**	**dB(A)**
ISO 9613-2	46.5	36.8	30.4	27.1	23.5	12.5	-	-	**29**
Wave Solution	46.8	39.4	30.9	24.8	25.5	14.2	3.4	-	**30**
Harmonoise	47.1	39.8	31.5	25.0	26.6	17.0	5.7	4.9	**31**

Comparison of the overall level results between Tables 5 and 6 indicates that an approximate 4 dB reduction results for propagation over absorbing ground compared to over hard ground, this reduction being seen in the results according to all three calculation. However, the Wave Method and Harmonoise results indicate that the addition of absorption for the ground surface produces a relatively lower attenuation in the 125 Hz frequency band, by approximately 3 dB, for the particular parameters chosen in this case to model the ground surface. The potentially sensitive nature of the results to the combination of receiver height and ground effect parameters is one of the reasons for the recommendation made in reference [38] that a receiver height of 4.0 m should be adopted, as this lessens the degree of sensitivity to the ground factor. Tables 7 and 8 summarise the results of repeating the calculations of Tables 5 and 6 but this time adopting a receiver height of 4.0 m instead of 1.5 m. The results indicate that adopting a receiver height of 4.0 m in ISO9613-2 calculation would have resulted in the same calculated level of 33 dB(A) for the case of hard ground (calculated sound levels using ISO9613-2 are independent of receiver height for the case of hard ground) but a level of 31 dB(A) for the case of mixed ground, as opposed to the level of 29 dB(A) calculated for a receiver height of 1.5 m.

Table 7. Octave band linear and overall A-weighted sound pressure level predictions at 1500 m separation distance in downwind conditions over hard ground at a receiver height of 4.0 m above ground level

Frequency, Hz	**63**	**125**	**250**	**500**	**1000**	**2000**	**4000**	**8000**	**dB(A)**
ISO 9613-2	46.5	41.0	35.4	31.1	25.3	14.1	-	-	**33**
Wave Solution	48.9	43.0	35.3	26.5	26.8	15.7	3.4	-	**33**
Harmonoise	48.6	43.9	35.4	23.3	25.3	16.5	5.3	4.9	**33**

Table 8. Octave band linear and overall A-weighted sound pressure level predictions at 1500 m separation distance in downwind conditions over grassland at a receiver height of 4.0 m above ground level

Frequency, Hz	**63**	**125**	**250**	**500**	**1000**	**2000**	**4000**	**8000**	**dB(A)**
ISO 9613-2	46.5	37.5	32.9	29.6	23.8	12.6	-	-	**31**
Wave Solution	46.2	37.4	30.2	31.3	22.2	13.9	3.4	-	**31**
Harmonoise	44.5	37.6	29.2	30.9	20.1	13.7	3.4	4.9	**30**

Overview of Other Hybrid Exact Methods

As an alternative to the use of the well established but relatively simplistic engineering methods, a group of hybrid numerically derived methods are available for evaluating the propagation of sound from industrial operations. At present the use of such methods is undoubtedly limited in practice to relatively complex investigations and not for practical application, although this situation may change in the foreseeable future with increased understanding being combined with increasing computational power [35]. The general principle of these methods is to solve the wave equation or Helmholtz equation so as to deduce the sound field generated by an acoustic source.

Although simply stated above, the procedure for solving the wave equation is generally difficult to implement due to the complexity of the atmospheric-acoustic environment. In fact, except for the very simplest boundary conditions and uniform media (which rarely occur in reality), it is not possible to obtain a complete analytical solution for either the wave or Helmholtz equation. As a consequence it therefore becomes necessary to use numerical methods to solve these equations. Several different types of solution for the sound field have evolved over the past thirty years or more: ray tracing provides a very graphic picture of the field, the Fast Field Program is accurate but computationally intensive and the Parabolic Equation is an approximation to the wave equation that has been solved using explicit and implicit finite difference schemes.

To demonstrate the potential use of these various methods, the following provides an overview of how they work, along with a brief review of the potential advantages and limitations. This includes an indication of situations where their use may or may not be appropriate. Consideration has been given to the Parabolic Equation (PE) method. It is noted that the review only examines the accuracy of the better known 2D models (in geometry not in field variable, as per the the Harmonoise 'reference model' used to validate the engineering method produced by Harmonoise [32], and does not include for the limited number of three-dimensional numerical propagation models which have been developed. The implicit assumption in such models is that a sound ray launched into a particular vertical plane remains in that plane over the entire transmission path. In the atmosphere, where variations of the sound speed profile in the horizontal may be negligibly small compared to vertical variations, this assumption is often reasonable, in which case a 2D radially symmetric model provides a satisfactory description of the acoustic environment.

Crank-Nicolson Parabolic Equation (CNPE)

In contrast to other engineering methods, the Crank Nicolson Parabolic Equation method is not restricted to systems with a homogeneous atmosphere or a homogeneous ground surface. In fact, the PE method in general is the only current technique that can handle range-dependent variations of environmental factors.

As with the other methods it is useful to gain an understanding of the basic theory. Essentially, the PE method is based on an approximate form of the wave equation. Depending on the approximation made to the wave equation, one can obtain either a wide-angle parabolic equation or a standard narrow-angle form such as the one shown below:

$$2ik_0 \frac{\partial u}{\partial r} + \frac{\partial^2 u}{\partial z^2} + k_0^2 \left(n^2(r,z) - 1\right) u = 0 \qquad [13]$$

where u is a variable proportional to the sound field. For more details about the derivation of this standard PE or other wide-angle PEs the reader is referred to Salomons [1] and the tutorial article by West Gilbert and Sack [30].

Different solution techniques exist to cope with the parabolic equation. Amongst them, the most popular are: The Crank-Nicholson PE (CNPE) method and the Green's Function PE (GFPE) method. Both methods make use of a rectangular grid in the rz plane. The source is represented by a starting function $p_c(r = 0,z)$ of the complex pressure amplitude at range r = 0. This starting function is extrapolated step-wise on the grid, in the positive r direction. An extrapolation step from range r to range r+Δr can be expressed as

$$p_c(r,z) \rightarrow p_c(r + \Delta r, z) \qquad [14]$$

By repeating this extrapolation step many times the complete field $p_c(r,z)$ of the complex pressure amplitude is computed.

The CNPE method was the first PE solution scheme introduced in atmospheric acoustics and essentially involves a finite-difference solution of a wide-angle parabolic equation. The horizontal and vertical grid spacings Δr and Δz in the CNPE method are limited to a maximum value of about $\lambda/10$, where λ is an average wavelength. Consequently, the number of grid points, and hence the computing time, increases with increasing frequency.

The GFPE method is less accurate than the CNPE method in situations with wide-angle propagation and large sound speed gradients. However, for most applications and particularly for many ground-to-ground propagation configurations, the GFPE method is sufficiently accurate. In the GFPE, the vertical grid spacing is also limited to about $\lambda/10$, but the horizontal grid spacing (or range step) may be chosen to be considerably larger, up to about 50λ. The consequent advantage of the GFPE over the CNPE method is that it can take considerably larger extrapolation steps (i.e. range steps), resulting in a more efficient and faster computation.

There are three distinct limitations to the various forms of PE methods. Firstly, PE algorithms have an unfortunate disadvantage in that they only give accurate results in a region limited by a maximum elevation angle α_{max}, as demonstrated in the figures below. The value of this angle oscillates from 10° to 70° or higher [1] depending on the angle approximation used in the derivation of the parabolic equation.

Secondly, the computation of a full frequency spectrum is not a straightforward process. The PE methods compute one frequency at a time. Thus the computing time for a complete spectrum can be considerable, particularly for frequencies above 600 Hz.

Finally, scattering by sound speed gradients in the direction back to the source is neglected by a parabolic equation. In other words, a parabolic equation is a one-way wave equation, taking into account only sound waves travelling in the direction from the source to the receiver. As the sound speed is usually a smooth function of position in the atmosphere this one-way wave propagation approximation is usually a good approximation. However, when turbulence is taken into account the backscattering limitation might become an effect to be considered.

Parabolic Equation: Example

The CNPE wide-angle formulation has been used in the following models over flat ground. Figure 18 displays examples for PE predictions of sound pressure levels for still and downwind conditions. The range of interest up to 5 km includes a homogeneous absorbing ground, which may be input as varying on the grid scale. However the results give some support to evidence for the apparent onset of curved sound paths in downwind conditions. The manner in which the sound paths refract downwards may have some impact on the implications for wind turbine noise propagation modelling where the source is elevated.

The PE method used for the example study below is a CNPE implementation, similar to the method described in West et al. [30]. A Gaussian sound source model approximating a point source located 80m above the ground was implemented with a logarithmic sound speed profile simulating the wind shear with a wind speed of 8 m/s at 10m height and such that sound speed at the ground is 340 m/s. The ground was characterised by a Delany and Bazley [17] empirical model adopting a flow resistivity of $\sigma = 250$ kNsm^{-4}. Solutions at the two frequencies of 63 Hz and 125 Hz are shown to illustrate the association of long distance sound propagation with downwind conditions compared to a neutral wind environment.

In Figure 18 (a)-(d) at constant height, just above ground level, there is little difference in overall decibel levels between sound propagation results at neutral conditions to that with downwind conditions up to a range 5 km.

As with many situations with sound propagation over long distances, there are significant effects of refraction. These effects arise either from temperature variations with height or non-constant wind speed profiles. It has previously been recognised from a ray-tracing perspective that this arises from additional ground reflected sound paths. Due to the specular reflections from the ground surface at around 100m from the source it is difficult to conclude whether sound paths directly from the sound source will have a downward curvature – it seems likely that curved sound paths may exist as a first

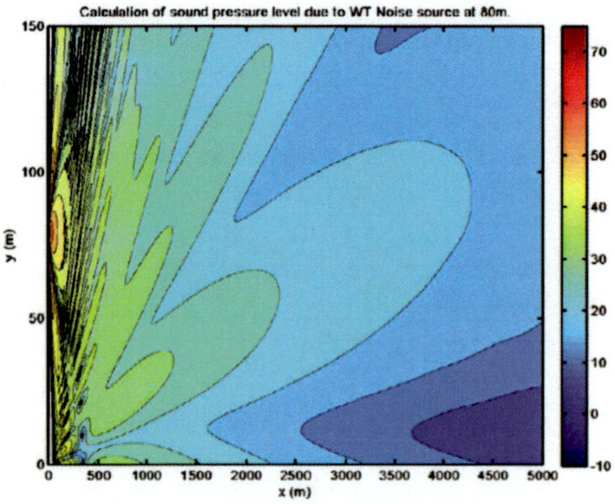

(a) Neutral conditions. f = 63 Hz

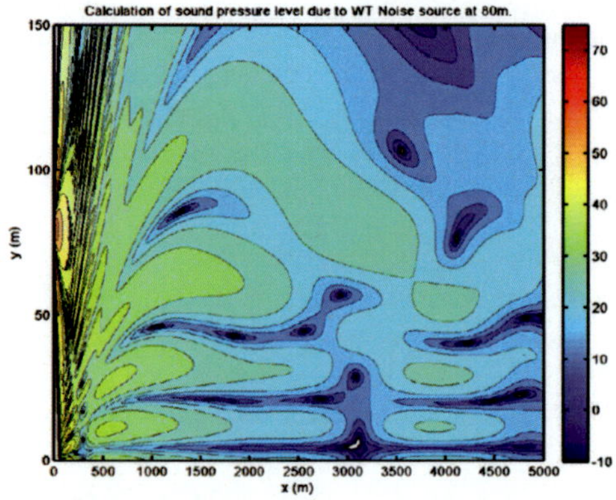

(b) Downwind conditions 8m/s at 10m. f = 63 Hz

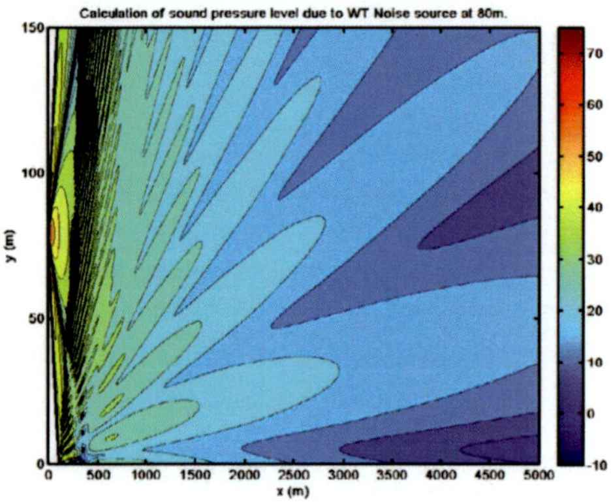

(c) Neutral conditions. f = 125 Hz

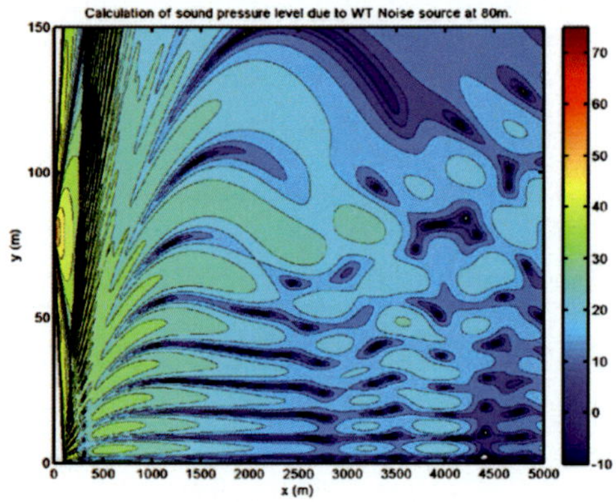

(d) Downwind conditions 8m/s at 10m. f = 125 Hz

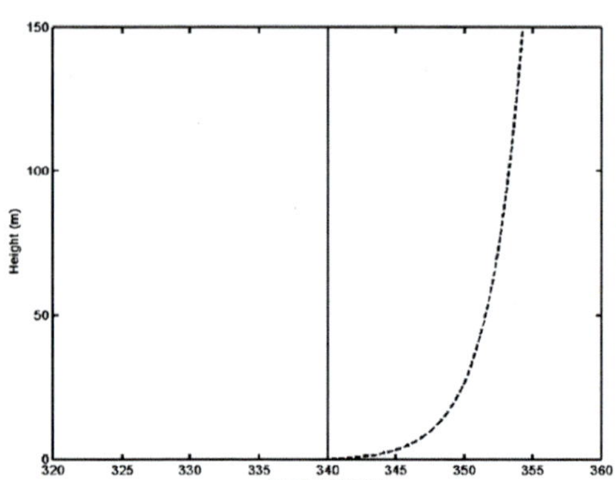

(e) Wind shear sound-speed profile in neural conditions (solid) and downwind (dashed).

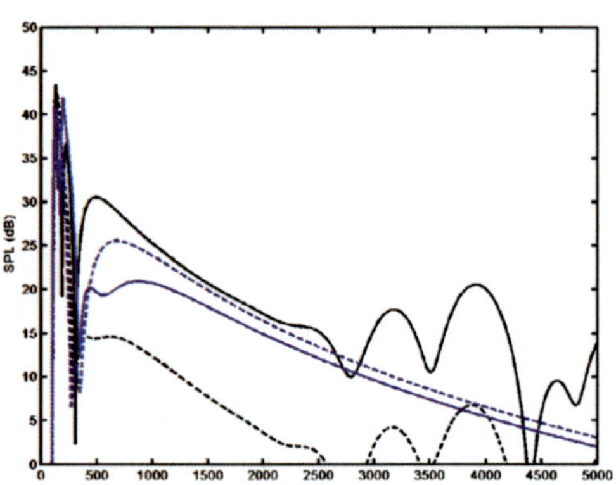

(f) Sound pressure levels at receiver heights 1.5m (dashed) and 4.0m (solid). Neutral (blue) and downwind conditions (black) 8m/s at 10m. f = 125 Hz

Figure 18. Sample PE predictions of transmission loss (a)-(d) illustrating numerical sound propagation results from an elevated source at 80m height in neutral and downwind conditions (8 m/s) at distances up to 5 km from source.

ground reflection at around 3.5 km to 4.5 km from the source, regions where sound levels around 20 dB to 40 dB are evident close to the ground surface. In comparison, for neutral conditions sound levels of 20 dB to 30 dB in the same region are apparent in Figures 18 (a) and (c).

In summary, the PE method allows for a precise description of the atmosphere as both the sound speed and the ground impedance can vary along the propagation path. The PE method gives accurate results at long distances but is limited in the near-field through the procedure's imposition of a maximum angle of elevation. The example presented here is for flat ground but can be extended to provide results for valleys or variable terrain cross-sections [1].

More complex models may also be used to validate and provide greater insight into the output of simpler models. An example of this can be found in the work reported by Heimann and Salomons [32] which was undertaken in support of the development of the Harmonoise model. The reported study involved the acquisition of detailed meteorological data over the course of a whole year. This data included wind speed, wind direction, humidity, pressure and temperature at multiple heights up to 100 m. The acquired meteorological data was then divided into separate contiguous six hour periods over the course of the year, with average meteorological parameters being calculated for each six hour period for subsequent use as inputs to a CNPE model. The overall A-weighted noise level was then calculated separately for each six hour period using the CNPE model at distances of 20 m, 200 m and 1000 m from a constant sound power output source. Thus variations seen in the output noise levels were limited to propagation effects. Figure 19 shows the results of the calculations assuming partly absorbing ground at a distance of 1000 m from the source.

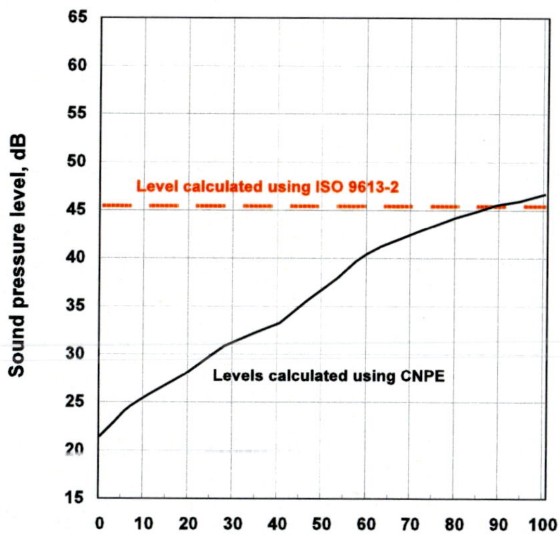

Figure 19. Percentiles of the calculated A-weighted sound pressure level at a 4.0 m height receiver located 1000 m from a 0.5 m height point source on partly absorbing ground. Results are shown both for the ISO 9613-2 model and for the CNPE method, the latter including the effects of varying meteorological conditions measured over the course of a year, as reported by Heimann and Salomons [32]

The results show a variation of calculated noise levels at the receiver over a range of approximately 25 dB(A). The upper levels relate to a positive sound speed gradient and the lower levels to a negative sound speed gradient. Also shown on the plot are the results of inputting the same source levels into the ISO 9613-2 model, again assuming partly absorbing ground. Rather than predicting a range of noise levels as a result of the varying meteorological conditions, the ISO 9613-2 model predicts a single noise level under the general condition of a positive sound speed gradient that promotes the propagation of sound from source to receiver. The difference in outputs between the two calculation procedures is evident from Figure 19, where the single level calculated using ISO 9613-2 intersects the range of levels calculated using the CNPE at approximately the 90th percentile.

Whilst the example provided here was not specifically related to wind turbines (the source height was 0.5 m and the receiver height 4.0 m), it nevertheless usefully illustrates the comparison between the targeted output of engineering methods of calculation such as ISO 9613-2 and the more flexible output of increasingly complex calculation procedures. A 2011 paper by Bolin and Karasalo [35] shows an example of calculations using a hybrid ray-tracing calculation method that accounts for varying atmospheric conditions, the ultimate output being the resultant long-term statistical distribution of noise levels from which the annual average L_{DEN} or L_{Aeq}, or indeed any other statistical index, may be calculated. A further example is provided by Andersson et al [39] where more complex models can aid in the understanding of observed effects, as discussed in greater detail in the following section.

MODEL VALIDATION

Having presented a general review of noise propagation calculation methodologies of varying degrees of complexity, this final section considers work undertaken to validate such models with specific regard to wind turbine noise measurements. Whilst many studies have reported similar validation measurements for other types of source, wind turbines exhibit certain characteristics that set them apart, thereby suggesting that dedicated validation measurements are required. Such characteristics include the height of the source, the extended physical size of the source, the wind speed dependence of the source sound output and the potential uncertainty of individual source levels on multiple wind turbine installations.

All noise propagation prediction models work on the basis that the source can be accurately characterised as an input to the model. However, all the foregoing issues can lead to uncertainty in the source characterisation. The certainty associated with noise propagation validation measurements based on measurements of wind turbine installations may, therefore, be encumbered by variable source effects. Potential issues additionally arise from the fact that wind turbine noise levels at distances of more than a few hundred meters are frequently within the range of the levels of background noise that may otherwise be encountered. This introduces particular difficulties when undertaking validation measurements. To illustrate this, Figure 20 shows a scatter plot of noise levels measured approximately 1 km from an operational wind farm containing ten wind turbines each in the 2 MW class. Each separate point represents a single measured $L_{A90,10min}$ noise level which has been plotted against the wind speed measured at the wind farm averaged over the same ten minute period.

Each of the individual measured noise levels includes contributions from both the wind farm plus any natural sources that may also have been present during the period. The heavy dashed blue line shows the wind turbine noise immission levels calculated using ISO 9613-2, modified as discussed previously. These predicted levels therefore equate to levels expected to occur under downwind conditions only. Under other conditions the wind farm noise immission levels would be expected to be lower. However, the predicted wind farm noise immission levels are embedded within the dominant range of noise levels actually measured. This is an indication that the scatter in the data is controlled primarily by the natural variability of the background noise environment.

In an attempt to overcome some of the aforementioned issues, reported validation studies for wind turbine noise have adopted two principle approaches: either to use an 'artificial' source of sound (by which is meant in this context a source which is not a wind turbine) located in an elevated position representative of turbine hub heights, or to use operational wind turbines but planning the experimental approach such that the ratio of wind turbine noise to residual noise is maximised, generally also basing results on the statistical analyses of large quantities of data rather than relying on individual measurements.

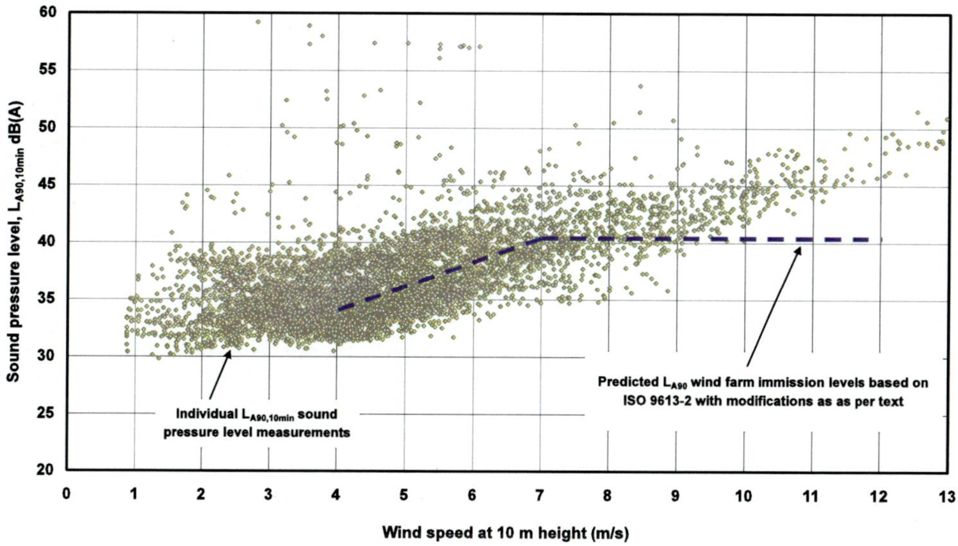

Figure 20: Scatter plot of $L_{A90,10min}$ sound pressure levels measured approximately 1 km from an operational wind farm and plotted against the wind speeds measured on the wind farm. The data covers an approximate three month long period. Also shown as the dashed line are the wind farm immission levels calculated in accordance with ISO 9613-2 with appropriate modifications for wind farm applications.

The advantage of using an artificial sound source is that its sound output can be accurately controlled. Provided a suitable source is selected, its output can also be set at levels higher than those radiated from wind turbines. As a consequence, higher signal to noise ratios may be achieved at larger distances than would be possible using wind turbines as the source. Furthermore artificial sound sources can be easily switched on and off to directly determine the influence of residual noise in the measured data. Thus, the measured noise levels can be more certainly associated with the specific source of interest, and any variations in these levels can be more positively ascribed to propagation effects. However, having devoted this chapter specifically to the consideration of propagation effects, it is important to acknowledge that such effects are just one element of the total 'source-propagation-receiver' model that needs to be considered in wind farm noise assessments. In applying the various available noise propagation models, the assumption is often made that each wind turbine radiates an 'apparent' sound power from a fixed point source located coincident with the hub of the turbine. The assumption is also commonly made that, at any given moment in time, the sound powers of all the individual turbines are equal, these individual sound powers being determined by an assumed wind speed distribution commonly being experienced across the rotors of all the turbines. In reality, however, the source is more complex. The actual sources comprise the potentially large radiating areas of the rotating blades with the associated directivity and convective effects, and with the turbines across a wind farm site possibly seeing differing air inflow conditions (and therefore potentially also radiating differing sound powers) over any given period of time. Clearly these factors may also affect the received noise levels in addition to any propagation effects. It is only through the study of noise radiated from actual wind turbines that the relative significance of such non-propagation related effects can also be established.

Starting with a review of validation measurements that have utilised artificial sound sources, the outcome findings of studies reported by Bass et al [33], Plosving et al [40] [41] and Andersson et al [39] are summarised here. The work reported by Bass et al [33] was undertaken in the early 1990s when turbines were of a smaller scale than today. As a consequence, the work focussed on measurements of sound propagation from an omnidirectional loudspeaker sound source elevated to heights of up to 30 m on a mast from which all measurements of meteorological conditions were also made. Measurements were undertaken on three different types of terrain classified as 'flat', 'rolling' and 'complex'. The source was continuously cycled on and off over periods of several weeks at each site and the resulting data filtered to correct for the effects of residual noise prior to its statistical analysis. Using this process useful data was obtained for source to receiver separation distances of up to approximately 700 m. The key findings of the study comprised:

- variability in received noise levels at large distances is primarily controlled by the vector component of wind between the source and the receiver;
- a positive component of wind speed blowing from source to receiver results in a small increase in received noise level of typically no more than 2 dB(A) when compared with neutral conditions with no identified increase with increasing wind speed;
- an increasing negative component of wind speed blowing from source to receiver results in a decreasing received noise level when compared to neutral conditions, although quantification of the maximum reductions other than an indication that it could be greater than 10 dB(A) was difficult due to corruption of the lower measured noise levels by residual noise;
- under downwind conditions the variability (and therefore also the predictability using more basic calculation techniques) of received noise levels is significantly lower than under upwind conditions;
- under downwind conditions the variability of received noise levels was found to increase with increasing source to receiver separation distance, a sample result being shown for the flat terrain site in Figure 21.
- where topographical features interrupted the direct line of sight between source and receiver the observed excess attenuations due to acoustic screening were limited to no more than 2 dB to 3 dB under conditions of a positive wind speed vector between source and receiver;
- where a valley existed between the source and receiver the calculated levels underestimated the measured noise level by approximately 2 dB to 3 dB;
- the use of practical engineering methods of noise prediction, and in particular ISO 9613-2, generally resulted in a good correlation between predicted and measured noise levels, with ISO 9613-2 predicting levels corresponding to the mean measured level plus one standard deviation under downwind conditions provided: that appropriate selections were made for the user selectable parameters in the ISO 9613-2; that the attenuations due to topographic screening were limited to no more than 2 dB; that a negative attenuation of 3 dB was accounted for in the case of propagation across a valley.

The experimental database obtained as part of the foregoing studies was later re-analysed by Alberola [34] who used more complex modelling techniques to identify the underlying causes of the variability in the received sound levels. Alberola concluded that for source to receiver separation distances of up to 300 m all variability in the received noise levels of the test dataset could be explained by the measured variability of the parameters included in the ISO 9613-2 model. However, at distances greater than 300 m the variability became dominated by meteorological factors not incorporated into the ISO 9613-2 model, most notably vector wind speed.

More recent work based on the use of artificial sound sources has been reported by Sondegaard et al [40] [41]. This work was aimed at validating the use of the Nord2000 model for elevated sound sources and involved the elevation of the source to heights of up to 50 m over flat terrain and 70 m over complex terrain. Useful data was achieved at source to receiver separation distances of up to 1500 m over flat terrain and 1000 m over complex terrain, with more attention being given to third octave band accuracy than in the earlier work reported by Bass at al [33]. The key outcome conclusions of the work were that Nord2000 is capable of predicting overall noise levels under downwind conditions with a standard deviation difference between predicted and measured noise levels of less than 1 dB(A). Nord2000 was also demonstrated to be capable of producing quite usable third octave band levels under these same downwind conditions, albeit to a relatively lower degree of accuracy than the overall noise level predictions. Unlike ISO 9613-2, Nord2000 is capable of explicitly incorporating the effects of meteorological conditions. Under upwind conditions the agreement between predictions and measurements was reduced, but the method still produced usable results provided it is recognised that a degree of conservatism is built in to the results in that the predicted levels are typically around 4 dB higher than the measured overall A-weighted noise levels with a standard deviation of approximately 2 dB. The point is additionally made that these greater discrepancies occur in an acoustical shadow zone where the overall noise levels are, in any event, generally low and more variable in practice.

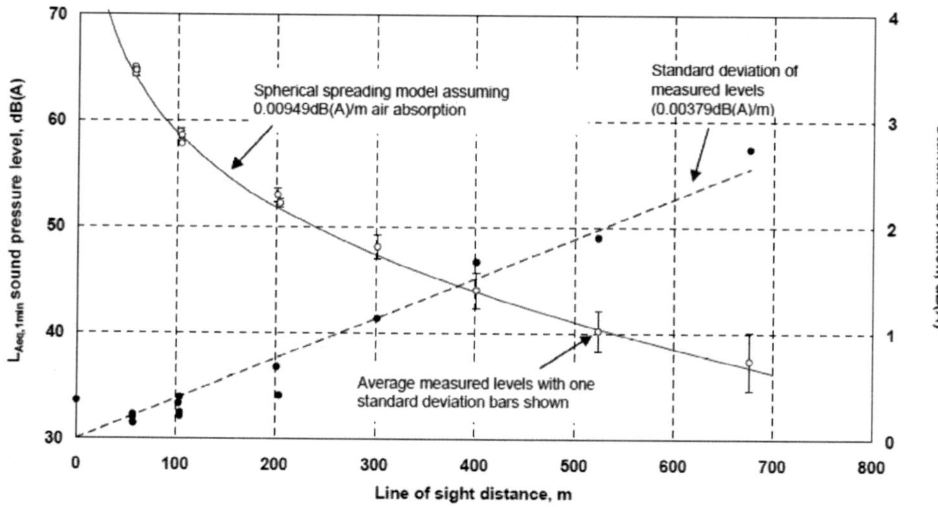

Figure 21. Results from the noise propagation study reported by Bass at al [33] showing, as the hollow white markers, the average sound pressure levels measured at a 1.5 m high receiver as a function of increasing distance over flat arable land from a 30 m high omnidirectional loudspeaker sound source. Also shown are the one standard deviation ranges in the measured data at each distance. The solid markers show the increasing standard deviation as a function of increasing distance, with the dashed line showing the best fit linear regression to this data. All results are limited to downwind propagation conditions only.

Another study reported by Andersson et al [39] studied the propagation of narrow band noise over water for much greater distances of approximately 10 km. The source characteristics comprised a loudspeaker and coupled resonator tube producing a tone at 80 Hz and a compressed air siren producing a narrow band frequency centred around 200 Hz but with a usable harmonic at around 400 Hz. The height of the source used for these tests was 30 m above the water while the receiver was located at a height of 1.7 m above local ground level approximately 750 m inland across flat ground with a typical height above sea level of 7 m. Meteorological data was acquired at lower levels of up to 90 m from an instrumented mast at the source location, and at higher levels from free flying balloons up to 3500 m at the receiver location. Using a parabolic equation model the predicted noise levels varied by about 40 dB between downwind and upwind propagation. However, these calculated differences were not reflected in the reported measurements, which more typically indicated differences of around 20 dB. The parabolic equation model was then used to demonstrate the influence of atmospheric turbulence in 'leaking' sound energy into the shadow regions that are otherwise predicted to exist under conditions of a negative sound speed gradient (i.e. upwind or temperature lapse), hence limiting the reductions that are observed in practice in acoustical shadow zones, as shown in Figure 22(a). The report also identifies the possible existence of a low level 'channel' in which sound becomes trapped closer to the surface, resulting in enhanced noise levels at larger distances. This effect occurs when the sound speed has a maximum at a relatively low level (approximately 200 m in the example reported, see Figure 22(b)) thus causing the sound to be refracted down within the channel. This result highlights the potential dangers of using 'general' engineering methods of calculation under specific circumstances in that such effects can not be modelled using such idealistic methods. The particular special case identified by Bolin is recognised in the scope of ISO 9613-2 which specifically warns that such conditions may result in higher sound pressure levels than predicted using the Standard.

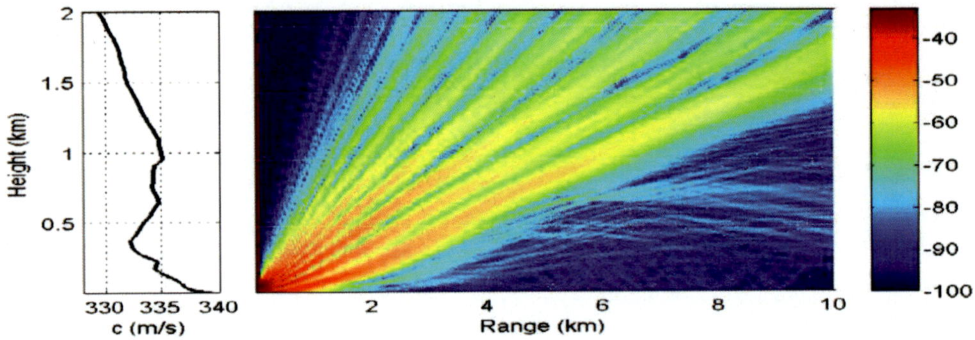

Figure 22(a). Sample result of transmission loss obtained by Andersson et al [39] for long distance propagation across water using a parabolic equation model. The figure shows the formation of an acoustical shadow zone under conditions of a negative sound speed gradient (i.e. upwind or temperature lapse) but with the 'spilling' of sound energy back into the shadow zone through the effects of turbulence. This turbulent scattering limited the attenuations actually observed in measurements reported in the same paper to around 20 dB. Exclusion of the effects or turbulence resulted in predicted attenuations of 40 dB or more.

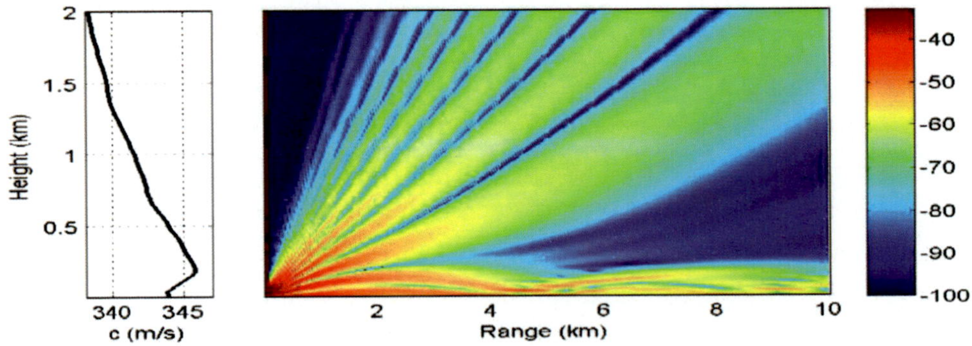

Figure 22(b). Sample result of transmission loss obtained by Andersson et al [39] for long distance propagation across water using a parabolic equation model. The figure shows the formation of a sound 'channel' under conditions of a positive sound speed gradient (i.e. downwind or temperature inversion) to a height of approximately 200 m with the sound speed gradient being negative above this height.

In terms of validation measurements undertaken on operational wind turbines, Bass et al [33] compared predictions using the ISO 9613-2 model with measurements undertaken at three operational wind farms. All three sites had more than 20 relatively large turbines rated at over 2 MW generating capacity each, with hub heights of 60 to 70 meters. The analysis first required careful considerations of the relevant wind speed reference (meteorological masts or turbine anemometers). Considerable data filtering was undertaken to minimise the influence of extraneous sources and to restrict analysis to conditions of downwind propagation only, as predicted by the ISO 9613-2 model. The analysis also focused on conditions where the signal-to-noise ratio was maximised. Despite these restrictions, the difficulty remained in isolating with 100% certainty the measurement contribution attributable to turbine immissions alone, but the comparison of measurements made at similar distances but with varying background environment allowed meaningful comparisons. The agreement with predictions was found to be quite acceptable in this context. Conservative results were observed when considering the use of commercial warranted sound power levels, or from the choice of pessimistic propagation parameters such as hard ground conditions, the latter conservatism being observed even in situations where the choice of such parameters would have been expected to provide a good representation of actual conditions. The conclusions of this work are in line with recommendations for best practice in the use of the ISO 9613-2 model outlined above [38].

A noteworthy finding of the studies just referenced [39] was that of the effect of turbine source levels. The general assumption made when predicting wind farm noise immission levels is that, for a given site wind speed, all the turbines on a multiple turbine wind farm will radiate the same sound power level. However, in practice significant variations in wind speeds can be experienced across the different turbines for any given site wind speed when this is determined, as is common practice, from measurements on a single meteorological mast. For one of the reported sites variations in wind speed across individual turbines translated to potential variations in sound power levels in excess of 2 dB around the sound power that would otherwise have been expected for that site wind speed. A standard deviation of +/- 0.7 dB in sound power levels was similarly found to exist across all wind turbines. Recalculating the wind farm noise immission levels using ISO 9613-2, but this time adopting the sound powers based on the actual wind speeds seen by each individual turbine as opposed to a common sound power across all turbines based on a single site wind speed reference, resulted in an improved agreement between predictions and measurements. It is also noteworthy that the variation in sound power levels across the wind turbines tended to result in lower rather than higher downwind noise immission levels. This effect resulted from those turbines lying closest to the receiver location also lying downwind of the other turbines, thus generally experiencing a lower wind speed than the site average. These closest located turbines therefore radiated a lower sound power than would be assumed based on the site averaged wind speed, with the consequence that the measured downwind noise immission levels generally were not increased above those predicted using a single site wind speed reference for all turbines. Notwithstanding this finding, the adoption of wind turbine sound power levels that include the test measurement uncertainty also indirectly compensates for potential variations in sound power levels across the turbines.

The studies by Plovsing et al [40] [41] were aimed at validating the Nord2000 model for wind farm applications. Extended measurements were undertaken to include a single operational wind turbine in addition to the previously discussed use of an artificial sound source. Signal to noise issues limited the maximum propagation distance at which useful results were obtained to 800 m downwind of the turbine. The outcome results of the study were that, whilst the use of the Nord2000 model could still be recommended as providing acceptable results, the agreement between measurements and predictions was less good than for the artificial sound source, with a measured standard deviation of 2.3 dB in the overall A-weighted sound pressure level measured under downwind conditions. Earlier work reported by Bass et al [33] also compared measurements and predictions at similar distances of up to 800 m downwind of the closest located turbine of a wind farm site comprising 42 turbines, these all being 300 kW machines with a hub height of 31 m and rotor diameter of 28 m. Similar results were obtained, with typical measured standard deviations in the overall A-weighted levels of around 2.5 dB at distances of up to around 500 m. At greater distances the corresponding standard deviations increased to over 5 dB, but this was identified as being due to corruption by extraneous sources of noise due to the low signal to noise ratio at these larger distances. The results of more recent studies on measured wind farm noise immissions have been reported by Forssen et al [27]. One of the key drivers of this work was to establish potential variability in practically experienced noise immission levels with a view to explaining some of the variability in reported subjective response to such noise. The work compared the results of measurements made 530 m from two 65 m hub height wind turbines against the output results both of engineering methods of calculation, including Nord2000, and also the output of a Crank Nicholson Parabolic Equation procedure, with the focus being on downwind propagation conditions. The study concluded that the use of engineering methods of calculation provided a good estimate of measured noise levels under downwind conditions, the agreement being to within one standard deviation of the measured levels. The study did further identify, however, that the variability in the received noise levels under downwind conditions at any given wind speed could not be fully explained by propagation factors alone and that variability in turbine emission levels may therefore also be a significant factor. Most recently Evans and Cooper [43] have reported a study into the use of several practical engineering methods, including ISO9613-2 and NORD2000, for wind farm noise predictions. The report concludes that engineering methods of calculation can result in good agreement with noise levels experienced in practice, but it also highlights the important effect that topography can have on measured noise levels.

As a final practical example, reference is made again to Figure 20 which showed the extent to which wind turbine noise may be embedded within the general range of environmental noise levels measured at typical residential separation distances from a wind farm. In order to better isolate wind farm noise from residual noise the data must be heavily filtered, with possible data rejection criteria including: removal of all periods with rainfall; removal of periods following rainfall when elevated noise from running watercourses may occur evident; use of data recorded during the night only; use of data recorded when the monitoring location was downwind of the wind farm only; use of data only when all turbines are operating normally; use of data at wind speeds for which the wind farm noise to residual noise ratio is likely to be highest; rejection of data when any other non-wind farm related noise sources were known to be significant, common examples being the early morning bird ‘dawn chorus’, agricultural activity and traffic noise.

By adopting the foregoing criteria, resultant datasets contain noise data points that generally correspond to those periods when the wind farm noise is at its highest (i.e. when all turbines are operating and the monitoring location lies downwind of the wind farm) and the background noise is at its lowest (i.e. when it is not raining and only during the night when human and animal activity has died down). Figure 23 shows the individual measured $L_{A90,10min}$ data samples remaining after the aforementioned data filtering exercise has been undertaken on the data previously shown in Figure 20. The results in this instance clearly indicate the effectiveness of the modified ISO 9613-2 model in predicting levels at the upper end of the range of the actual levels measured under downwind conditions. It should be noted that, despite the heavy data filtering that has been undertaken in order to try and isolate those periods when wind farm noise dominates the overall measured $L_{A90,10min}$ noise levels, it can not be concluded that any individual data point lying above the predicted worst case wind farm noise immission level curves is still not the consequence of residual noise, from whatever source other than the wind farm, that may have occurred in the affected measurement period. For this reason, when assessing the results of wind farm noise immission measurements the use of some form of averaging is generally advised as reliance on individual data points may lead to erroneous conclusions. In this regard the solid curve in Figure 23 shows the best fit third order polynomial regression curve derived from the measured data points.

CONCLUDING REMARKS

Having presented an overview of various environmental sound propagation calculation methodologies of gradually increasing complexity, the question may now reasonably be asked as to which is the

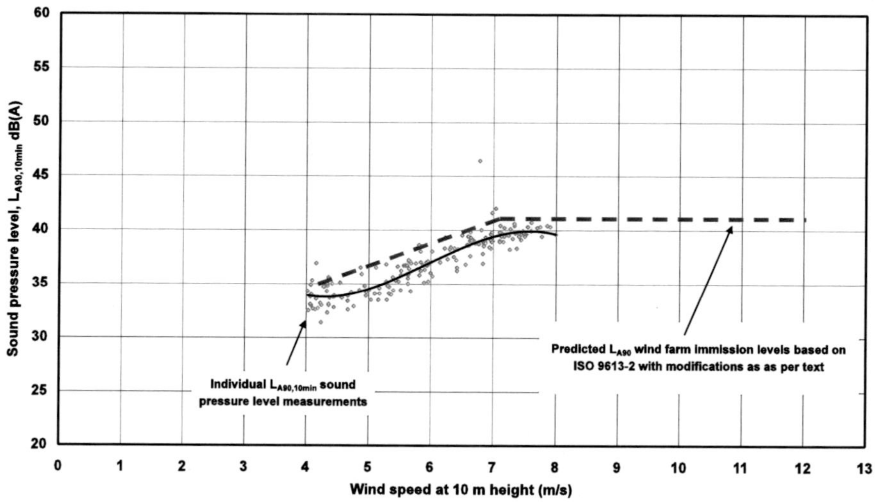

Figure 23. Scatter plot of $L_{A90,10min}$ sound pressure levels measured approximately 1000 m from an operational wind farm and plotted against the wind speeds measured on the wind farm. The data has been filtered to isolate as best as possible the contribution of wind farm noise only from the total measured levels previously shown in Figure 20. Also shown as the dashed line are the wind farm immission levels calculated in accordance with ISO 9613-2 with appropriate modifications for wind farm applications, as discussed previously.

preferred methodology for wind farm noise. The answer to this question lies in the introduction to this chapter, where it was highlighted that the choice of model should be guided by the use to which the model's outputs are to be put. For this reason it is not possible to recommend a single 'best' model. Sample issues that should be accounted for when selecting the appropriate model for a given situation include:

I) Is it required to predict noise levels based on a simple 'worst case' basis?

II) Is it required to predict noise levels for one or more specific sets of meteorological conditions and, if this is the case, then does the model allow the specification of the necessary parameters?

III) Is it required to predict overall A-weighted noise levels or are band limited noise levels, or even tonal noise levels, required?

IV) If the model requires the detailed specification of multiple input parameters, are the values of these parameters known to the required degree of accuracy?
N.B. It should be remembered that the output of 'accurate' or even 'exact' models are only as good as their input data: inaccurate input data may give rise to outputs that are no better than much simpler models, possibly even misleading the user into reaching erroneous conclusions that aren't actually a reflection of the noise model but are instead solely a function of the input data.

V) What resources, including computational power, are available to undertake the modelling exercise?

VI) What is the ultimate use of the results and what would the financial, social or other possible consequence be of the potential over or under prediction of the levels?

Hybrid methods are generally based on solutions of the wave equation, ranging from simplified analytical solutions (simple ray tracing) through to direct solutions such as the wave solution over absorbing ground and large-scale numerical solutions such as the Parabolic Equation. They possess the

Table 7. Comparative guide to properties of different sound propagation prediction models

Characteristic	Engineering			Hybrid modelling methods			
				Approximate / analytical		Numerical	
	NORD2000	**HARMONOISE**	**ISO 9613**	**Ray tracing**	**WAVE**	**PE**	**GFPE**
Computing time (Present)	Fast	Fast	Fast	Fast	Fast	Slow	Medium
Accuracy (Relative)	High	High	Poor	Medium	Exact	Very good	Good
Ideal frequency range	All freq	All freq	All freq.	High frequency	All	Low frequency	Low and mid frequency
Range-dependent conditions	NO	NO	NO	NO	NO	NO	YES
Shadows and caustics	NO	NO	NO	NO	NO	NO	NO
Elevated sources	YES	YES	YES	YES	YES	YES	NO
Met Conditions	YES	YES	~	NO	NO	YES	YES
Banded freq	YES	YES	YES	NO	NO	NO	NO

ability to provide highly accurate representations of propagation effects for individual frequencies in certain conditions, albeit with some assumptions. Such methods have provided the basis for the 'reference model' used to validate the engineering method produced by Harmonoise. They have also been successfully used to assess the parametric sensitivity of received noise levels and to isolate and better understand individual propagation effects.

On the downside, the techniques are generally computationally intense and are most commonly employed only for 2D predictions except where radial symmetry has been assumed. Furthermore, the methods are not widely available within common commercial software and they involve fairly complex computational procedures to implement. It should also be recognised that there are different ranges of accuracy, relating to limited source radiation angle, optimum frequency ranges, non-homogenous ground conditions, and topographical complexity. As discussed above, the accuracy will be in reality limited by the imperfect knowledge of the full range of atmospheric and ground conditions along the propagation path.

The brief review contained herein indicates that numerical methods have many strengths, mainly in accuracy, but also some weakness, mainly in practical application. It is reasonable to conclude that none of the examined methods is capable on its own of handling all possible environmental conditions, frequencies and transmission ranges of interest in practical applications. One method will be more appropriate than another for a particular problem scenario, and thus selection of the best method must be situation specific.

As a rough preliminary guide to demonstrate how the most appropriate method might be selected, an attempt has been made to compare in relative terms how well the properties of the various modelling methods would deal with important procedural characteristics. For comparison purposes the practical engineering method ISO 9613-2 has also been included. A summary is provided in Table 7.

In conclusion, for most practical applications engineering methods, provided they are applied correctly within their appropriate scope of applicability, provide reasonably accurate predictions. However, what determines 'reasonable accuracy' should be interpreted within the context of the natural variability of environmental sound fields, a feature which occurs in practice both for general sources as well as wind turbines and also within the context of the use to which the output results of any modelling exercise are to be put.

REFERENCES

[1] E. M. Salomons, *Computational atmospheric acoustics*, 2001, Kluwer Academic Press.

[2] K. Attenborough, K. M. Li and K. Horoshenkov, *Predicting outdoor sound*, 2006, Taylor & Francis.

[3] D. Bies and C. Hansen, *Engineering noise control; theory and practice*. Spon Press 2009.

[4] T.F.W. Embleton, G.J. Thiessen and J.E. Piercy. *Propagation in an inversion and reflections at the ground.* 1976, J. Acoust. Soc. Am. 59:278-282.

[5] D.G Albert, *Observation of acoustic surface waves propagating above a snow cover*, Proc. Fifth Symposium on Long Range Sound Propagation, 10-16, 1992

[6] S. Bell. A beginner's Guide to Uncertainty in Measurement. GPG(011),. National Physical Laboratory, August 1999.

[7] Review and analysis of published research into the adverse effects of industrial noise, in support of the revision of planning guidance. FINAL REPORT MARCH 2004, DEFRA Ref. NANR 5 Bernard F Berry, Berry Environmental Ltd in association with Nicole Porter Independent Consultant.

[8] Imagine – Improved Method for the Assessment of Generic Impact of Noise in the Environment. State of art review, Deliverable 2 of the Imagine Project, Document identity: IMA10TR-040423-AEATNL32, 14 October 2004

[9] HARMONOISE, FINAL TECHNICAL REPORT Final public version. Document identity: (HAR7TR041213AEAT04public.doc), 1 March 2005

[10] HARMONOISE Reference Model Task 2.4, Choice of basic sound propagation models. Deliverable 14 of the Harmonoise project, HAR24-TR021018-TNO10.doc. 12 December 2002.

[11] HARMONOISE Reference Model, Description of the Reference model Deliverable 16 of the Harmonoise project, HAR29TR-041118-TNO10.doc, 22 December 2004

[12] HARMONOISE Technical Report - Deliverable 18, Harmonoise WP 3 Engineering method for road traffic and railway noise after validation and fine-tuning, HAR32TR-040922-DGMR20, 20 January 2005.

[13] HARMONOISE, Validation of the Harmonoise models, Deliverable 21, HAR28TR-041109-TNO11.doc, 24 February 2005

[14] European Commission Working Group Assessment of Exposure to Noise (WG-AEN), Position Paper, Good Practice Guide for Strategic Noise Mapping and the Production of Associated Data on Noise Exposure, Version 1, 5 December 2003.

[15] ISO 9613-1. Acoustics of sound during propagation outdoors – Part 1: Calculations of the absorption of sound by the atmosphere.

[16] ISO 9613-2. Acoustics of sound during propagation outdoors – Part 2: General method of calculation.

[17] M. Delaney and E. Bazley *Acoustical properties of fibrous absorbent materials*. 1970. Appl. Acoust. 3(2): 105-116.

[18] NordTest Method NT Acou 107 Approved 2001-05. Acoustics: A Framework for the Verification of Environmental Noise Calculation Software

[19] G. A. Daigle and M R. Stinson Attenuation of Environmental Noise: Comparing the Predictions of ISO 9613-2 and theoretical models. Internoise 2005.

[20] Benoit GAUVREAU, Bertrand LIHOREAU, Michel BERENGIER, Philippe BLANC-BENON, Isabelle CALMET Estimating long-term representative SPL in complex environments using a coupling method based on acoustic (MW-WAPE) and micrometeorological (SUBMESO) numerical predictions., Internoise 2005

[21] Fabrice JUNKER, Benoit GAUVREAU, Michel BERENGIER, Philippe BLANC-BENON, Cora CREMEZI-CHARLET Qualitative and quantitative study on relative influence of physical parameters affecting long-range sound propagation., Internoise 2005.

[22] Igor ROZMAN, Ale_ GLOBEVNIK and Lilijana KUHELJ UNCERTAINTY IN NOISE MAPPING OF INDUSTRIAL SOUND SOURCES., Internoise 2006.

[23] Richards. J M VARIATION OF ISO 9613, CONCAWE AND THE JOINT NORDIC METHODS OF PROPAGATION CALCULATION,., W.Kellogg Limited, Newman. M, Campus Marine A.S, Institute of Acoustics Conference Paper, Vol. 25. Pt 3. 2003

[24] The Uncertainty of Sound Pressure Levels calculated with Noise Prediction Programs, Wolfgang Probst, Ulrich Donner, ACCON GmbH. www.datakustik.de/articles/Uncertainty.pdf

[25] J. Nilsson. Sound propagation from wind turbines – A comparison of measurements and calculations at the forest site Ryningsas. MSc. Thesis. Faculty of Engineering, Lund University, 2009.

[26] G. B. Jonsson. A comparison of two numerical models for outdoor sound propagation – Harmonoise and Nord2000. MSc thesis. Acoustic Technology, DTU, 2007.

[27] J. Forssen, M Schiff, E Pedersen, K Waye. Wind turbine noise propagation over flat ground : Measurments and predictions. Acta Acustica 96. 2010

[28] Naturvårdsverket 2009. Ljud från vindkraftverk, reviderad utgåva av rapport 6241, Koncept 26 feb 2009. NV dnr 382-6897-07 Rv

[29] K. Bolin. Wind turbine noise and natural sounds – masking,propagation and modeling. PhD thesis. Department of Aeronautical Engineering. Kungl Tekniska Hogskolan. TRITA-AVE-2009:19. 2009.

[30] M. West, K Gilbert and R Sack. A tutorial on the parabolic equation (PE) model used for long-range sound propagation in the atmosphere. 1992. Applied Acoustics (37) 31-49.

[31] A. D. Pierce, *Diffraction of sound around corners and over wide barriers*. J Acoust Soc of Am, 1974, 55, 941-955.

[32] D. Heimann and E. M. Salomons. *Testing meteorological classifications for the prediction of long-term average sound levels*. Applied Acoustics 2004 (65) 925-950.

[33] J. H. Bass and A. J. Bullmore and E. Sloth. *Development of a wind farm noise propagation prediction model*. Final report. Contract CEC/DTIJOR3-CT95-0051. 1998.

[34] J. Alberola. *Predicting variability in environmental noise measurements* PhD thesis. ISVR 2004.

[35] K. Bolin and I. Karasalo. *Wind turbine noise exposure in a complex terrain.* 2011. Fourth International Meeting on Wind Turbine Noise.

[36] J. Adcock, A. Bullmore, M. Jiggins and M. Cand, *Wind farm noise predictions – the risk of conservativsm.* 2007 Second International Meeting on Wind Turbine Noise, Lyon, France.

[37] A. Bullmore, J. Adcock, M. Jiggins and M. Cand. *Wind farm noise predictions and comparison with measurements* 2009 Third International Meeting on Wind Turbine Noise, Aalborg, Denmark.

[38] Bowdler, Bullmore, Davis, Hayes, Jiggins, Leventhall, McKenzie *Prediction and assessment of wind turbine noise – agreement about relevant factors for noise assessment from wind energy projects.* 2009. 34(2) Proc IOA (Acoustics Bulletin).

[39] B.L. Andersson, K. Bolin, A Cederholm and I Karasalo *Long distance sound propagation over a sea surface* 2009 Third International Meeting on Wind Turbine Noise, Aalborg, Denmark.

[40] B Sondergaard and B. Plovsing *Prediction of noise from wind farms with Nord2000 Part 1* 2009 Third International Meeting on Wind Turbine Noise, Aalborg, Denmark.

[41] B. Plovsing and B Sondergaard *Prediction of noise from wind farms with Nord2000: Part 2* 2009 Third International Meeting on Wind Turbine Noise, Aalborg, Denmark.

[42] C. F. Chien and W. W. Soroka *A note on the calculation of sound propagation along an impedance surface.* Journal of Sound and Vibration 1980, 69. 340-343.

[43] T. Evans and J. Cooper Comparison of measured and predicted wind farm noise levels and implications for assessments of new wind farms. *Proceedings of Acoustics 2011*, Conference of the Australian Acoustical Society, November 2011

Chapter 4. Wind Turbine Noise at the Receiver

Bo Søndergaard (Denmark), Grontmij Acoustica.
Bo.Sondergaard@grontmij.dk

INTRODUCTION

If there were no receivers there would be no problems with noise from wind turbines and this book would never have been written. When wind turbines are erected in remote areas, as sometimes happens, there are no receivers. However, most wind turbines and wind farms are built in the vicinity of local settlements such as small villages, individual residences, farm houses and so on. Because of this the authorities have set up criteria for noise limits with the intention of protecting neighbours from the wind turbine noise. The degree of protection is a political decision based on what is considered to be an acceptable balance between the need for the project and the number of people that are affected and the degree to which they are affected. This is dealt with in detail in Chapter 8.

Most noise criteria are based on free field noise levels only, taking into account the influence of the ground, but not the local geometry of buildings at the receiver. Almost no noise criteria include indoor noise levels, largely because of the difficulties in measuring inside occupied properties. But the low frequency part of the noise, which can be dominant at large distances because the higher frequencies have been selectively attenuated by air absorption, becomes even more dominant internally, because higher frequencies are again selectively attenuated by building structures. It is generally accepted that upwind wind turbines of a standard design do not emit audible infrasound, but an investigation made by Hayes-Mackenzie [1] concluded that low frequency noise (20 to 200Hz) was measurable and could be rated audible according to ISO 226 [2]. Comparing C-weighted and A-weighted noise levels, especially in-doors is considered to be an inadequate way of rating the low frequency component of a noise but can be used as an indicator of whether a more detailed analysis should be made. Although most noise criteria are based on external free field measurements, the noise perceived at – neighbouring properties can be different due to reflections, and other local conditions. This chapter will deal with local phenomena that influence the level and characteristics of wind turbine noise at the neighbouring residences, such as reflections and noise insulation of buildings. Audibility and masking from background noise will also be treated. Some model calculations are made for illustration. As the noise spectra of wind turbines do not change significantly with size [3] the conclusions based on the spectral content of the noise are valid for most turbines.

LONG RANGE PROPAGATION

When noise propagates from a wind turbine or wind farm to the receiver position, it is attenuated due to spherical spreading and is influenced by air absorption, ground and vegetation absorption, screening from terrain and the meteorological conditions. The details of this mechanism can be seen in detail in chapter 3 but, in summary, the general formula for noise propagation can be written as:

$$L_{Aeq} = L_{WA} + \Delta L_d + \Delta L_a + \Delta L_r + \Delta L_s + \Delta L_v + \Delta L_g \qquad [1]$$

where

- L_{eq} is the A-weighted noise level at the receiver
- L_{WA} is the A-weighted sound power level
- ΔL_d is the correction term for distance (divergence), spherical spreading is used as the source is modelled as a point source.
- ΔL_a is the correction term for air absorption
- ΔL_g is the correction term for ground effect
- ΔL_r is the correction term for reflection
- ΔL_s is the correction term for screening
- ΔL_v is the correction term for vegetation

All of these correction terms, except the correction for distance, are frequency dependent and the most used prediction model ISO 9613-2 [4] works in 1/1-octave bands from 63 Hz to 8 kHz.

For wind turbines the most important terms are the correction for distance which ensures a general lowering of the noise level over the entire spectrum, the correction for air absorption, which reduces the noise at all frequencies but at a higher rate at the higher frequencies and the correction for influence of the ground, which is a bit more complex. Air absorption coefficients are shown in Figure 1.

Tempera-ture	Relative humidity	Atmospheric attenuation coefficient α, dB/km							
		Nominal midband frequency, Hz							
°C	%	63	125	250	500	1 000	2 000	4 000	8 000
10	70	0,1	0,4	1,0	1,9	3,7	9,7	32,8	117
20	70	0,1	0,3	1,1	2,8	5,0	9,0	22,9	76,6
30	70	0,1	0,3	1,0	3,1	7,4	12,7	23,1	59,3
15	20	0,3	0,6	1,2	2,7	8,2	28,2	88,8	202
15	50	0,1	0,5	1,2	2,2	4,2	10,8	36,2	129
15	80	0,1	0,3	1,1	2,4	4,1	8,3	23,7	82,8

Figure 1. Air absorption coefficients from ISO 9613-2.

The complex effect of the ground has been discussed in detail in the previous chapter. In practical applications it is common to use either octave or one-third octave bands of sound rather than narrow bands. Some of the models available using such bandwidths and used commonly for wind turbine noise are also discussed in the previous chapter. By way of example the graph below shows the ground effect calculated for a typical turbine and receiver configuration by Nord2000 and ISO 9613 together with the measured results. As can be seen the excess attenuation varies with frequency but not in a relatively simple fashion as with air absorption.

As a consequence of the correction terms, the noise spectrum changes when the noise travels from the source to the receiver. There is a slight shift towards low frequencies mainly due to air absorption which is most efficient at higher frequencies. This is often used as an argument that wind turbine noise is specifically low frequency in nature. This is not the case as the sound power spectrum of a wind turbine is broad band and has its most significant part at frequencies from 200 Hz to 2000 Hz. In Figure 3 A-weighted noise spectra predicted at distances from 300 m to 10 km are shown and it is clear that the important part of the spectrum is changing towards the lower frequencies. At the shorter distances it is the frequency range 200 Hz to 2 kHz which is most important and at 5 km it is the frequency range 125 Hz to 500 Hz which is dominant. But at these distances the levels are rather low as noise at all frequencies is attenuated equally with the distance due to spherical spreading. The predictions are made for a typical spectrum for a large wind turbine with a hub height of 90 m using Nord2000, flat farmland and a downwind wind speed of 8 m/s. At 300 m the A-weighted noise level is 45 dB.

For a large wind farm (68 wind turbines, 140 MW) with nearest receiver around 1 km and a cross section of the wind farm of 2 to 3 km, the overall spectrum at the receivers will be shifted further towards the lower frequencies as the sum of the noise from the farthest wind turbines is not negligible. An example of this is included in Figure 2. There are no clear rules as to when the most distant wind turbines can be ignored. It is dependent on the geometry of the wind farm. In a square layout the most distant wind turbines are in a group while in a line array with the residence at the end of the line the

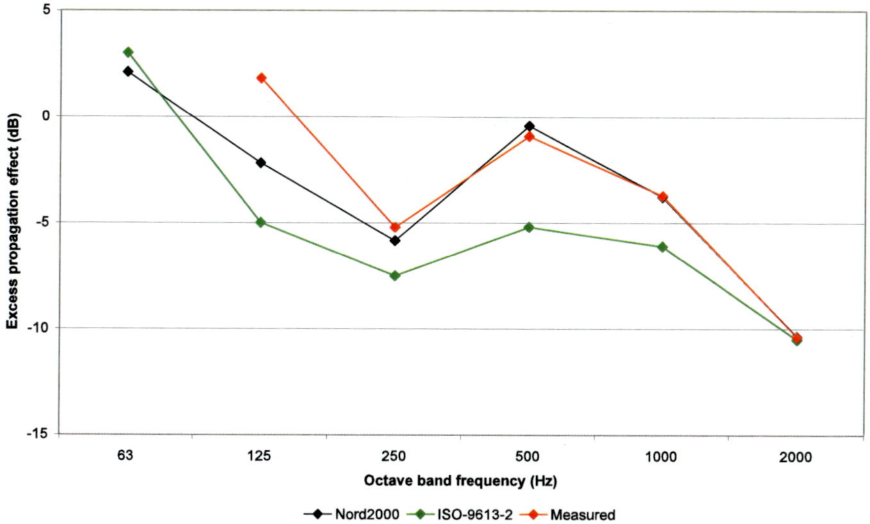

Figure 2. Ground correction (Excess propagation effect) predicted by Nord2000 (black), by ISO 9613-2 (green), and measured (red). Downwind, distance 1500 m, source height 50 m, and receiver height 2 m

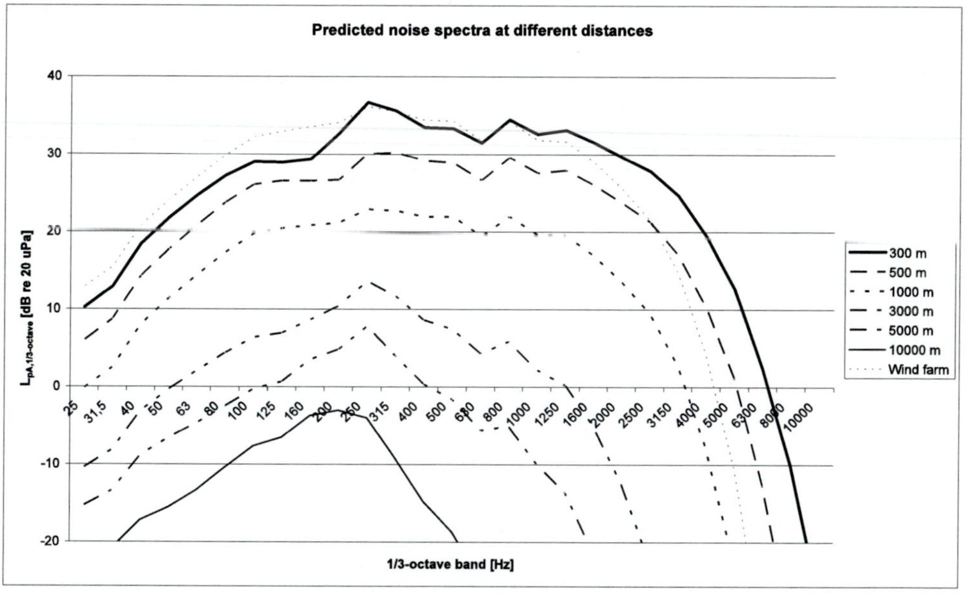

Figure 3. Predicted out-door noise spectra at different distances from a single multi MW-wind turbine. Included is the noise spectrum for a large wind farm giving a noise level of 45 dB(A).

noise is dominated by the nearest wind turbines. In both cases it is assumed that there are similar propagation situations over the entire wind farm.

It is important to distinguish between active/passive stall regulated wind turbines and pitch/pitch-rpm regulated wind turbines even though the general spectrum of the these wind turbines do not deviate significantly. For stall regulated wind turbines the noise increases considerably with the wind speed at wind speeds above 8 m/s, which is normally taken as the reference wind speed, while pitch regulated wind turbines have a maximum noise level at around 8 – 10 m/s. This is illustrated in Figure 4. Large wind farms with stall regulated wind turbines can actually be heard at larger distances and thus at lower frequencies, than wind farms with pitch regulated wind turbines at high wind speeds. Almost all new MW wind turbines are pitch controlled with a well known maximum sound power level. This design is better for controlling the noise and the load.

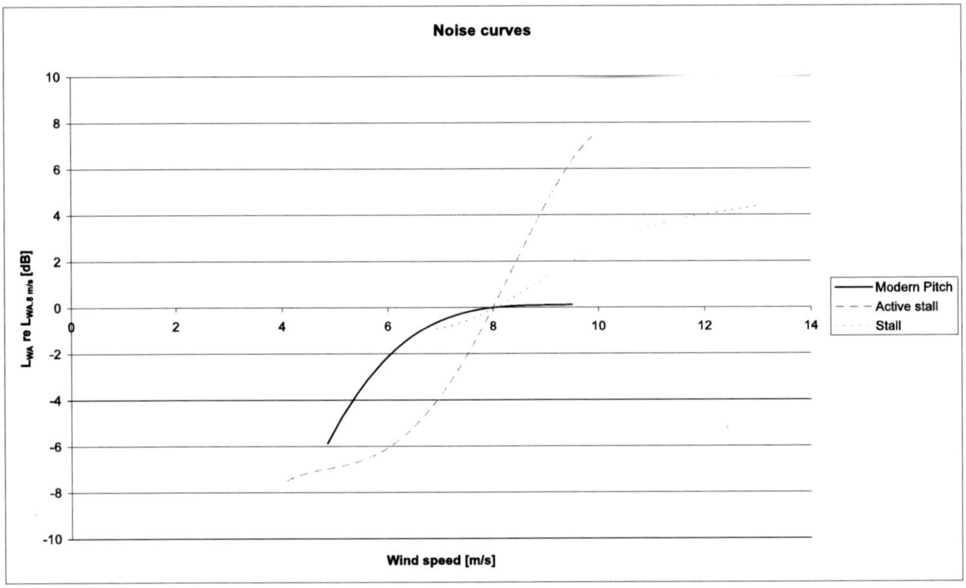

Figure 4. Noise curves for 3 types of wind turbines. The curves are normalised to pass through 0 dB at 8 m/s for comparison, but are from actual measurements.

FACTORS AFFECTING EXTERNAL NOISE

Steep mountainsides with bare rock and forest edges close to the receiver can in principle contribute with reflections. Not many prediction models include mountain sides or forest edges as sources of reflections but we know from studying echoes that these reflections actually occur. In Nordtest method NT ACOU 099 [6] dealing with prediction of shooting noise the reflection coefficient of a partly wood covered mountain slope is estimated to be 0.015 and for a dense forest to 0.025 corresponding to losses of 18 and 16 dB per reflection. Shooting noise and noise from wind turbines is similar in the sense that it involves large distances and most often the same type of area. However the spectrum of shooting noise is probably slightly more high frequency than wind turbine noise and the losses in the case of turbine noise could be less than those in the shooting noise method. The size of the losses however gives a good indication why reflections from mountain sides and forest edges usually do not contribute to the overall noise level. If, however, the direct noise path is screened, for example in a mountainous area and the reflected path is not, reflections can be significant if the transmission path lengths are comparable.

Local factors that mainly influence the noise level at receivers are the buildings at a receiver itself or neighbouring residences. Criteria are based on free-field principles. Reflections from the buildings at the residence such as the façade effect are excluded in the noise predictions and only reflections from neighbouring buildings are included. From a receiver perspective this can seem strange as the noise in an enclosed yard can be influenced by single or multiple reflections. If you stand within 1 m of a façade pointing in the direction of a wind farm the noise you hear will be increased due reflections.

The amount of reflected sound is dependent on the size of the reflecting surface, angle of incidence, reflection coefficient of the surface and frequency/wavelength of the noise as is illustrated in the excerpt and figure from the prediction model ISO 9613-2 [4] in Figure 5 and Figure 6. The reflection coefficient describes the amount of energy that is preserved in a reflection. E.g. a reflection coefficient of 1 describes a lossless reflection, 0.8 means a loss of 1 dB per reflection and 0.5 is a loss of 3 dB per reflection. 0.8 is valid for most buildings

The reflections from an obstacle shall be calculated for all octave bands for which all the following requirements are met:

- a specular reflection can be constructed, as shown in figure 8;
- the magnitude of the sound reflection coefficient for the surface of the obstacle is greater than 0,2;
- the surface is large enough for the nominal mid-band wavelength λ (in metres) for the octave band under consideration to obey the relationship

$$1/\lambda > \left[2/(l_{\min}\cos\beta)^2\right]\left[d_{s,o}d_{o,r}/(d_{s,o}+d_{o,r})\right] \quad \ldots (19)$$

where

λ is the wavelength of sound (in metres) at the nominal midband frequency f (in hertz) of the octave band $\left(\lambda = \dfrac{340\ \text{m/s}}{f}\right)$;

$d_{s,o}$ is the distance between the source and the point of reflection on the obstacle;

$d_{o,r}$ is the distance between the point of reflection on the obstacle and the receiver;

β is the angle of incidence, in radians (see figure 8);

$l_{\min}$ is the minimum dimension (length or height) of the reflecting surface (see figure 8).

Figure 5. Excerpt from ISO 9613-2 [4] showing the necessary conditions for reflections to occur.

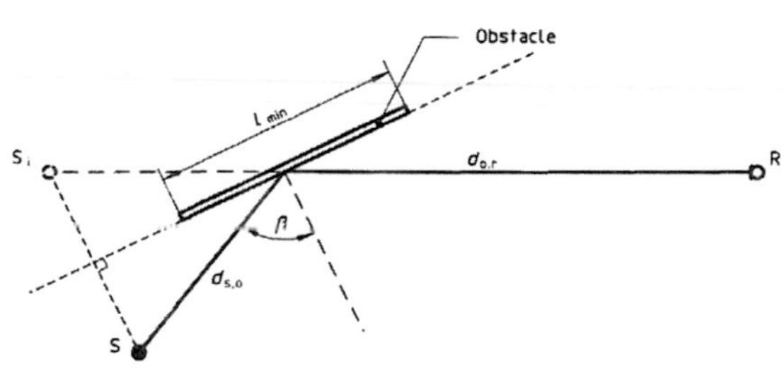

NOTE — A path $d_{s,o} + d_{o,r}$ connecting the source S and receiver R by reflection from the obstacle exists in which β, the angle of incidence, is equal to the angle of reflection. The reflected sound appears to come from the source image S_i.

Figure 6. Figure from ISO 9613-2 [4] showing the geometry of specular reflection.

If the receiver is approximately 5 m from a building façade with a height of 6 m (2-storey) and assuming an angle of incidence of 0° only frequencies above about 100 Hz will be reflected. At larger distances or other angles of incidence this lower limit is shifted to higher frequencies. At shorter distances the frequency of the lower limit is shifted downwards.

Model calculations have been made for a rectangular house, an L shaped house and a U-shaped house including reflections in the three figures below. In Figure 7 to 9 contours of the noise from a wind turbine are given relative to the free-field noise predicted without the houses present. The

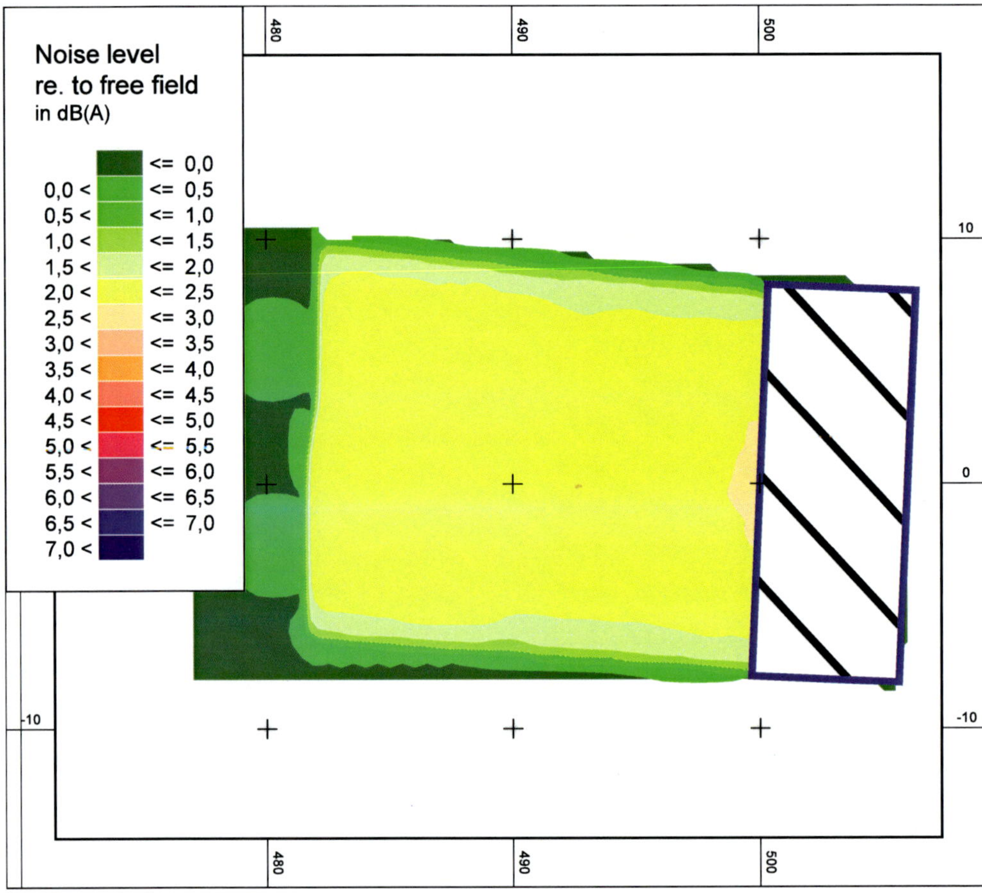

Figure 7 Model predictions of the noise in front of a façade pointing in the direction of the wind turbine. In a small area just in front of the façade the noise is increased by almost 3 dB relative to free-field. In a larger area the noise is increased by 2 – 2.5 dB. Distances are in meters.

predictions are made for a wind turbine with a hub height of 90 m at a distance of 300 m.

It can be seen that the noise is increased relative to the free-field noise level by almost 3 dB in a small area just in front of the façade for a rectangular house and up to 5 dB for L- and U-shaped houses. Therefore it is important that free-field noise limits allow for such situations. Note that when setting up monitoring systems, positions influenced by reflections should be avoided if the noise criteria are based on free field conditions.

It is important to note that these model calculations shows what happens to the noise level, but do not give any information on the dynamic character of the noise. If the noise has a periodic variation which is prominent, instead of a steady noise, the perception of the noise can be very different from the free-field situation.

INDOOR NOISE

The internal noise level is dependent on the noise level at the façade and roof of the residence, the noise insulation of the façade and roof including windows and, to some extent, the room dimensions. In most cases only the low frequency part of the noise is important indoors. Noise standards rarely specify

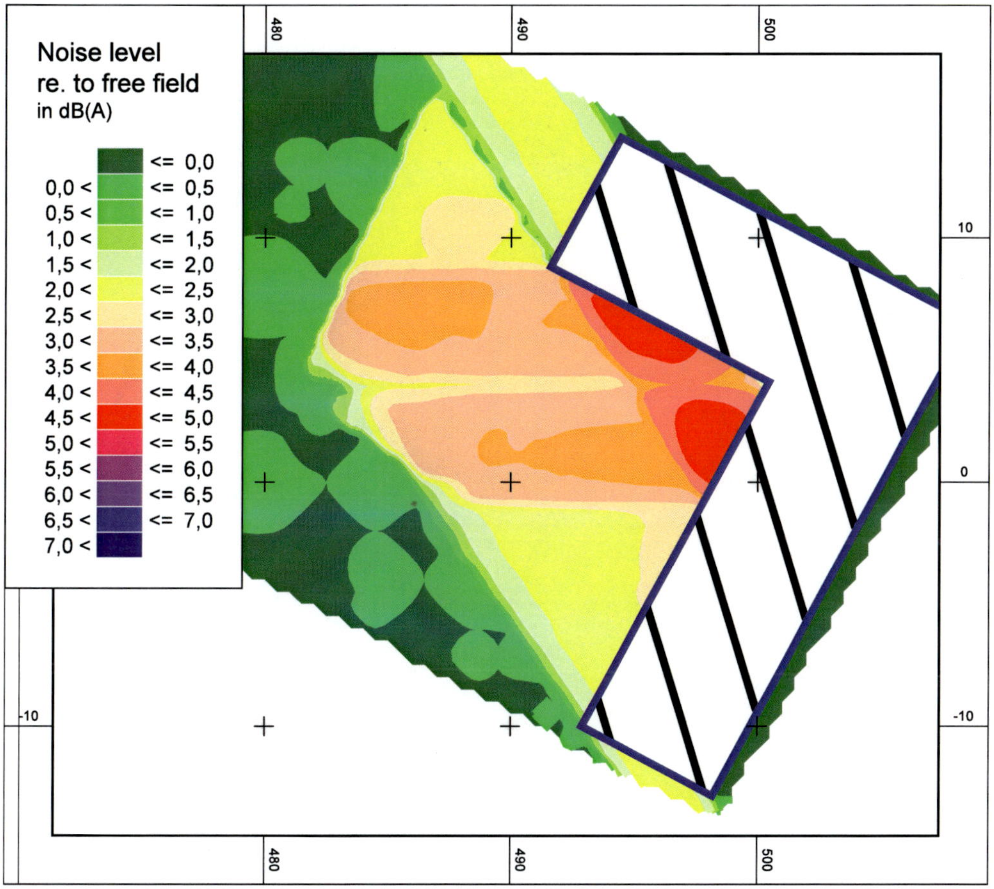

Figure 8. Model predictions of the noise in front of the facades of an L-shaped building. The wind turbine is positioned 300 m to the left. In a small area just in front of the façade the noise is increased by almost 5 dB relative to free-field. In a larger area the noise is increased by more than 2dB. Distances are in meters.

internal noise levels. This is because there are practical problems with assessing noise internally, in particular that the level varies significantly in different parts of a room due to reflections and standing waves (known also as room modes).

In [3] and [7] a model for prediction of the low frequency noise at the neighbours is suggested and includes prediction of the indoor noise. The method is based on the knowledge of the insertion loss of a building at frequencies below 200 Hz. At these frequencies the insertion loss is used instead of the transmission loss values R_w which demands knowledge on room characteristics like reverberation time. The method suggested for measuring the insertion loss is based partly on EN ISO 140-5 [8] and partly on the indoor noise measurement method suggested by Pedersen et al [9]. Pedersen suggests that microphones should be put within centimeters of a 3-dimensional corner position close to the ceiling or the floor, see Figure 10, and at least 4 corner positions should be used.

The method recommended by the Danish Environmental Protection Agency [10] suggest the use of a microphone position at a distance of 0.5 -1.0 m to a 2-dimensional corner approximately 1.0 – 1.5 m above the floor and 2 microphone positions representative of the use of the room - that is at least 0.5

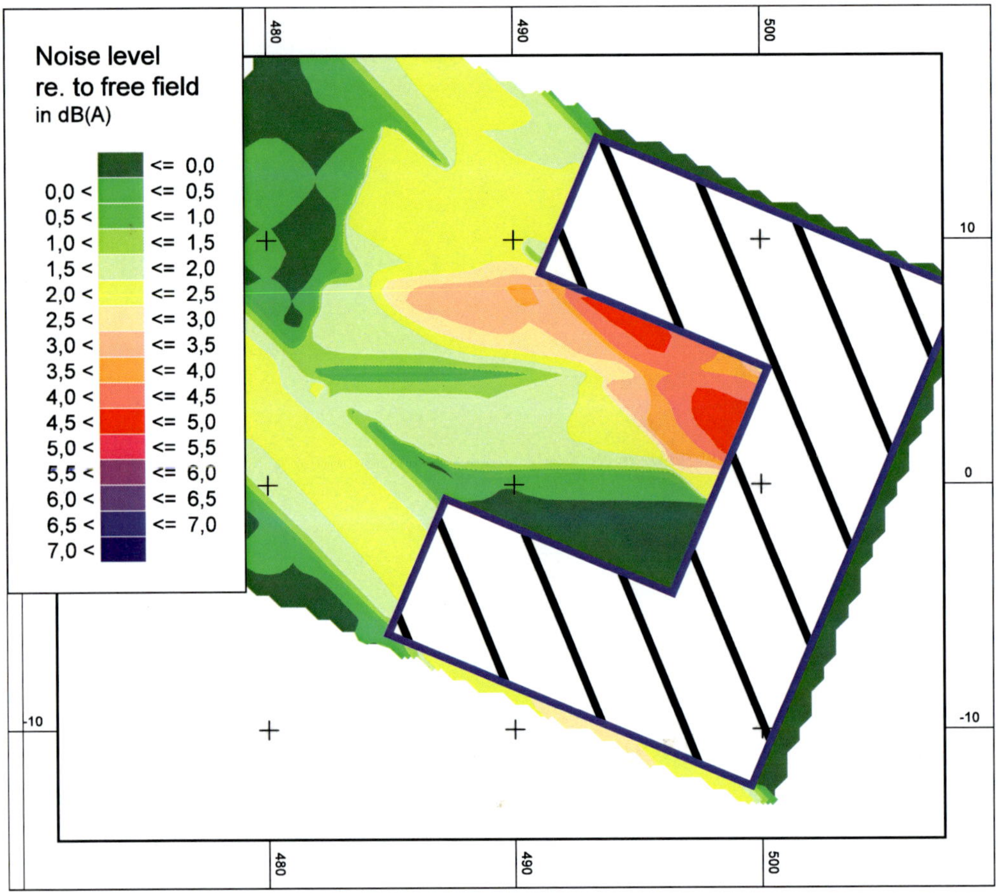

Figure 9. Model predictions of the noise in front of the facades of an U-shaped building. The wind turbine is positioned 300 m to the left. The situation is more complex than for the L- shaped building due to both reflections and screening. In a small area just in front of the façade the noise is increased by almost 5 dB relative to free-field. In a larger area the noise is increased by more than 2 dB. Distances are in meters.

m from walls and heavy furniture and 1 – 1.5 m above the floor. Hoffmeyer and Jakobsen [11] compare the two methods and show that the 3D-corner method gives higher noise levels and thus lower noise insulation in the frequency range from 63 Hz to 200 Hz. This is shown in Figure 11. There is at the moment a discussion whether the 3D-corner method is actually representative of the highest levels perceived by resident or if it is an overestimation. Brunskog et al [12] conclude after reviewing the two methods and the arguments behind them that:

"There simply is no simple, practical and objective method that can guarantee to give the maximum sound pressure level that people are exposed to in ordinary rooms. It may be possible to develop an objective method based on scanning, but such a method will undoubtedly be more complicated and time-consuming than the method specified in ISO 16032 and in the Danish and Swedish guidelines. Alternatively the diagonal method suggested by Moorhouse & Ramandorai 2006 may have potential if a suitable correction can be established – but this remains to be seen. However, it would seem to be

Figure 10. Example of a 3D corner position. From [9]

more reasonable to make use of information provided by the people who actually occupy the room. It is clearly a good idea to use corner positions in rectangular rooms if it is known that a single mode is dominating. Unfortunately, it is more likely that the sound field is composed of several modes, and in this case there is no general relation between the sound pressure levels in corners and the maximum levels that occur at a distance from all surfaces. Moreover, in rooms with walls of lightweight material corner positions can easily lead to significant overestimation because of evanescent waves.

The ongoing controversy between the protagonists of various measurement procedures is essentially political, since the very definition of the 'target' varies. Thus, perhaps the most pressing matter is to reach consensus on what the target is. It would also seem to be important to investigate whether it is possible for a person with normal hearing to point out regions in a room with a particularly high sound pressure level of low-frequency noise."

As the differences in insertion loss determined by these methods are large at the important frequencies where the wind turbine noise have a significant amount of energy it is necessary to be aware what the target is, as Brunskog puts it. The Danish EPA have decided for now to keep the target values and the method already described. The Danish target values are $L_{pA,LF}$ = 20 dB and L_{pG} = 80 dB.

In the UK Moorhouse et al [13], [14] and [15] have made a review of existing standards of low frequency noise in combination with laboratory studies and noise measurements in the field. From this they suggest a criteria curve in 1/3-octave bands from 10 Hz to 160 Hz which the noise should not be allowed to exceed in any 1/3-octave band. The criteria curve is accompanied by the instructions below:

Measurement should be taken with the microphone in an unoccupied room where the complainant says the noise is present. (Note that the person taking the measurements may not be able to hear the sound).

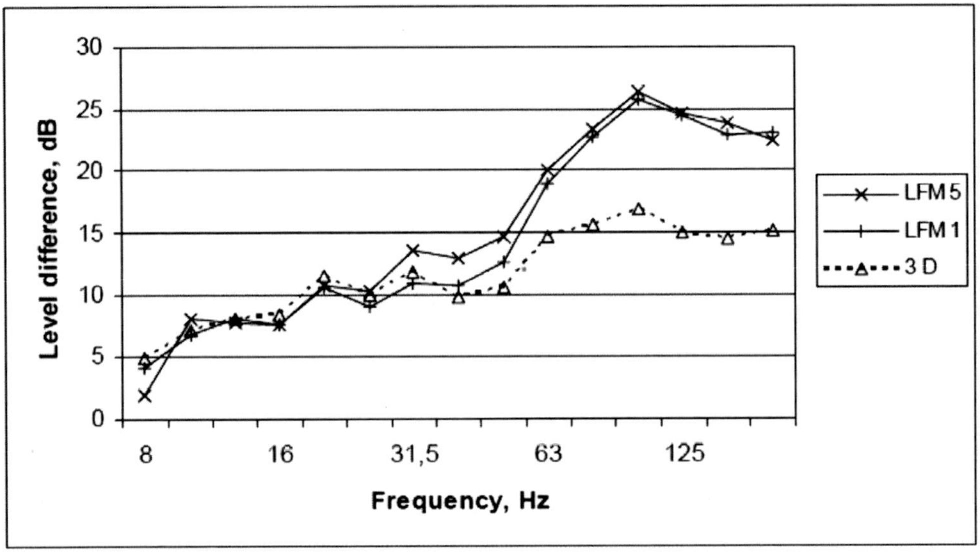

Figure 11. Insertion loss of a building determined according to the Danish EPA method (LFM 5 and LFM 1) and according to the 3D-corner method suggested by Pedersen. From [11]

Record Leq, L10 and L90 in the third octave bands between 10Hz and 160Hz.

If the Leq, taken over a time when the noise is said to be present, exceeds the values in Figure 12 it may indicate a source of LFN that could cause disturbance. The character of the sound should be checked if possible by playing back an audio recording at amplified level.

If the noise occurs only during the day then 5dB relaxation may be applied to all one-third octave bands.

If the noise is steady then a 5dB relaxation may be applied to all one-third octave bands. A noise is considered steady if either of the conditions a. or b. below is met:

a. L10-L90 < 5dB

*b. the rate of change of sound pressure level (Fast time weighting) is less than 10dB per second**

where the parameters are evaluated in the one-third octave band which exceeds the reference curve values Figure 12) by the greatest margin.

** For a meter capable of storing short term Leq the rate of change is $(L_1-L_2)/t_s$ where L_1 and L_2 are subsequent values of the level and t_s is the time for each sample (should be less than 0.1s). For simpler instruments it should be possible to estimate the rate of change from the depth and speed of fluctuations judged by eye. For example, if there are 2 fluctuations per second with a difference of 6dB from peak to trough then the total change in a second is 24dB (two up, two down, each 6dB). The rate of change would therefore be at least 24dB if the level changes smoothly, and more than this if it changes irregularly or suddenly.*

The method does not specify measurement positions but only states that the measurements should be made in an unoccupied room, where the residents have stated that low frequency noise is present. The discussion about measurement positions indoors is important and can shift the noise spectra by 10 dB or more at frequencies above 63 Hz.

Even though the low frequency noise is perceived through the ears like noise at higher frequencies, there are some differences in the sensitivity to the noise at low frequencies compared to other

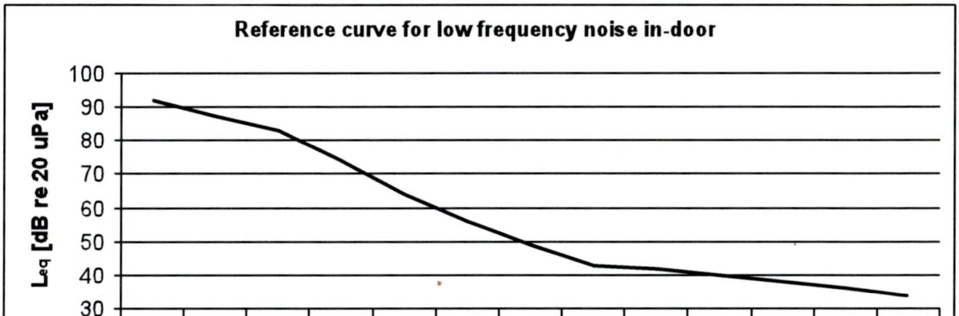

	10	12.5	16	20	25	31.5	40	50	63	80	100	125	160
dB	92	87	83	74	64	56	49	43	42	40	38	36	34

1/3-Octave band [Hz]

Figure 12. Reference curve for indoor noise from [13]

frequencies. This is illustrated in the equal loudness contours from ISO 226-2004 [2] shown in Figure 13. Also included in the figure are curves for the predicted noise of a wind turbine outside and inside a house when the level outside the house is 45 dB(A) (blue), 40 dB(A) (green) and 35 dB(A) (red) The curves use the data from [11] for the noise reduction of a house. The predictions are made with Nord2000 giving the best results for the ground reflections at low frequencies.

The figure shows that at low frequencies the interval between the equal loudness contours is smaller than at higher frequencies. This so-called "narrowing of the dynamic range" is very marked in the infrasound range which is not included in Figure 13. E.g. a 10 dB change at 1 kHz and a change of 4 dB at 20 Hz are perceived as equally loud.

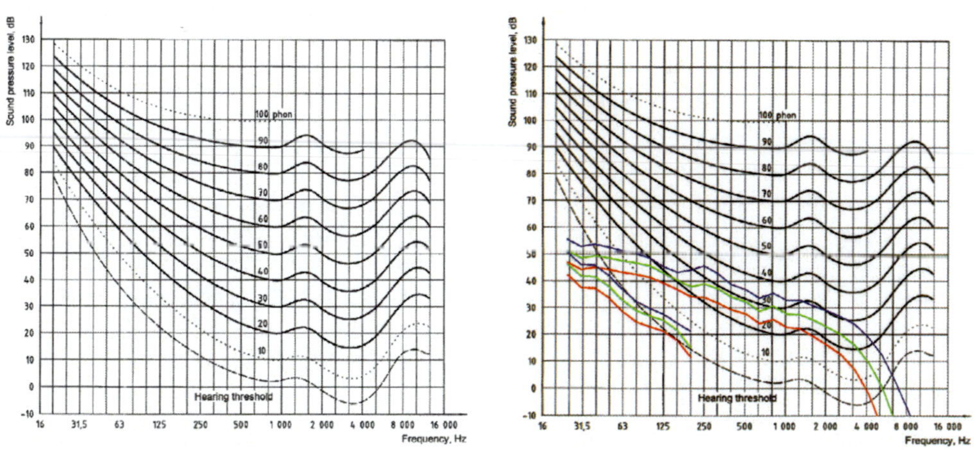

Figure 13. Equal loudness contours from ISO 226-2004. Curves representing wind turbine noise at a predicted level of 45 dB(A) (blue), 40 dB(A) (green) and 35 dB(A) (red) outside and inside a residence are included.

As there are some variations between the individual hearing thresholds (of the magnitude +/- 5dB), this means that low-frequency sound or infrasound that is inaudible to one person may be clearly audible or even rather loud to another, (especially in the infrasound range).

It is often stated in relation to wind turbine noise that even small changes in the level of low frequency noise therefore can lead to strong reactions, but as shown in the curves, the wind turbine spectra are crossing the equal loudness contours, indicating that it is the higher frequencies where there is almost no compression of the dynamic range at which the wind turbine noise is most audible. Strong tones can occur at lower frequencies causing adverse reactions.

AUDIBILITY AND MASKING

Pedersen has developed a procedure for evaluation of the audibility for low frequency sound which is described in this section. [16]

Comparing spectra directly with equal loudness contours or the hearing threshold is not the best method to evaluate the audibility of broad band low frequency noise. Different results will apparently be obtained depending on the bandwidth in which the noise is measured. The left hand graph of Figure 14 shows the same noise spectrum measured in various bandwidths compared with the hearing threshold, indicating that a simplistic comparison cannot be made, Pedersen et al, have proposed the following method to overcome this using critical bands instead and introducing the hearing threshold (HT) as a weighting of a measured spectrum.

First the hearing threshold (HT) is applied as a weighting of the measured spectrum independently of the frequency resolution. This is shown in Figure 14 where the values in the left hand graph are weighted in this way to produce the curves in the right hand graph.

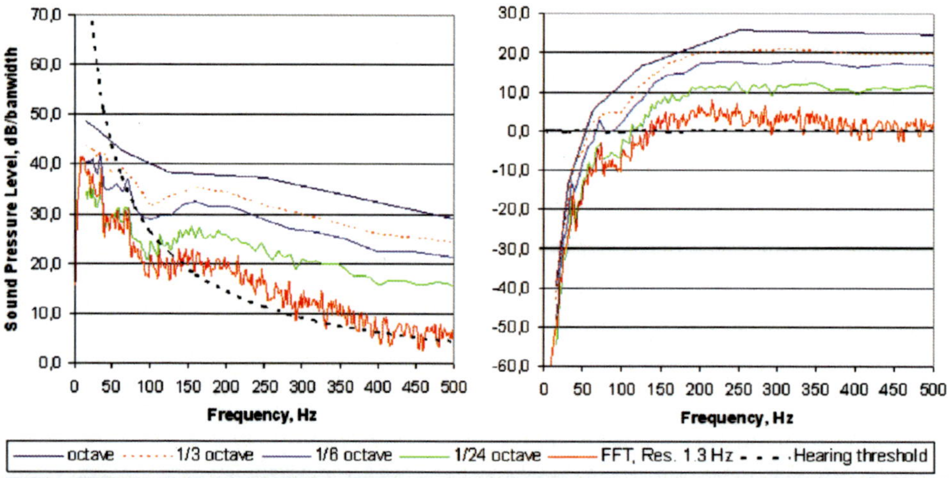

Figure 14. Spectra with different resolution of the noise from a 1.3 MW wind turbine, referred to a distance of 280 m (measured at 70 m). The abscissa is the level per effective analysis bandwidth. The hearing threshold is also shown at the graph. The A-weighted sound pressure level of the noise is 33 dB. The right side shows the corresponding HT-weighted spectra. From [16]

The next stage is that the resulting spectrum is energy added for each critical band to get the total HT-weighted level per critical band. The relevant frequencies are shown in Figure 15.

From the spectra in the right part of Figure 14 the total HT weighted level per critical band is calculated to give the curves in Fig 4.17. For this purpose the energy in a number of bands as shown in Figure 15 has to be added.

The energy addition is made according to the following formula:

$$L_{crit.band} = 10 \cdot \log(\sum_{i=1}^{n} 10^{\frac{L_{HT,i}}{10}}) \qquad [2]$$

Where $L_{HT,i}$ is the HT-weighted level of the i'th frequency band.

It will be seen that the lowest critical band includes both a low frequency and an infrasonic region. This subdivision is only conventional and there are no physical, physiological, or psychological reasons to maintain it in the above calculations.

Critical band	0-100 Hz	100-200 Hz	200-300 Hz	300-400 Hz	400-500 Hz
Octaves	0-63	125	250	250	500
1/3 octaves	0-80	100-160	200-250	315-400	500
1/6 octaves	0-90	100-180	200-285	320-400	450-506
1/24 octaves	0-98	101-196	201-293	301-390	402-492
FFT analysis	0-99	100-199	200-299	300-399	400-499

Figure 15. Centre frequencies in Hz for bands to be energy added from the HT weighted spectra to get the total HT-weighted level per critical band. Only analyses with a resolution of 1/3 octaves or better should be used. This "definite" definition of the placement of the critical bands is for "pragmatic" reasons only. The critical bands of hearing are not centred at any specific frequencies but can be placed at any frequency. From [16]

A procedure that calculates the critical band level as a continuous function of frequency would be more in line with the fact that the critical band may be centred at any frequency, but the procedure described may be seen as a first attempt to develop an overview of the matter.

The low frequencies of the wind turbine noise will be regarded to be below the hearing threshold, if the critical band levels found from the HT-weighted wind turbine spectra is less than 0 dB.

The example illustrated in Figure 14 and Figure 16 shows that the method is relatively insensitive to the analysis bandwidth except for the 1/1-octave band analysis.

Assessment of Masking [16]

In the preceding section, a method was defined for comparing the wind turbine noise in quiet surroundings with the hearing threshold. In practice there is always some background noise, at least from wind in vegetation and buildings, so the wind turbine noise may not be audible due to masking, even if the HT-weighted critical band levels are above the hearing threshold.

Assuming that no tones are present in the noise, the wind turbine noise is broadband and varying as is the wind-induced noise from vegetation or buildings. In this case the following principles can be used:

1. In simple cases the spectra of the wind turbine noise and the background noise can be compared directly provided that the spectra are measured or referred to the same frequency weighting and analysis bandwidth and that the bandwidth is 1/3 octave bands or less:
   a. The wind turbine noise will be masked if the levels in all analysis bands of the wind turbine noise is more than 2 dB below the levels of the background noise.

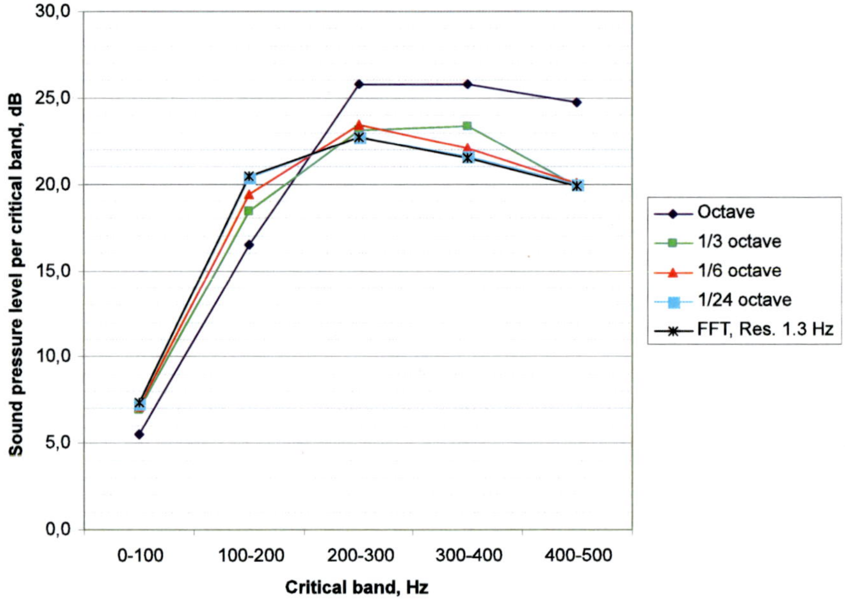

Figure 16. The critical band levels computed from HT weighted frequency analyses of the noise from the 1.3 MW reference wind turbine shown in Figure 14. The different curves are the result of calculations based on analyses with different analysis bandwidths

b. If the levels of all analysis bands of the total noise is less than 2 dB above the background noise the wind turbine will be masked.

If this is not the case, the wind turbine noise may be masked even if the levels of some analysis bands of the wind turbine exceed the levels of the background noise. In this case the following rule applies:

The wind turbine noise is masked if the levels of the critical bands of the A-weighted wind turbine noise are more than 2 dB below the levels of the critical bands of the A-weighted background noise.

The energy sum of the levels of the two lowest critical bands of the A-weighted wind turbine noise equals in practice $L_{pA,LF}$.

For unstable wind conditions another problem exists: For a specified average wind speed (1 minute averaging time) short term variations of the wind speed (for example gusts of wind of duration less than 10-15 seconds) cause variations of the levels of both the wind turbine noise and of the background noise. The variations of the A-weighted or the octave band levels measured with time weighting F are in the magnitude of 5-10 dB. Due to horizontal and vertical difference in location of the wind turbine rotor and the vegetation generating the natural background noise, these level variations are not synchronous in time. This means that the maximum values of the wind turbine noise may occur at times where the background noise is at minimum and vice versa.

If we want to find out if the low frequencies from the wind turbine is below the hearing and masking threshold at all times for a specified average wind speed, then the maximum levels (with time weighting F) of the wind turbine noise should be compared to the minimum levels (expressed perhaps as the L_{95}) of the background noise using the same criteria described above.

Normally such information is not available in the measurements so, as a first approximation, we have to rely on the procedures for stable conditions.

NOISE FROM VEGETATION AND WAVES

Masking of the wind turbine noise by wind noise in vegetation was for a long time thought to occur, especially at higher wind speeds, because the spectrum of the wind noise is similar to that of a wind turbine. However, the periodic and deterministic nature of wind turbine noise makes it more audible through the more stochastic wind noise than originally expected.

One reason for this could be that, in the beginning, many wind turbines were stall regulated and at high wind speeds the wind turbine noise increased at a larger rate than the wind noise. Masking was never expected at low wind speeds. The wind turbines of today, and probably also of the future, will be regulated in a way to reduce the load on the turbine at high wind speeds and most probably also the noise. If this is the case it may be relevant to look at masking in more detail.

In [17] Bolin has described a semi-empirical model for prediction of noise from vegetation modelling the wind profile and turbulent wind field in the vegetation. Predictions are compared with measurements giving reasonable agreement as can be seen in Figure 17.

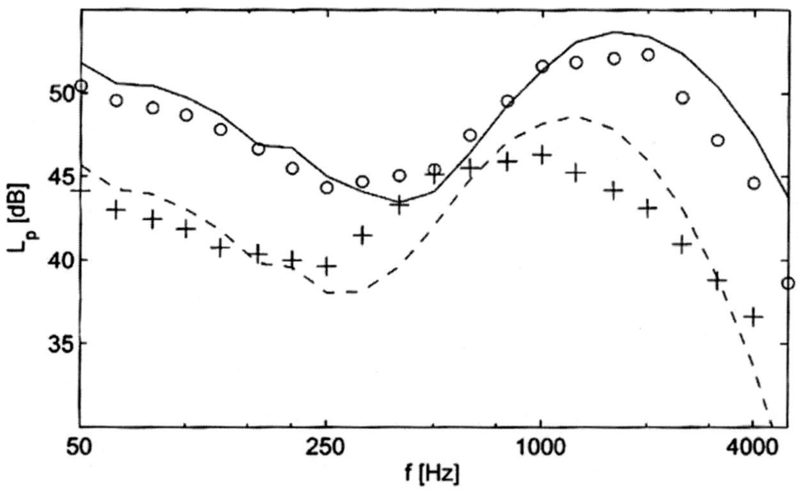

(a) Site 1 at U=8.5 m/s (o) measurement and (—) simulation. U=6.1 m/s (+) measurement and (- - -) simulation.

Figure 17. Measurements and predictions of noise from vegetation. From [17]

A model for noise from breaking waves have also been described and it seems that what is called the breaker type (how the wave breaks on the shore or in the water) and the wave height are the dominant effects. Comparison of predictions and measurements show good agreement. See Figure 18.

Tests of the masking abilities of vegetation noise have been made in laboratory experiments concluding that the model described by Moore and Glassberg tends to overestimate the masking ability. See Figure 19 and Bolin(17)

OTHER NOISE CHARACTERISTICS

Apart from the noise level and the spectrum, which are averaged parameters, the noise of turbines at neighbouring properties also has dynamic characteristics that influence people's perception of it. This chapter concludes with a review of these characteristics.

Tonality

Tonality is an important part in the assessment of noise and from many investigations it has been

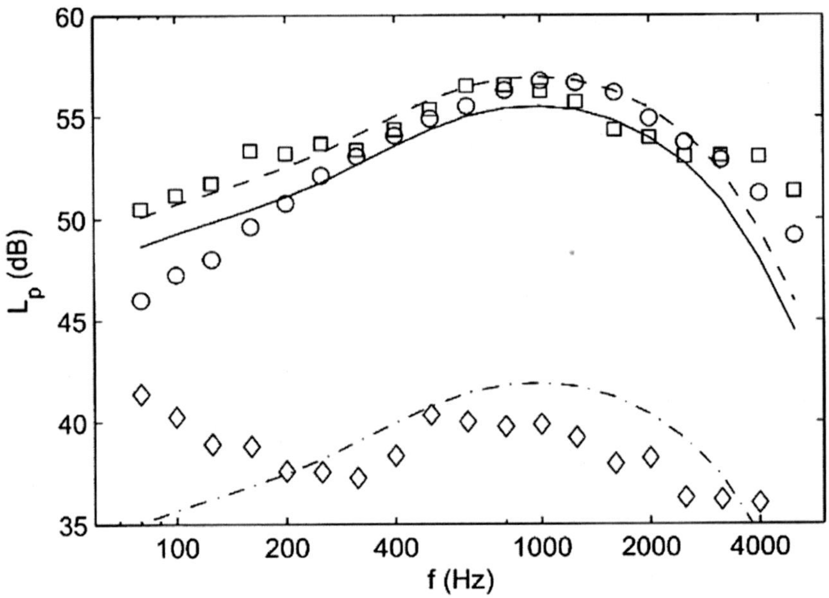

(a) Measurements are shown as markers and prediction as lines for site 6 (◇,- · -), site 7 (○,—) and Site 8 (□,- - -).

Figure 18. Measurements and predictions for noise from waves. From [17]

documented that noise that includes prominent tones are more annoying than the same noise without the tones. Wind turbines have many rotating parts including gears, generator, fans which can generate tones. Noise from the transformer and frequency converter in modern turbines can also be significant.

If tones are rated clearly audible it is normal in most jurisdictions to apply a penalty to the level of the noise when it is assessed. The penalty may be different from country to country. There are also several methods describing procedures for assessing the tonality of noise. Nearly all the methods are based on narrowband spectra, FFT-spectra, with a line width around 2 Hz allowing for the finer details in the noise to be seen.

The most documented method is the method described in annex C of ISO 1996-2:2007 [18]. The spectra used are A-weighted and linearly averaged over a period such as 1 minute. A tone is identified as a peak in the spectrum where the lines are separated into noise, tones and noise pause. This is shown in Figure 20.

The tonal audibility is calculated as:

$$\Delta L_{ta} = L_{pt} - L_{pn} + 2dB + \log\left[1 + \left(\frac{f_c}{502}\right)^{2.5}\right] dB \qquad [3]$$

Where

L_{pt} is the total sound pressure level of the tones in the critical band
L_{pn} is the total sound pressure level of the masking noise in the critical band
f_c is the centre frequency of the critical band, expressed in hertz

And the penalties K_t suggested in the standard are given as

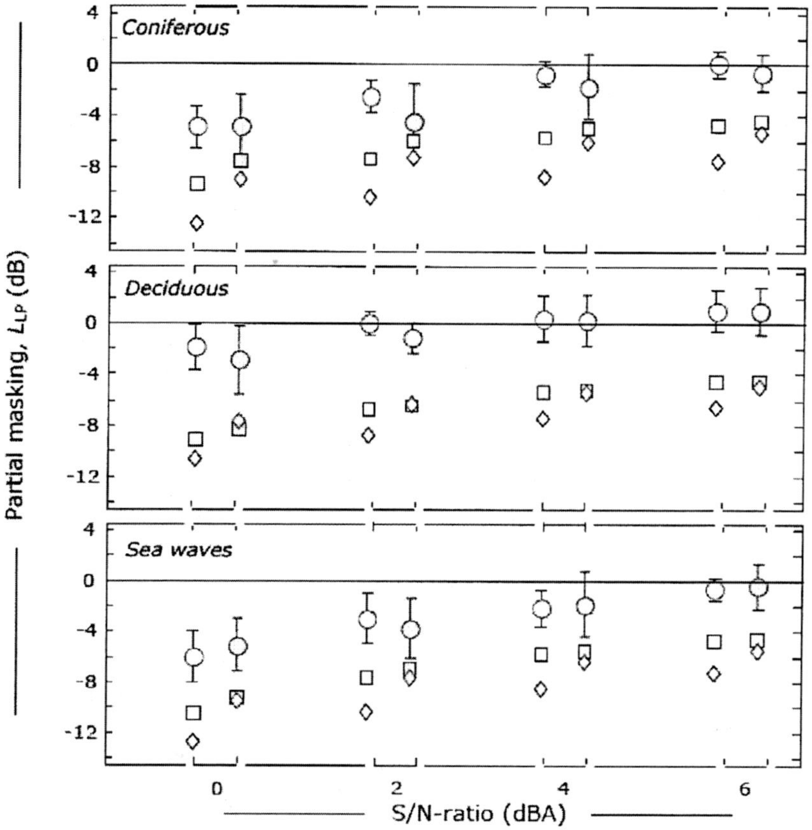

Figure 4: Level differences between equally loud WT noises heard in the presence of masking sound and heard alone, as a function of S/N ratio. Results are shown for coniferous trees (upper diagram), deciduous trees (middle diagram) and sea waves (lower diagram), separately for single WT noise (open symbols) and multiple WT noise (filled symbols). Error bars indicate the 95 % confidence interval around the mean. Model predictions of Moore's and Glasberg's model are indicated with circles and diamonds, respectively.

Figure 19. From [17]

For 10 dB < ΔL_{ta} in accordance with Equation 3:
K_t = 6 dB
For 4 dB < ΔL_{ta} ≤ 10 dB in accordance with Equation 3::
$K_t = \Delta L_{ta}$ - 4 dB
For ΔL_{ta} < 4 dB in accordance with Equation 3::
K_t = 0 dB
K_t is not restricted to integer values.

The DIN 45681 standard [20] is very similar to the ISO 1996-2:2007 standard and also includes suggestions for penalties.

The ETSU-R-97 [19] method is based on similar principles to the ISO standard, though the method for separating the lines in the spectrum is slightly different. However as the tones in wind turbine noise are considered to be varying in nature the level of the tone is determined from a large number of rms-weighted spectra with an integration time between 0.29 to 0.4 seconds over a 2 minute period, as the

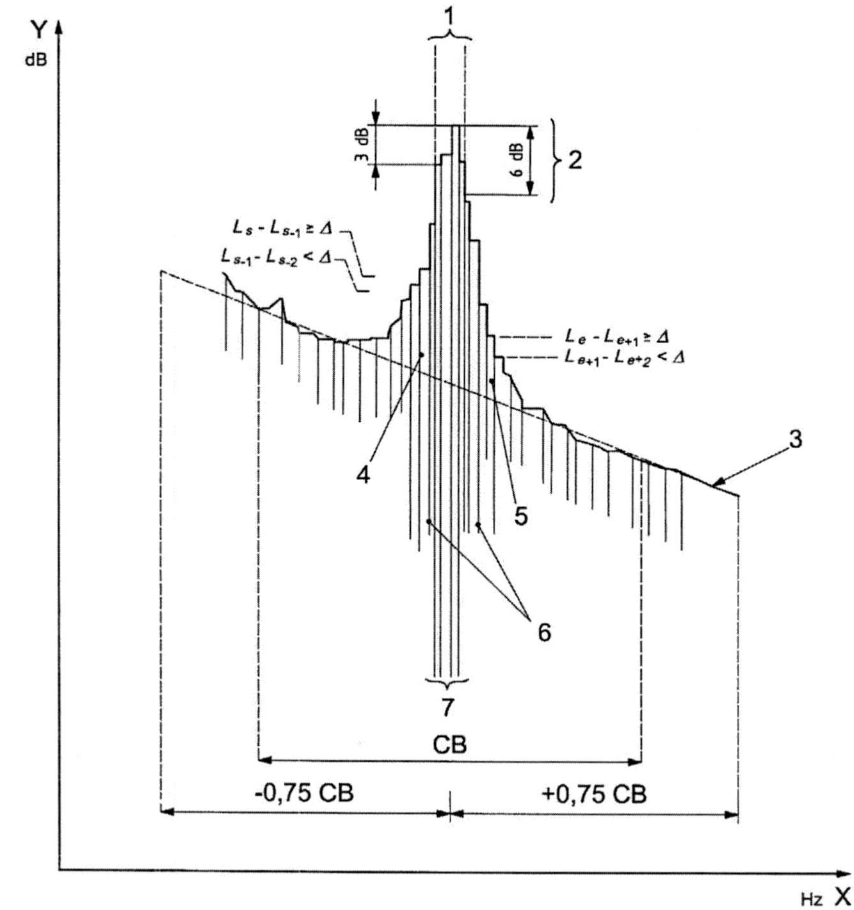

Key

1 tone when the 3 dB bandwith is less than 10 % of the critical band
2 tonal energy
3 linear regression line of the noise level
4 noise-pause start
5 noise-pause end
6 neither tone nor noise
7 tone
CB critical band

Figure C.2 — Definitions of tones, noise and noise pause (neither tone nor noise). Δ is the tone-seek criterion and is normally chosen as 1 dB

Figure 20. Classification of lines in a spectrum. The parameters f and 0.75 are selectable in the method. From ref. [18]

arithmetic average of the 10% highest values. The level of the noise in the critical band is determined from an rms-weighted spectrum for the entire 2 minute period.

The method in IEC 61400-11 [21] differs from the other methods in the way the lines are divided into the 3 categories. This method is devised to give a simpler and easier interpretable and programmable method. The method has a tendency of giving slightly lower tonalities than the ISO-method. This method is only used in the near field of the wind turbine and does not include penalties. In practice the tonality is determined for 12 10-second periods and averaged to a tonality for a 2 minute period.

In annex D of ISO 1996-2:2007 [18] there is a simplified method based on 1/3-octave bands. This method should be considered indicative of whether prominent tones are present only and not for detailed analysis. It is included in the BWEA-standard for small wind turbines, but measurements using this method are unlikely to be accepted in other countries.

Impulsivity

Impulses are another parameter which is documented to increase the annoyance of the noise. A method for rating the prominence and annoyance of impulsive sound are described in Nordtest method ACOU112 [22]. The method is based on the onset rate and the level difference as the prominence P given by:

$$P=10\cdot\log(\text{onset rate [dB/s]}) + 2\cdot \log(\text{level difference [dB]})$$

This means that a sudden onset of a constant noise like for a compressor can give a penalty. The penalties are given as:

$K_i=1.8\cdot(P-5)$ dB for $P>5$
$K_i=0$ dB for $P\leq 5$

Amplitude Modulation

Other parameters like amplitude modulation are not yet investigated to a degree where it is possible to make an objective assessment. Field investigations and listening tests are necessary to give a relation between a measured amount of amplitude modulation and the corresponding annoyance. There is some further discussion of this in the next chapter.

REFERENCES

[1] Hayes Mackenzie Partnership Ltd "The measurement of low frequency noise at three UK wind farms" Contract number W/45/00656/00/00 for the Department of Trade and Industri 2006.
[2] ISO 226-2004: Acoustics Normal equal-loudness-level contours.
[3] Søndergaard B. Low Frequency Noise from Large Wind Turbines – Summary and Conclusions form Measurements and Methods. AV 140/08. DELTA April 2008
[4] ISO 9613-2. Attenuation of sound during propagation outdoors – Part 2: General method of calculation
[5] B. Søndergaard, Plovsing B. Noise and Energy Optimization of wind farms - Validation of Nord2000. DELTA Acoustics report AV1236/09,2009.
[6] Nordtest Method NT ACOU 099 edition 2. Shooting Ranges: Prediction of noise.
[7] Hofmeyer D. Søndergaard B. Low Frequency Noise from Large Wind Turbines – Measurements of Sound Insulation of Facades. AV 1097/08. DELTA April 2008
[8] EN ISO 140-5:1998, Acoustics – Measurement of sound insulation in buildings and of building elements – Part 6: Field measurements of airborne sound insulation of facade elements and facades, 1998.
[9] Pedersen, S., Møller, H. and Persson, K.: "Indoor measurements of noise at low frequencies – problems and solutions". Journal of Low Frequency Noise, Vibration and Active Control, Vol. 26, (2) 2007, 249-270
[10] Jakobsen, J.: "Danish guidelines on environmental low frequency noise, infrasound and vibration". Journal of Low Frequency Noise, Vibration and Active Control, Vol. 20 (3) 2001, 141-148.

[11] Hoffmeyer D. Jakobsen J. "Noise insulation of dwellings at low frequencies" Journal of Low Frequency Noise, Vibration And Active Control Vol. 29 (1) 2010, 15-23

[12] Brunskog J. et al. A note on measurement of low-frequency noise in rooms. 14th International Meeting on Low Frequency Noise and Vibration and its Control Aalborg, Denmark 9 – 11 June 2010.

[13] DEFRA report contract no NANR45, 2005 Proposed criteria for the assessment of low frequency noise disturbance

[14] DEFRA report contract no NANR45, 2005 Procedure for the assessment of low frequency noise complaints

[15] DEFRA report contract no NANR45, 2005 Field trials of proposed procedure for the assessment of low frequency noise complaints

[16] Pedersen, Torben Holm "Low Frequency Noise from Large Wind Turbines – A procedure for evaluation of the audibility for low frequency sound and a literature study." AV 1098/08. DELTA April 2008

[17] Bolin K. "Wind Turbine Noise and Natural Sounds – Masking, Propagation and Modelling" Doctoral Thesis, KTH, Stockholm Sweden 2009.

[18] ISO 1996-2:2007: Acoustics – Description, measurements and assessment of environmental noise – Part 2: Determination of environmental noise levels

[19] ETSU – R – 97: The assessment and rating of noise from wind farms

[20] DIN 45681 Acoustics – Detection of tonal components and determination of a tone adjustment for the assessment of noise immissions

[21] IEC 61400-11:2002 ed 2.1: Wind turbine generator systems – Part 11: Acoustic noise measurement technique.

[22] Nordtest Method NT ACOU 112 "Acoustics: Prominence of impulsive sounds and for adjustment of LAeq"

Chapter 5. Amplitude Modulation

Frits van den Berg (Netherlands) and **Dick Bowdler** (UK).
fvdberg@ggd.amsterdam.nl. dick@dickbowdler.co.uk

In Chapter 6 we will see that 75% of the people who live near a (Dutch) wind farm think that the variation in sound level (swishing, lashing) is a proper description of wind turbine noise. Other descriptions were found adequate by less than 25% of the respondents. The UK Government published research in 2007 that showed that complaints had been received in about 20% of all wind farms in the UK. As Van den Berg pointed out [1], in all but one of the cases where a description of the sound was available (13 out of 14 cases), the sound was described with one or more words indicating a regular variation in level. In the remaining case the question was not resolved as no description was given.

Studies indicate that people are more sensitive to modulated noise than to steady noise and so it seems that AM is a major factor in perception of wind turbine noise. Laboratory investigations of wind turbine sound have shown that the variation in level is an important factor in noise annoyance (see pages 134 et seq). The variation is at the rate of the blade passing frequency (=BPF: for a three-bladed turbine three times the rotation frequency) and is technically denoted as Amplitude Modulation or AM. The AM level or modulation depth is the difference between the highest and lowest level.

AM AND 'SWISH'

The presence of 'swish' near a wind turbine is a well known phenomenon. We know that swish is the type of AM which is ubiquitous near to a turbine. It is due to the directivity of trailing edge noise and Doppler amplification as described in Chapter 2. It is a mid-frequency (400 to 1000 Hz) phenomenon and because of the cardiod directivity of the noise source it falls off rapidly up and down wind, but transmits over longer distance at 90 degrees to wind direction. However, in this latter direction the absolute level is significantly lower, so overall it seems that swish does not generally transmit over distances more than a few hundred metres at levels that are likely to be a significant problem to neighbours.

So what is the AM that troubles residents and why is AM a problem at some sites? How often does it occur and at how many sites? Apparently residents are not bothered by this 'normal' swish, but another form of AM which has been called "thump". Van den Berg observed the difference between "thump" and "swish". As he says in one paper [2]: "In the wind park the turbines are audible for most of the (day and night) time, but the thumping is not evident, although a 'swishing' sound . . . is readily discernible" and later "At night the turbines cause a low pitched thumping sound superimposed on a broadband 'noisy' sound . . ." In his thesis Van den Berg remarks: "Our experience at distances of approx. 700 to 1500 m from the Rhede wind farm, with the turbines rotating at high speed in a clear night and pronounced beating audible, is that the sound resembles distant pile driving. When asked to describe the sound of the turbines in this wind farm, a resident compares it to the surf on a rocky coast. A resident living further away from the wind farm (1200 m) likens the sound to an 'endless train'. Another resident near a set of smaller wind turbines, described the sound as that of a racing rowing boat (where rowers simultaneously draw, also creating a periodic swish). On the website of MAIWAG, a group of citizens from villages near four wind farms in the south of Cumbria (UK), the sound is described as 'an old boot in a tumble dryer', and also as 'Whumph! Whumph! Whumph!'. Several residents near single wind turbines remarked that the sound often changed to clapping, thumping or beating when night falls: 'like a washing machine'. It is common in all descriptions that there is noise ('like a nearby motorway', 'a B747 constantly taking of') with a periodic fluctuation superimposed. In all cases the sound acquires this more striking character late in the afternoon or at night, especially in clear nights and downwind from a turbine."

On p 52 of a report by Hayes McKenzie [3] it says, in relation to a wind farm site where significant AM was present: "However, the presence of high levels of modulation at Site 1: Location 1 is associated with wind direction and the inappropriate aerodynamic conditions seen by the closest three wind turbines to the dwelling." Hayes goes further in the description of the problem in his evidence to the court in the case of nuisance brought by complainants at the same site where he stated that "the source of modulation may be related to wind shear by which I refer to changes of wind speed and direction at different heights above ground level. I consider it likely that wind shear effects caused by the topography of the site and (for specific wind directions) wake effects caused by turbines upwind of the turbine exhibiting the noise may cause the direction of the wind at some points on the arc of the blades to be different from that measured by the turbine anemometer at the hub height of the wind turbine. Thus the blades at these points in the arc may not be fully pointed into the wind which may result in increased aerodynamic noise in the frequency region where amplitude modulation has been measured."

Measurements of AM

Few measurement results have been published in scientific journals. Apart from early studies on now obsolete types of wind turbines (such as the downwind types with strong impulsive noise) only one peer reviewed article was found. Van den Berg published measurement results from the Rhede wind farm, a 3 km line of ten 1.8 MW, 98 m hub height, pitch regulated wind turbines and seven similar turbines on a line almost parallel to the first. [4, 5] Amplitude modulation depth, the difference between the peak and trough of the sound level varying at the rate of the BPF, was usually –at a time when AM was audible- up to 5 or 6 dB and never more than 9 dB. As figure 1 shows (taken from ref. v), the change in sound level corresponding to this AM occurred at frequencies from about 200 Hz to several kHz when close to the turbines. Simultaneously at the dwellings this occurred at a lower frequency range: between 50-100 Hz and one kHz. At one location (Figure 1, panel A) the infrasound level is almost as high at the dwellings as closer to the turbines. Thumping at residences west of the Rhede wind farm was most pronounced in clear evenings and nights with an eastern wind. At those times the wind at ground level was often very weak. Similar spectral differences were measured at another location with a single 45 m high wind turbine (see same references)

Di Napoli [6] measured AM near five 600 kW stall regulated wind turbines on a 500 m hill at distances up to two km. He found that AM was most pronounced downwind from the turbines with level variations up to 7 dB, though more usually about 3 dB. Upwind AM was not as strong and in a crosswind direction (where one expects the normal swish) it was still weaker and often absent. The AM was weakest close to the turbines and well audible (and easy to measure) up to 2 km in the downwind direction. His results also show that the thumping was not prominent close to the turbines, similar to Van den Berg's observation that it was not as clearly audible close to a turbine. Di Napoli also showed that the sound from this wind farm was clearly amplitude modulated in the infrasound region (6-10 Hz): level variations here were up to and over 10 dB.

A recording made by Bowdler (available at www.dickbowdler.co.uk) close to a turbine shows normal swish changing in character to what is recognised by complainants as "thump". An analysis has been made by Carlo di Napoli and is shown in Figure 2 below.

The waveform starts as swish and half way through audibly changes to thump. The modulation of low frequencies plays a minor role to begin with when the subjective sound is swish, then it begins to rise so that it has a significant role in the overall sound. In thump the low frequencies also have a modulation depth approaching 10 dB compared with the swish of 5 or 6 dB. However, it should be remembered that this is measured close to the turbine and such modulation depths may not propagate over large distances. The other feature to note is that there is a slight phase difference between the high pass and low pass sound.

The third-octave frequency analysis of swish and thump from this recording is shown in Figure 3. Swish, from the first third of the recording, is the black line and thump, from the last third, the blue line. This shows clearly how this example of thump is dominated by the three frequency bands of 125, 160 and 200 Hz compared with 800 Hz in the case of swish.

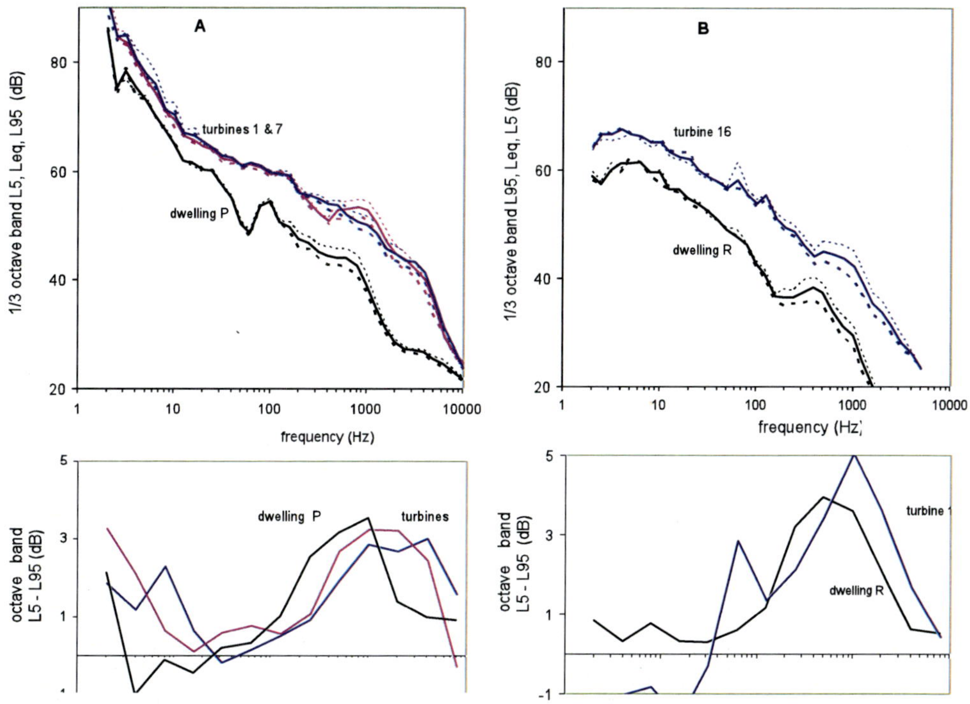

Figure 1. measured wind turbine sound spectra near the Rhede wind farm (see text) with dwelling P at 750 and R at 625 m from the closest turbine; upper panels: thick lines: 1/3 octave band Leq near windturbines and dwellings; dotted lines: Leq of all samples with resp. 5% highest (thin dotted lines) and 5% lowest (thick dotted lines) values of broad band LA ; lower panels: difference between Leq of 5- and 95-percentile octave band levels

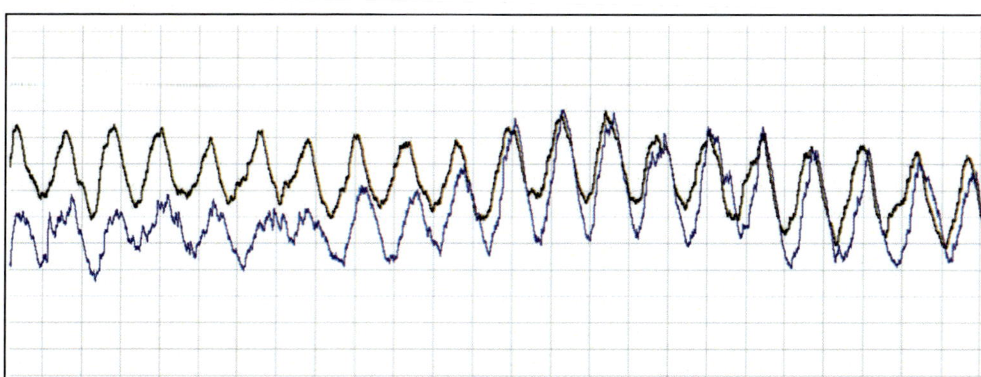

Figure 2. Trace of swish turning to thump recorded close to turbine. The graph shows sound level in dBA on the vertical axis with a 2 dB grid and time on the horizontal axis on a 1 second grid. The lower curve in blue shows the low pass (1-210 Hz) level and the upper curve in black shows high pass (240 Hz-2 kHz).

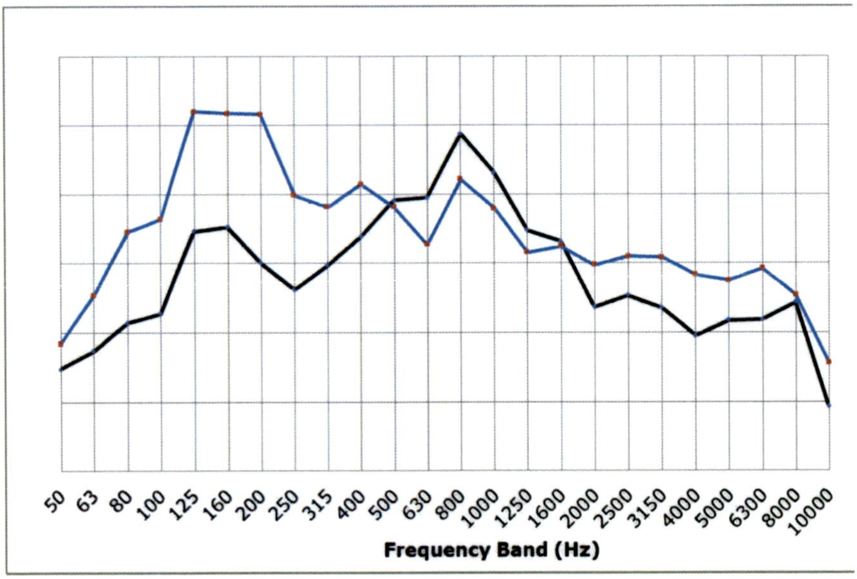

Figure 3. Frequency analysis of swish and thump taken from measurements shown in Figure 2. The vertical scale grid spacing is 5 dB.

Vos and Houben made a large number of sound recordings at distances of 200 to 850 m from the closest turbine in one wind farm. [7] From all recordings (without disturbances from other noise sources) the AM frequency and depth were determined. They found that in about 1 in 5 recordings the modulation depth was between 2 and 3 dB or higher. A depth of 2 dB or more is considered to be audible AM.

CAUSES OF AM

So there is evidence that another form of AM, not the normal 'swish', transmits over longer distances downwind. In principle there would appear to be three main possible contributory reasons for AM being heard at a distance. The first is a different source noise mechanism from swish and in particular one that has a different directivity pattern so that the sound is transmitted downwind. The second is propagation effects moderating turbine sound. The third is a local meteorological or topographical situation that makes audibility more likely – such as housing in a sheltered valley. Of course it is possible that such AM is due to more than one cause. For example we could speculate that, in conditions of high wind shear there might be a different source mechanism at work because the angle of attack of the blades changes significantly. Whilst this may not be enough in itself to create a problem, the high shear might mean that the turbines are running at full sound power but the ground level wind which generates background noise is low so that the AM is more audible than expected. But, as yet, this is no more than speculation.

Van den Berg based his explanation of 'thump' on two factors: it is related to a stable atmosphere (that is: an atmosphere where the temperature does not decrease with height as it usually does in daytime) and to modern, tall turbines. When the atmosphere turns stable (usually at sundown, when not fully overcast), the difference in the wind speeds at the blade tips in their lower and higher position increases. This leads to a regular variation in the thickness of the trailing edge boundary layers (at both sides of the rotating blades). Calculations showed that this causes periodic variations in the sound power level.

Van den Berg also considered directional wind shear or the change in wind direction with height (wind veering), but according to calculations this could not lead to significant AM. Richarz et al [8] included 'diffraction': for an observer downwind from a turbine the blade sound is (partly) blocked when the blade passes the mast, resulting in a sudden, small decrease of sound level. They showed that this also resulted in very weak AM.

A second mechanism proposed by Van den Berg is the interference of the sound from several wind turbines. [6] Wind turbine noise AM is not a constant and strictly regular variation. This may be due to variations in wind speed (and hence sound power) and propagation effects, but it could also be due to interference of the sounds from different turbines. If we picture the AM pulses from one turbine arriving at the listener's ear as a pulse train, then from a wind farm there are several such pulse trains. Of course those from the nearest turbines are loudest, and pulse trains from turbines at increasing distance from the listener will be less relevant. When the turbines rotate almost synchronously there will be a period of some length when the pulses arrive at the same time. Then the peaks and troughs coincide and the modulation level is highest. Some time later they will be out of phase again with less overal level difference. Near synchronicity is not a rare event, but in fact very probable in a stable atmosphere. [4]

Di Napoli noticed the same AM variation when doing his measurements. [6] Figure 4 is taken from his paper and shows how the modulation depth changes from less than 1 dB, presumably when the AM is not synchronous, to a maximum of (in this case) 4 dB. Though this interference does not increase the AM level[1] (relative to a single turbine), the change in AM level could make the sound more conspicuous when it draws attention every time AM becomes more prominent.

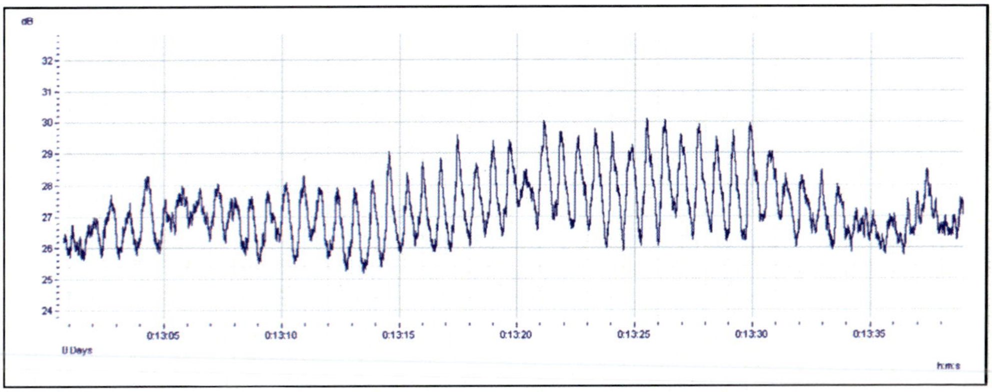

Figure 4. Sound level in dBA at 2 km downwind from 5 turbine wind farm (see text) on a 510 m hill. Wind speed 8 m/s on hill, no wind at measurement location (230 m height). Vertical axis is dBA with grid lines at 1dBA. Horizontal axis is time – Total 40 minutes

The increase in infrasound level as noted by Di Napoli and Van den Berg cannot be explained by the source mechanisms described in chapter 2 (trailing edge noise and inflow turbulence noise) as these sources are insignificant in the infrasound frequency range. The only known wind turbine infrasound source is thickness sound. [9, 5] As van den Berg mentioned in a footnote of his thesis (p. 34), pulses at the blade passing frequency (BPF) yield a spectrum that is a convolution of the BPF and its harmonics, and the pulse spectrum. When the pulse has a length t_{pulse}, its spectrum will have a maximum at $1/t_{pulse}$. When BPF ≈ 1 Hz and t_{pulse} ≈ 1/8 s, then the BPF harmonic at 8 Hz may be most prominent. Thus a sudden and repeated loading of a blade, perhaps because of (added) stall at a certain position of each blade, could explain the change in infrasound level[2].

[1] Although Van den Berg proposed an amplification in level as a result of interference, in a discussion about this Oerlemans correctly argued that with N equally loud (for the observer) turbines, both the maximum and minimum level increases with 10*log(N), so there is no increase in modulation depth.

[2] Although the infrasound is measurable it is explained in Chapter 6 that it has not been demonstrated that the infrasound part of wind turbine noise can be perceived except when very close to a turbine

Ongoing Research

As this book is being finalised a number of organisations are investigating amplitude modulation and some of these produced papers at the Fourth Conference on Wind Turbine Noise in Rome in April 2011. RenewableUK, the trade and professional body for the UK wind and marine renewables industries has commissioned research by a consortium of consultants and universities, including an investigation into the mechanisms causing AM and dose response tests and the results are expected in Autumn 2011. The consortium consists of Hoare Lea Associates, Southampton University, Robert Davis Associates, National Aerospace Laboratory in the Netherlands and Salford University. More information about the conclusions of the RenwableUK research project will be available on their website www.renewable-uk.com when the project is complete. What follows here is largely a summary of the preliminary findings of the consortium.

In the creation of trailing edge noise as described in Chapter 2, the boundary layer over the blades changes thickness due to changes in the angle of attack. When the boundary layer thickens the peak frequency moves down. So when the angle of attack increases, the swish heard near to the turbine increases in level and the peak frequency lowers. However, the directivity of the sound source remains cardiod and so, whilst the effect may be significant in the near field it would not be expected that the change in character would make this swish become significant in the far field.

With a further increase in angle of attack, parts of the blade may stall locally as flow separation takes place at a point along the blade where the angle of attack has increased too much. This results not only in a further lowering of the peak frequency, but also an increase in sound pressure level. Furthermore the directivity pattern changes to a dipole and so the predominant direction of propagation will be upwind and downwind. If the wind speed were the same at all points across the rotor then this stall would either occur all round the blade path or not at all and so there would be no AM. However, the normal situation is that wind speed increases with height so that, if the blade pitch remains constant, stall would happen first at the top of the blade trajectory but not elsewhere. This would mean that there would be a pulse of noise in the upwind and downwind directions as each blade passed its highest point. One of the significant points about this possible causal mechanism is that it would become more likely when wind shear is high, as has been suggested by a number of people.

All the above effects can occur in relatively stable atmospheric conditions with relatively little atmospheric turbulence. However, turbulence may also result from local topographical or meteorological features or from the wake of other turbines. This means that inflow turbulence noise may increase as well as the increase in trailing edge noise caused by changes in the local angle of attack as blades rotate. Inflow turbulence noise is also a dipole source and so would radiate sound upwind and downwind in the same way as stall noise. If pulses of noise created by either mechanism are caused by such turbulence at specific parts of the rotation of the rotor and periodically, AM may be created.

Propagation effects may also modify the character of AM in some circumstances. As explained in Chapter 4, air absorption shifts the spectrum of the noise to a lower frequency because high frequencies are absorbed more than low frequencies. Also, because the sound insulation of dwellings is less at low frequencies, there is a shift of spectrum when noise is heard indoors. The shift of spectrum due to air absorption is probably too small to be significant in most cases though the shift from outside to inside is more significant.

There are some propagation effects that theory suggests may not only moderate AM (such as the interference explained above) but might actually cause it. One of these is an upwind effect. The reason sound is not heard over long distances upwind is because of the effect of wind shear. This has the effect of bending sound waves upwards and away from the ground so that beyond some distance there is a sharp reduction of sound level. This can be seen in Figure 10 of Chapter 3. However, the higher the noise source the greater the distance at which it can be heard. So with turbine noise it is possible that, at say 600 metres distance, the sound of the turbine blade at the top of its trajectory could be heard, but not the sound at the bottom or half way down. So a form of AM could be generated in this way. It is also possible that at certain frequencies the ground effect (interference of the direct sound and the sound reflected at the ground) could reduce turbine noise at the bottom of the trajectory more than at the top or vice versa. This could cause a form of AM in certain frequency bands.

In line with other results (see chapter 6) the preliminary results of pilot dose response tests in the Renewable UK study seem to show that the only significant factors in people's adverse perception of AM of a particular periodicity are the modulation depth and the level, and possibly the spectral content of the noise although the effect of all these may be moderated by masking from background noise. It is already well known that the greatest annoyance caused by modulated sound is at a periodicity of 4 Hz and that annoyance decreases either side of this. It does not appear from the preliminary tests that the skew in signal shape (proportionally different rise and fall times) or the width of the pulse are significant in human perception.

REFERENCES

[1] F. van den Berg (2009), Why is wind turbine noise noisier than other noise?, *proc. Euronoise 2009*, Edinburgh

[2] G.P.van den Berg. "Effects of the Wind Profile at Night on Wind Turbine Sound", *J. Sound Vib.* 277 (4-5), 955-970 (2004)

[3] The Measurement of Low Frequency Noise at Three UK Wind Farms" produced for the DTI in the UK

[4] G.P. van den Berg, "The beat is getting stronger. the effect of atmospheric stability on low frequency modulated sound of wind turbines", *J Low Freq Noise Vib Active Control* 24 (1), pp. 1-24 (2005)

[5] G.P. van den Berg, "The sounds of high winds" (doctoral thesis), University of Groningen (2006)

[6] C. Di Napoli, "Long distance Amplitude Moudulation of Wind Turbine Noise:, *proc. 4th Int. Meeting on Wind Turbine Noise*, Rome, 2011

[7] J. Vos. M.M.J. Houben (2010), "Analyse van opnamen van windturbinegeluid" ("Analysis of wind turbine sound recordings", in Dutch), report TNO-DV 2010 C014, TNO

[8] W. Richarz, H. Richarz, "Wind turbine noise diagnostics", *proc.3rd Int. Meeting on Wind Turbine Noise*, Ålborg (2009)

[9] S. Wagner, R. Bareiss, G. Guidati (1996), "Wind turbine noise", Springer, Berlin

Chapter 6. Effects of Sound on People

Frits van den Berg (Netherlands), GGD Amsterdam Public Health Service.
fvdberg@ggd.amsterdam.nl

Sound is of major importance to people. Sound is one of our links with the surrounding world (others are light, odours and the taste and feel of that world). If loud enough, sound can impair our hearing, even when the sound may be enjoyable. But softer sounds can have a profound effect too: it can lift our spirits or cause irritation or terror, depending on the sound as well as the listener.

The audio-cortex, the part of our brain that processes incoming sound, is always active and always 'listening in'. The signal is, perhaps unconsciously, constantly interpreted, essentially to decide whether it is 'friend or foe'. If not important but friendly or neutral, such as muzak, the refrigerator or a ticking clock, we may not even 'hear' it, at least not consciously. But when it is unfriendly or interferes with our activities (e.g. reading or resting) the sound is brought to conscious attention. Then a sound may irritate even if we could appreciate it at other times.

The Nature of Sound and Noise

Sound[1] can be defined as an objective quantity that can be measured. The usual physical measure of the level of sound is the decibel. For young, normal hearing people the hearing threshold is approximately 0 dB (zero decibel) and the level where sound hurts is approximately 130 dB. When people grow older, the hearing threshold usually deteriorates and can be 20 or 30 dB or worse. Human hearing is most sensitive at frequencies between approximately 500 and 10000 Hz. At lower, as well as higher frequencies, the hearing threshold increases rapidly. As a result we cannot hear very high frequencies (ultrasound, over 20000 Hz) or very low frequencies (infrasound, below 20 Hz). At both ends of the range there is no sudden limit: even infrasound can be heard, and then felt, if it is at very high levels. To mimic human hearing, a frequency dependent correction is applied to physical sound pressure levels. This frequency correction or weighting has been dubbed A-weighting and the A-weighted decibel or dB(A) has become the standard for sound levels relevant to human hearing.[2] In fact it is correct and should be applied only at low to medium levels of sound, but this detail is usually lost in its application.

Loudness is the perceived 'strength' of the sound. When a sound level increases by 10 dB it sounds twice as loud.[3] A 20 dB increase is thus four times as loud, a 30 dB increase eight times as loud. However, at very low frequencies, growth of loudness is more rapid. Here a doubling of loudness at 20Hz corresponds to a 5dB change in level.

Environmental sound levels range from about 20 dB(A) in very quiet environments up to about 80 dB(A) in very busy streets (or a violent storm). At both ends of this range there is most agreement about

[1] That is, if one defines sound as a physical disturbance travelling at the speed of sound. Another widely used definition refers to sound as an audible phenomenon. The definitions often, but not always concur. Examples of sound where they do not concur are masked sound (physically present but not audible), tinnitus (audible, but often not physical sound) or 'wind rumble' (turbulent air movements at the ear, not travelling at the speed of sound).

[2] Except C-weighting (B has become derelict) that is still used at very high sound levels and/or (sometimes questionably) at very low frequencies

[3] Unfortunately this is different for the physical energy present in sound: a doubling in sound energy corresponds to a 3 dB increase of the sound level, a tripling to 5 dB; 10 dB extra means ten times as much sound energy.

the respective quiet- or noisiness. In between there is less agreement: moderate sound levels can be annoying to some, but neutral or even agreeable to others. For example: for a road traffic noise level of 60 dB(A) on dwelling facades, 21% of the inhabitants are annoyed of which 8% are seriously annoyed. But there are also 79% that are not or only mildly annoyed. Apparently the sound as such can evoke rather different reactions.

Noise[4] is often defined as unwanted sound and therefore has a subjective quality as its evaluation depends on the person experiencing that noise. This of course applies to a wide range of sounds: loud music may be heaven for some but intolerable for others, just as a tranquil landscape can be an area to relax or a place of boredom. What may be –at moderate sound levels- more important than just level is the source or nature of the sound and its appropriateness at that time and place. Here we will use 'noise' when referring to sound as a (possibly) adverse factor, whereas 'sound' will be used as a neutral description of a physical phenomenon.

Adverse Health Effects of Noise

The definition of health in the Preamble to the Constitution of the World Health Organization (WHO) is: "Health is a state of complete physical, mental and social well-being and not merely the absence of disease or infirmity". An adverse health effect can thus be defined as a negative impact on the physical, mental and social well-being of a person or group of persons. This description is also given in the WHO Guidelines (2000), as well as a more precise definition of adverse health effects of noise: "changes in the morphology and physiology of an organism that result in impairment of functional capacity, or an impairment of capacity to compensate for additional stress, or increases the susceptibility of an organism to the harmful effects of other environmental influences". A number of such adverse effects are distinguished (WHO, 2000):

- hearing impairment
- interference with speech communication
- cardiovascular and physiological effects
- sleep disturbance
- mental health effects
- effects on performance
- effects on residential behaviour and annoyance

For most effects the WHO guideline value is the lowest level where an adverse health effect has been shown to occur. This may depend upon the environment where the effect occurs: workers on a construction site or spectators of a movie are less vulnerable to disturbance by noise than children learning at school or people trying to sleep. Hearing impairment only occurs at high sound levels: for this effect the guideline values range from 70 dB(A) for public indoor and outdoor areas to 100 dB(A) for a loud event a few times a year. Low guideline values apply to sound levels within bedrooms and hospital wards: 30 dB(A). A limit value set by an authority may be higher than a guideline value as the limit is the result of balancing social, economic, environmental and health considerations.

Wind turbines are a relatively low level source (compared to, *e.g.*, a nearby busy road or airport), so the high level effects do not occur. This will be dealt with from page 131 onwards.

Personal and Social Factors

Sound can have a direct physiological impact, such as being startled by a sudden loud sound or by causing hearing impairment as a result of long lasting exposure to loud sounds. It can also have a psychological impact that eventually may lead to physiological effects. For example: noise may disturb task performance and the stress caused by continued disturbance and/or annoyance could lead to

[4] It is also, in a broader sense, used for a meaningless component accompanying a signal that carries information

cardiovascular effects. Psychological effects depend on how the listener appreciates the sound, on the mood and activity of that person, and a person's individual traits and social environment. In reality, body and mind are of course not separate entities and any (adverse) impact will have a physiological as well as a psychological effect where the one may or may not be the result of the other.

Several individual factors have a clear influence on annoyance from a noise source. Most important are fear in relation to the noise source and sensitivity to noise. For rail, road and aircraft noise Miedema and Vos (1999) have related these factors to the reported annoyance, using a database containing the response to noise surveys from nearly 50000 subjects. The results could be expressed as an increase or decrease in annoyance due to each factor, or as an equivalent increase or decrease in noise level. Fear associated with the noise source was found to have a significant and large impact on annoyance. When dividing respondents into three groups, the difference between the group with lowest and highest fear was equivalent to a difference in sound level (Ldn)[5] of up to 19 dB. Although the effect of fear was found for all three transportation modes, few persons associated high fear with railroads. The authors point out that the added annoyance could be associated with actual fear of that source or it could be related to a predisposition of respondents to noise annoyance and fear.

A second important factor is noise sensitivity. Again dividing respondents into three groups, the difference in annoyance between groups with lowest and highest sensitivity was equivalent to a difference in sound level (Ldn) of approximately 11 dB. Age also had a relation with noise annoyance, but less pronounced than fear or noise sensitivity: at the same sound level younger and elderly people are less easily annoyed than people of intermediate age. Education level, occupational status, the number of persons in a household, home-ownership, economic dependency on the noise source and use of the noise source (car, train or airplane) each had a small or very small effect on annoyance. There was no relation between noise annoyance and gender.

At a workshop of professional acousticians Flindell and Stallen (1999) reviewed non-acoustical factors that may have an effect on noise annoyance. Such factors are not related to any characteristic of the sound as such, *e.g.* level, frequency or variation over time. The list given by Flindell and Stallen (1999) is:

- predictability: the perceived likelihood of future changes in noise level;
- (perceived) control, either by the individual or by others, e.g. the noise source owner;
- trust and recognition that impacts are recognised by authorities;
- the ability of people to speak to and be listened to by authorities;
- general attitudes, such as awareness of economic and social benefits of the noise source, awareness of noise control action costs, fear of crashes, etc;
- personal benefits, such as employment in relation to the noise source;
- compensation, such as noise insulation or house purchase schemes;
- sensitivity to noise: noise sensitive people are more vulnerable;
- home ownership: home owners might be concerned about the value of their property;
- accessibility to information: this can affect the extent to which authorities are perceived to be taking an interest in the noise exposed community;
- understanding; attitudes can be based on irrelevant or incorrect information.

Some of these factors have been investigated in relation to wind turbine sound and will be described later in this chapter.

ANNOYANCE FROM WIND TURBINE SOUND

Wind turbines are a relatively new source of community noise. In general there is wide public support for wind energy, but residents are more aware of possible drawbacks when a wind project is planned

[5] Ldn is the average of the day and night time sound level after applying a penalty of 10 dB to the nighttime level

in their neighbourhood. As in many countries the number of wind turbines is increasing, the number of exposed residents is also increasing. In many projects the impact on the landscape is an important topic of public debate. Visibility of the wind turbines and their appropriateness in the neighbouring landscape are important issues that depend on the attitudes, motives and views of the different parties involved. According to Johansson and Laike (2007), and in line with earlier research results, people who felt that the wind turbines would have a negative visual effect on the landscape and thought that the turbines would decrease its recreational value, were more likely to oppose wind turbines. Of course, residents may also worry about the noise they will be exposed to in the future, about possible damage when part of a blade may fly off, or about the possible loss of property value because of the presence of a wind farm. And their views may be influenced by opinions on the ability of wind energy to help solve the energy problem or on the balance of advantages and disadvantages of wind energy for the parties involved.

After construction, when the wind turbine or wind farm is in operation, visual impact and noise are the most frequently reported problem (Devine-Wright, 2005). Here we will focus on post-construction noise only. Research has shown that annoyance is the dominant negative effect of noise on residents. There is an indication that sleep disturbance may be a second important effect, but there is as yet little information available. Sleep disturbance will be discussed in the next section. First we will consider noise annoyance in relation to wind turbines.

Annoyance in Relation to Immission Sound Level

Five studies have been published on wind turbine noise annoyance: 1) a study carried out in Denmark, the Netherlands and Germany by Wolsink *et al (*1993); 2) a Danish study by Pedersen and Nielsen (1994); 3) a study in the Swedish countryside by Pedersen and Persson Waye (2004); 4) a second Swedish study in seven different areas by Pedersen and Persson Waye (2007); 5) a study in the Netherlands by Pedersen et al (2009). The size of wind turbines has increased over the almost two decades of time covered by these studies. The 1994 Danish study included 16 wind turbines of up to 150 kW nominal power with hub heights up to 33 m high, while the most recent Dutch study focussed on wind turbines of at least 500 kW and ranging up to 3 MW with towers up to (and some over) 100 m high. The first four studies included mostly single wind turbines, the last study focused on wind turbine farms (at least two wind turbines of at least 500 kW each within 500 m from one another) that have become more common since the early studies. Here we will discuss the results of the three latter studies as they address wind turbines with dimensions that have become most common in the present century.

The three studies in Sweden and the Netherlands are surveys investigating the response to the sound present at respondents' dwellings as a result of one or more wind turbines close to the dwellings. The dose in all studies is the calculated, *A*-weighted equivalent sound level[6] at the respondent's dwelling under specific meteorological conditions: a wind speed of 8 m/s at 10 meter height (4 to 5 Beaufort) and, implicitly, a neutral atmosphere[7]. The response is the degree of annoyance as stated by the respondents in the survey questionnaire. These surveys were conducted in the summer half year. The Dutch survey was presented as a survey about the residential living environment and also dealt with road traffic noise and the perception of other environmental factors such as odour,

The first Swedish study was conducted in the year 2000 in a municipality in the south of Sweden in a flat and mostly agricultural landscape (Pedersen and Persson Waye, 2004). It included respondents exposed to levels of predicted wind turbine noise ranging from just below 30 dB(A) to just over 40

[6] an equivalent sound (pressure) level menas that the sound is averaged over time, so variations of the level within the averaging period are neglected

[7] in a neutral atmosphere no (significant amount of) heat is added by insolation to the near-ground air or lost due to cooling of the ground after sundown; a neutral atmosphere thus occurs when the sky is heavily clouded and/or when the near-ground air is mixed very well with higher altitide air by a strong wind

dB(A) from wind turbines of 150 to (predominantly) 600-650 kW nominal electric power. The second study, carried out in 2005, was conducted in seven areas in Sweden with turbines with nominal power of at least 500 kW (Pedersen et al, 2007). The areas differed with respect to terrain type (flat/complex) and with respect to degree of urbanization (rural/built-up). Here respondents were exposed to levels of wind turbine noise ranging from 30 dB(A) to just over 40 dB(A).

The Dutch study, carried out in 2007, included residents living within 2.5 km of a wind turbine of 500 to 3000 kW nominal power and at least one other turbine within 500 m from the first (Pedersen et al, 2009). This study investigated a wider range of noise levels (24 to 54 dB(A)) so as to obtain data for a dose-response relation over the range of noise levels occurring in practice. The survey included all of the Netherlands and three types of flat area: rural, rural with a main road and built-up. The questionnaire was based on the Swedish questionnaire and presented as a questionnaire concerning the respondents' living environment, now including questions on road traffic noise as well as wind turbine noise.

In the 2000 study 513 questionnaires were sent to residents of which 351 (68%) were completed and returned. In the 2005 study 1309 questionnaires were sent of which 754 (58%) usable questionnaires were returned. In the final 2007 study 1948 questionnaires were posted and 725 (37%) usable questionnaires were returned; a non-response analysis found no significant difference in annoyance due to wind turbine noise between respondents and non-respondents. Thus, the three studies together were sent to 3770 subjects of which 1830 (49%) were returned.

The results of these three studies have been collected by Janssen et al (2008, 2009) and presented as generalized dose-response curves for wind turbine sound where response was separated in the percentage of respondents being either annoyed or highly annoyed when being indoors and outdoors respectively. The results, using data from Janssen et al (2008) are presented in Figure 1. Here respondents with economic benefits from wind turbines have been excluded.

When the individual study results are compared to the aggregated results (Figure 1), it appears that the percentages of respondents that are (highly) annoyed *outdoors* is somewhat higher in the 2007

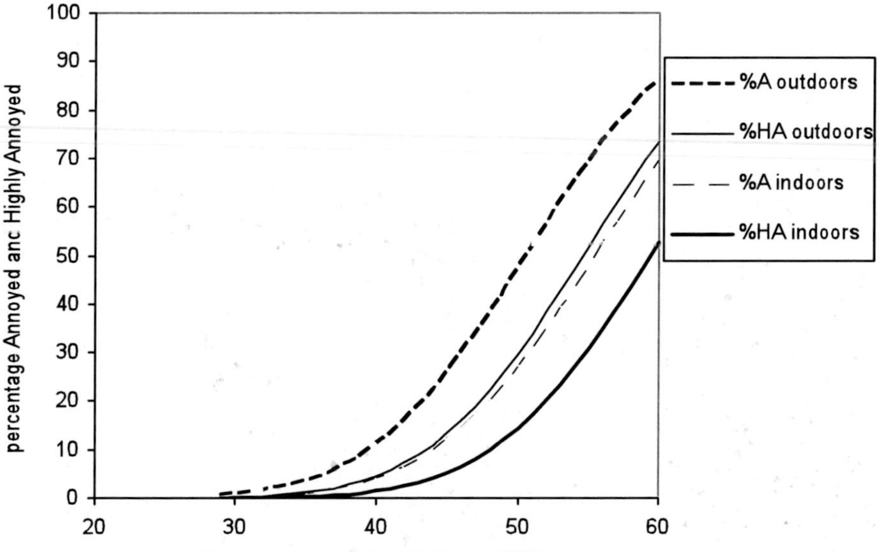

Figure 1. Percentages of respondents that were Annoyed (broken lines) or Highly Annoyed (solid lines) when being either outdoors (thin lines) or indoors (thick lines); data from Janssen et al (2008)

Dutch study and still higher in the 2000 Swedish study but, to balance, lower in the 2005 Swedish study. The percentages of respondents being (highly) annoyed *indoors* are somewhat higher in both the 2007 Dutch study and the 2000 Swedish (both yielding nearly identical results) but clearly lower in the 2005 Swedish study. Investigation of a number of acoustical and non-acoustical factors shows that the response is related to the sound character, to the attitude to and benefits from wind turbines and to the visibility of the wind turbines.

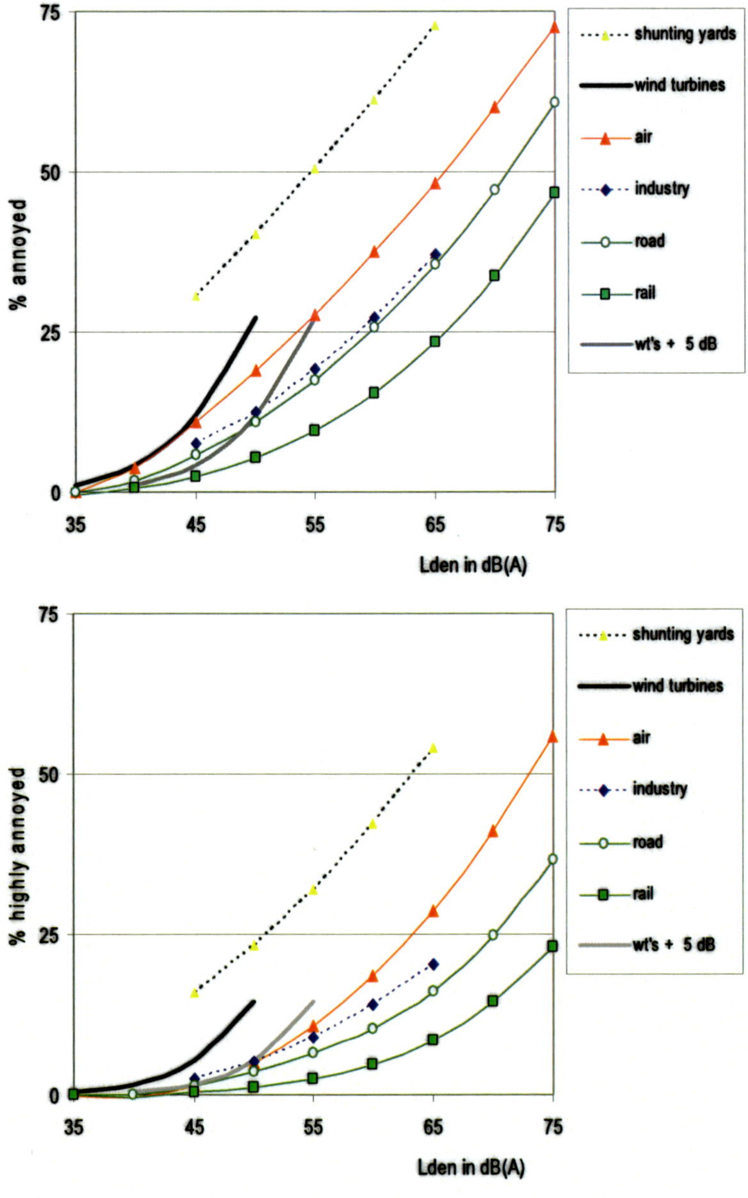

Figure 2. Percentages of respondents that are annoyed (upper panel) of highly annoyed (below) when exposed indoors to noise from wind turbines or other sources

Comparison of Noise Annoyance from Wind Turbines and Other Noise Sources

When comparing the dose-response curves for wind turbine noise to similar curves for other noise sources, wind turbines appear to be a relatively annoying noise source. Dose-response curves for indoor annoyance versus outdoor sound level have been determined for the dominant transportation modes (road, rail and air transport) and for industrial sources, including railway shunting yards, by Miedema and Oudshoorn (2001) and Miedema and Vos (2004). The results are shown in Figure 2, where the curves for indoor annoyance from wind turbine noise are the same as in Figure 1.

Figure 2 shows that wind turbine noise is more annoying than the other sources with the exception of shunting yards. There are several acoustical and non-acoustical factors that could explain this relatively annoying character. If the sound character is taken into account and a 5 dB penalty is added to the sound level of wind turbines because of the presence of amplitude modulation, the curves for wind turbine noise are within the range of the main sources at L_{den} levels below 50 dB(A) as shown by the grey curves in Figure 2.

INFLUENCE OF ACOUSTICAL FACTORS

Perception of Amplitude Modulation

People that live not far from a wind farm give descriptions of wind turbine sound that show that the variation in sound level (swishing, beating, thumping, etc.) apparently is an important characteristic of wind turbine noise. Several publications list a number of such descriptions (*e.g.* Van den Berg, 2006; Frey and Hadden, 2007), but from these it is not clear how widespread this phenomenon may be.

In the Dutch survey residents living within 2.5 km of a wind farm were asked to describe the sound made by the wind turbines (Van den Berg, 2008; Pedersen *et al*, 2009). A number of descriptions were given (and the possibility of giving another, personal description -which was hardly used). The response is shown in figure 3. Most respondents found 'swishing' or 'lashing' a proper description, regardless of whether they found the sound annoying or not. Rustling also appears to be a description shared by residents irrespective of their being annoyed or not, though it is mentioned rather less often. Other descriptions apply to fewer respondents and are far less equally shared by those who were or were not annoyed.

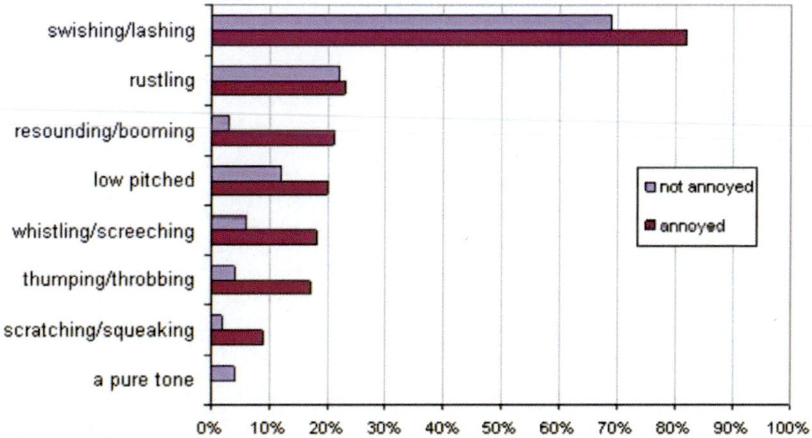

Figure 3. Percentage of respondents in Dutch survey that think the description fits the wind turbine sound they hear at home

Hayes (2006) performed sound measurements in dwellings close to three different wind farms and concluded that amplitude modulation was perceptible within the dwellings. He concluded that this could be a cause of (stronger) disturbance, rather than a low frequency character which had been suspected to be the main cause of disturbance. As a follow-up in a further study Moorhouse et al (2007) investigated whether this phenomenon could be established at other wind farms by asking Environmental Health Officers in the UK whether they could confirm the presence of amplitude modulation (AM) using a short questionnaire. This study concluded that in four cases out of 27 AM appeared to be a factor in the complaints, in fourteen it was not, and in eight it was considered marginal. However, the evidence for these conclusions could not be found in the completed questionnaires. An alternative conclusion from this survey is that AM figured in nearly all cases where a description was given (Van den Berg, 2009). In 13 out of 14 cases where complaints had been lodged and a description of the sound was available, the sound was described with one or more words indicating a (regular) variation in level or AM. In another 13 cases the question was not resolved as no description was given. In one case only, the descriptions did not fit AM.

Fluctuations are perceived as such when the modulation frequency is less than 20 Hz. Human sensitivity for fluctuations is highest at $f_{mod} = 4$ Hz, which is the frequency typical for rhythm in music and speech [Zwicker et al, 1999] and for frequencies of the modulated sound close to 1 kHz. For wind turbines a typical modulation frequency is 1 Hz, modulating the trailing edge sound that itself has dominant frequencies at approximately 500 to 1000 Hz. Fluctuation strength can be expressed as a modulation factor, that is: the percentage of modulation relative to the highest perceptible fluctuation strength (100%), or as an absolute value in the unit vacil. Physically, the modulation factor mf is defined as the change in sound pressure amplitude due to modulation, relative to the average amplitude. The sound level difference ΔL associated with an amplitude modulation (AM) factor mf is: $\Delta L = 20 \cdot \log((1+mf)/(1-mf))$.

For an AM pure tone as well as AM broad band noise, absolute fluctuations strength is zero until $\Delta L \approx 3$ dB, then increases approximately linearly with modulation depth up to a value of 1 vacil. 1 vacil equals the absolute fluctuation strength reference value: a 60 dB, 1 kHz tone that is 100% amplitude-modulated at 4 Hz [Zwicker et al, 1999].

For a broad band noise level L_A the fluctuation strength F_{bb} can be written as [Zwicker et al, 1999]:

$$F_{bb} = \frac{5.8 \cdot (1.25 \cdot mf - 0.25) \cdot (0.05 \cdot L_A - 1)}{(f_{mod}/5\text{ Hz})^2 + (4\text{ Hz}/f_{mod}) + 1.5} \text{ vacil} \qquad [1]$$

For $\Delta L < 9$ dB a good approximation (± 5%) is: $mf = 0.055 \cdot \Delta L$. Using typical values for wind turbine noise of $f_{mod} = 1$ Hz and $L_A = 40$ dB(A), equation 1 can be written as $F_{bb} = 1.31 \cdot (mf - 0.2)$ vacil or:

$$\Delta L < 9\text{ dB:} \quad F_{bb} = 0.072 \cdot (\Delta L - 3.6) \text{ vacil} \qquad [2]$$

With measurement results from one site Van den Berg (2005) showed that when ΔL rises from 3 dB, presumably a maximum value for a daytime (unstable or neutral) atmosphere, to 6 dB, mf rises from 17% to 33% and fluctuation strength F_{bb} from zero to 0.17 vacil. For a (possibly maximum) value of $\Delta L = 9$ dB, mf is 50% and $F_{bb} = 0.39$ vacil. For higher average sound levels these values are higher; *e.g.* for $L_A = 50$ dB(A), these fluctuation strengths would be 0.26 and 0.59 vacil, respectively.

Thus, human hearing is relatively sensitive to wind turbine sound fluctuations. A modulation depth of at least 3 dB at a modulation frequency of approximately 1 Hz can be easily picked up by human listeners, possibly because it is not unlike loudness modulation in speech. A similar modulation frequency is used in warning signals (such as beeps from trucks in reverse gear, alarm clocks) and in musical rhythm. The evaluation of a sound that can be picked up easily will (also) depend on its character: if it is unwanted, this characteristic will aggravate its disturbing character.

Annoyance from Amplitude Modulation

The influence of the character of wind turbine sound on its evaluation and impact has been investigated in laboratory and field studies. Persson Waye and Öhrström (2002) investigated the psychoacoustic characteristics of wind turbine sound in a laboratory setting with naive listeners (students not used to wind turbine sound). Of sounds recorded from five different turbines, the most annoying sound was described as 'swishing', 'lapping' and 'whistling', the least annoying as 'grinding' and 'low frequency'.

Legarth (2007) played recordings from a number of wind turbines in different conditions to listeners in a home like laboratory room. The fluctuation strength (determined by the modulation depth) was determined for the frequency band 350 – 700 Hz where the fluctuation was strongest. This fluctuation strength correlated very well with the swishing character perceived by the listeners. Lee et al (2009) found in laboratory listening tests that a sound with a higher modulation depth at blade passing modulation frequency was considered more annoying by listeners than sound with less modulation.

From the first Swedish survey amongst residents exposed to a level of at least 30 dB(A) of wind turbine sound, Pedersen and Persson Waye (2004) concluded that annoyance with wind turbine sound was correlated with descriptions of the sound characteristics. One in three respondents mentioned swishing and this was strongly correlated to annoyance (correlation coefficient 0.72). Whistling and pulsating/throbbing were mentioned by respectively 26% and 20% of the respondents, but were less strongly correlated (0.64 and 0.45, respectively). Other descriptors (resounding, low frequency, scratching/squeaking, tonal, lapping) were ever less often mentioned and had lower correlation coefficients (0.49 to 0.26).

Similar results were found in the Dutch survey: respondents that characterized the sound from wind turbines as swishing or lashing were more likely to be annoyed by the sound. Compared to respondents who were not annoyed by the sound, respondents that were rather or very annoyed by wind turbine sound more often characterized the sound as thumping/throbbing, whistling/screeching, scratching/squeaking and/or resounding.

To obtain a better understanding of residents living close to wind turbines Pedersen et al (2007) interviewed them and found they felt irritated because of the intrusion of the wind turbines in their homes and gardens, especially because of the swishing sound, the blinking shadows and the constant rotation.

Other studies also have shown that a sound that is amplitude modulated is not only more easily perceptible, but is also more annoying than an unmodulated sound of similar frequency content and level. Bradley (1994) investigated fluctuations in laboratory experiments. He presented wideband noises to nine subjects that were similar to a reference spectrum, but with an elevated level at 31.5 Hz and at higher frequencies gradually sloping towards the reference sound. The low frequency part could be amplitude modulated at modulation frequencies of 0 (no modulation), 0.25, 0.5, 1, 2 and 4 Hz and modulation depths of 10 and 17 dB (*i.e.* mf is 52% and 75%, respectively). Subjects were asked to adjust the overall level of these to be equally annoying to the unmodulated reference spectrum (a broad band sound with a slope of 5 dB per octave). For modulation frequencies of 0.5 to 4 Hz the adjustments varied with modulation depth. At 17 dB modulation depth an extra attenuation of about 4 dB resulted in equal annoyance with the unmodulated sound. Overall, at modulation frequencies of 0.5 to 4 Hz the 'equal annoyance correction', that is: the attenuation to be applied to the average level of the modulated sound so the annoyance is equal to that of the reference sound, could be expressed as $K*\Delta sd$. Here K is a constant for which a value equal to 4.5 fitted the majority of the data points. Δsd is the mean difference between the standard deviations of the sound pressure levels of the modulated sound and the unmodulated reference sound, averaged over the octave bands from 31.5 to 250 Hz. Values for Δsd are not given by Bradley (1994) but are in the order of 1 dB. The difference between L10 and L90 for the modulated sounds was 3 to 5 dB for each octave band between 31.5 to 250 Hz.

Later Moorhouse et al (2007, 2009) also presented steady and fluctuating low frequency sounds to eighteen subjects in a laboratory study and obtained similar results. They found that if the level of fluctuation (determined as the difference between L10 and L90) was 3 to 5 dB or higher, subjects found it equally acceptable as a steady sound at a 5 dB higher level.

Dittrich et al (2009) investigated the effect of sound level fluctuations by presenting sounds with a 2 dB or 4 dB standard deviation (sd) to thirteen subjects. The sounds consisted of nine contiguous broad band noises each one lasting 0.1 second, followed by 0.5 seconds silence and a second series of nine 0.1 second broad band noises. The levels of the 0.1 second sounds were randomly selected, but their total standard deviation was restricted to either 2 or 4 dB and the overall average (Leq) level was kept constant. A total of 300 test sounds, with random sequences of the 0.1 second sounds, was presented to each subject. The modulation did not influence the loudness estimates of the sounds, but it did influence the annoyance estimate: the annoyance score for the sound with 4 dB sd was significantly higher than the sounds with 2 dB sd (annoyance scores, normalized to a scale of 0 to 1, were 0.66 and 0.52, respectively). Thus, this study shows that noises with the same average level but with different modulation depths are judged to differ in annoyance but not in loudness.

Finally, Vos et al (2010) presented AM sounds to 32 subjects in a home-like laboratory room who were asked to score the degree of annoyance imagining they would hear this for a prolonged time at home. The sound was a noisy sound with a spectral content equal to that of road traffic sound and with a modulation frequency of 1 Hz and modulation depths of 0, 6 or 12 dB, all at a level of 32 dB(A) as measured at ear position. The difference in response to a modulated or unmodulated sound was highly significant, but the annoyance score did not differ significantly for the 6 and 12 dB modulation depth. The increase in annoyance corresponded to an increase of level of the unmodulated sound of 10 dB.

Thus, the laboratory studies show that amplitude modulation of a broad band sound, either with or without a strong low frequency component, increases its annoyance but not its perceived loudness. The effect on annoyance may occur when modulation is perceptible and perhaps does not increase substantially with increasing modulation depth. As the degree of modulation was quantified in different measures in these studies, the effect of modulation depth cannot readily be established. However, it appears that the increase in annoyance due to modulation corresponds to an increase of the level of the unmodulated sound of at least 3 dB.

Low Frequency Character of Wind Turbine Sound

Wind turbines produce a substantial amount of low frequency sound as shown by Wagner et al (1996), Van den Berg (2006a) and others. However, as the sensitivity to sound depends on the frequency of sound, the impact of a sound does not correspond directly to its physical (unweighted) sound level. Human hearing is relatively insensitive at low frequencies which thus compensates for physically higher levels at these frequencies. As the A-weighting mimics the frequency dependency of human hearing at low to moderate loudness, A-weighting should be a correct estimate of the loudness of a sound at low to moderate level (this loudness corresponding to the level of a tone at 1000 Hz of 40 dB).[8] Nevertheless, a number of researchers think that low frequency sounds may be more disturbing than 'normal' (higher frequency) sounds (see, *e.g.*, Leventhall, 2003). A laboratory study by Vos (2010) suggests that this is true when considering the effect of *outdoor* noise on a person being indoors. As a building façade reduces high frequency noise more effectively than low frequency noise, a low frequency sound impinging on the façade must be of lower level than a high frequency sound in order to arrive at the same indoor sound level. The World Health Organization states that even though for noises with strong low frequency components the A-weighted levels have been criticized as not being accurate indicators of the disturbing effects, the possible inaccuracy in prediction is small and the inaccuracy due to the variability of people is usually more important (WHO, 2000). This is supported by Hünerbein et al (2011, who conclude from a laboratory study with tonal sounds and recorded wind turbine sounds that tones at lower frequencies are not more annoying than tones at higher frequencies, if the tone prominence is the same. In practical situations therefore the limitations of A-weighted measures may not be very important. However, this conclusion applies to low frequency tones and may

[8] Other frequency weightings apply at higher levels: the B-weighting is best when the loudness is moderately high (corresponding to a sound level of 70 dB at 1000 Hz), the C-weighting when loudness is high (100 dB at 1000 Hz).

not apply to broad band low frequency sound without tonal components. Pedersen (2008) proposed to determine the loudness of low frequency sound by comparing the level of all sound in each critical band to the hearing threshold. For frequencies above 500 Hz a 1/3 octave band spectrum can be used, as here a critical bandwidth is nearly equal to a 1/3 octave bandwidth. For frequencies below 500 Hz the critical bandwidth is approximately 100 Hz and thus spanning a broader frequency range than a 1/3 octave band. As a result a 1/3 octave band spectrum will underestimate the audibility and loudness, and increasingly so at lower frequencies. Nevertheless, a narrow band (such as a 1/3 octave band) spectrum is still useful to identify frequencies where the sound may be relatively loud. Using this method, Pedersen showed that for a 1.3 MW wind turbine at 280 m distance the LF sound may be audible, but the sound at higher frequencies is dominant.

Figure 4 shows the spectral distribution of A-weighted sound levels at two distances from a wind turbine having maximum, average and minimum sound power. The spectral distributions are taken from the WINDFARMperception project (Van den Berg et al., 2008) and represent all existing wind turbine types in the Netherlands in 2008 with a maximum electric power production between 250 and 3000 kW. These spectral distributions are similar and the averages per octave band are the same within 1 dB when compared to somewhat earlier measurements of Søndergaard et al (2007), who collected data of sound power levels down to 25 Hz from wind turbines with a maximum electric power production between 75 and 2000 kW.

Figure 4 shows the average, maximum and minimum octave band sound levels for a receiver at 300 m and 1500 m distance of a single wind turbine of maximum, average and minimum sound power level (108, 102 and 94 dB(A), respectively). These receiver sound levels are calculated from the wind turbine sound power level, taking into account spherical spreading of the sound over half a sphere and atmospheric attenuation for each octave band as given by ISO9613-2. They can be seen to be very similar to those described in the chapter 4. The conclusion is that the noise impact of wind turbines is dominant at frequencies of approximately 100 to 1000 Hz, not at low frequencies. This is still true when applying the method proposed by Pedersen (2008; see above) to evaluate audibility.

Pedersen and Møller (2010) have shown that with increasing electric power of wind turbines sound power increases somewhat faster and the low frequency part of the sound (10 – 160 Hz) still faster. The differences are small (up to 3 dB) but are expected to grow with the development of more powerful wind turbines. Therefore, the low frequency component of wind turbine noise may become more important.

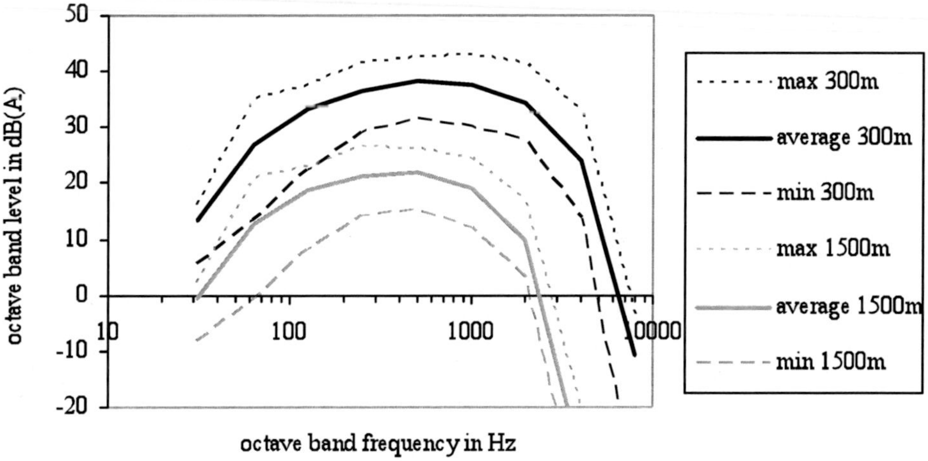

Figure 4. Sound immission levels in octave bands at 300 and 1500 m from a single wind turbine; maximum, average and minimum levels due to all turbine types present in the Netherlands in 2008

Infrasound is at the very lowest end of the audible spectrum, at frequencies of 20 Hz and below. It is audible only when the physical level is high when compared to the usual environmental sounds at higher frequencies. Infrasound frequencies are not shown in Figure 3 as infrasound levels usually are not measured and therefore not reported. An important reason for this apparent deficiency is that infrasound is rarely audible in the living environment. Also, infrasound is difficult to measure correctly because of the presence of air turbulence that creates pressure variations on a microphone (and on the human ear) that is not easily distinguishable from proper sound at the same frequencies. Infrasound, especially due to atmospheric sources, is certainly present in the environment, but because of the high hearing threshold at infrasound frequencies (at least 78 dB) most of this is not audible.

Jakobsen (2005) reviewed all published measurements of infrasound from wind turbines and concluded that at a distance of 100 m all measured infrasound levels from upwind wind turbines were below the normal hearing threshold. This included wind turbines with a maximum electric power production between 50 and 4200 kW. Downwind turbines (with the rotor downwind from the mast) produced most infrasound, and for the most powerful infrasound producer this could be audible up to 1 km. This high level of infrasound –produced in pulses whenever a blade passed behind the mast- was an important reason that downwind turbines were not developed further and upwind turbines became the standard. According to Jakobsen (2005) the infrasound level of upwind turbines is, at a distance of 100 m from a single turbine, 25 dB or more below the average hearing threshold. Thus, Jakobsen concluded that infrasound, or even low frequency sound, is an unlikely explanation for annoyance from wind turbine sound. The usual A-weighted sound level is a more likely candidate to explain -at least part of- the annoyance as this 'normal' sound exceeded the Danish noise limit in five out of six cases in his review. This view was supported by Van den Berg (2004) who concluded that infrasound due to the passing wind turbine blades must be considered inaudible, but the blade passing frequency does modulate well audible, higher frequency sounds and thus creates periodic (AM) sound that could be an important cause of annoyance. The fact that a modulation frequency is also expressed in Hz may have led to confuse this with infrasound frequencies.

Several authors have suggested that low frequency noise may be an important feature of wind turbine noise with regard to its impact on residents. Frey and Hadden (2007), Pierpont (2009) and Alves-Pereira and Castelo Branco (2009) have linked infrasound or low frequency sound from wind turbines to health effects experienced by residents. They assume that low frequency sound (including infrasound) can have an effect at levels below the (normal) hearing threshold. Salt and Hullar (2010) have shown that the differences between outer and inner hair cells in the inner ear could explain different frequency dependent sensitivities in human hearing. They also proposed that this could mean that infrasound can have effect at inaudible levels. Though this could perhaps explain specific auditive disorders, there is as yet no evidence that adverse effects occur for people without medical complications when exposed to infrasound levels present in common environmental sounds. Salt and Lichtenhan (2011) added that, as animal experiments have shown, the presence of high frequency sound may inhibit the perception of infrasound. Vice versa, the absence of high frequency sounds can lead to the perception of infrasound. If this is also true in humans, this could explain a sound perception indoors when higher frequency sounds are inaudible due to sufficient attentuation. It could perhaps also explain the presence of 'phantom sounds' that, according to several studies, people hear when put in a quiet room (van den Berg, 2006). However, as yet it has to be demonstrated that such an explanation holds in practice.

The view that inaudible infrasound can have (adverse) effects on people is not supported by studies investigating the effect of low levels of low frequency sound. Takahashi (2009) has shown that the threshold for body vibrations induced by low frequency sound (20 to 50 Hz) is higher than the hearing threshold. This was investigated with seven subjects with normal hearing. The vibrations were most often felt at the head and to a lesser degree at the chest or abdomen. Here the sound was produced by loudspeakers in the test room which is the usual procedure to determine low frequency hearing thresholds (for higher frequency thresholds head phones are commonly used). Therefore, if a person would sense the effect of low level low frequency sounds otherwise than by hearing, this could still be determined with the usual procedures to establish low frequency sensation thresholds. An indication that the ear is the most sensitive organ to perceive sound, including low frequency sound, was given much earlier by Yamada et al (1983) who investigated low frequency sensation thresholds of deaf subjects. These subjects could perceive the sound with their body, but only at levels of at least 40 to 50 dB above the hearing threshold of normal hearing subjects. The experience of chest vibrations for deaf subjects was comparable to subjects with normal hearing. Finally, in their evolution human beings have been exposed to low frequency pressure fluctuations from turbulence in wind. These fluctuations can be heard as 'wind noise' (where the auricle may also induce extra turbulence) and can be felt on the skin. Expressed as a (sound) pressure level, for wind velocities of 4 m/s (3 Beaufort) and higher the one-third octave band levels of these fluctuations are 60 dB or higher at frequencies below 50 Hz (Van den Berg, 2006b). It seems plausible that this would have the same effect as exposure to noisy (non-tonal) low frequency sound/infrasound.

Thus, as yet there is no evidence that inaudible (low frequency) environmental sounds can have somatic effects. To guide further research with respect to environmental health effects, it is important to have clear indications to the contrary rather than speculations.

Influence of Night Time Exposure

Sound levels from modern, tall wind turbines are not lower at night compared to daytime, at least not in the temperate climate zone. This is because at rotor height the wind does not abate at night as it often does closer to the ground. In fact, at 100 m hub height the long-term average wind turbine sound power level is somewhat (0.5 dB) higher at night than it is in daytime (Van den Berg, 2008). In most cases this slightly higher level adds perceptually to the already higher contrast with the lower ambient levels at night due to less near-ground wind, less traffic and less human activity. This is supported by the response to one of the questions in the Dutch survey, probing the circumstances for increased audibility of wind farm sound (Van den Berg et al, 2008). In Figure 5 the results are shown: as expected more respondents find the wind farm more audible when the wind blows from the wind farm towards the dwelling or when the wind is stronger. In contrast to the usual abatement of near ground wind only a minority of residents close to a wind farm (22%) find the wind farm less audible at night. According to 40% of respondents the wind farm is louder at night and for 38% there is no clear difference between night and day in this respect. Thus, for most respondents wind turbine sound does not abate at night.

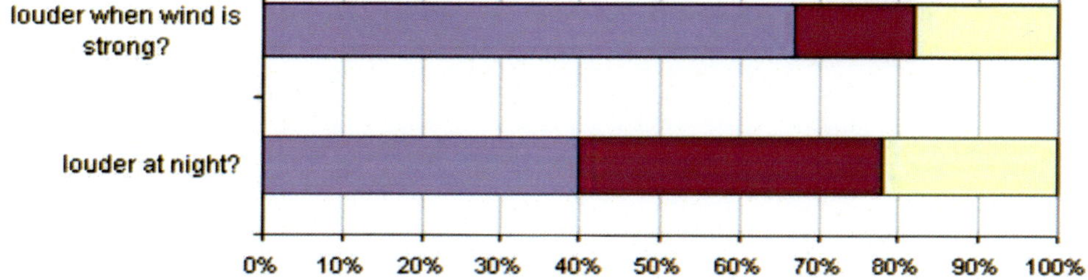

Figure 5.

INFLUENCE OF NON-ACOUSTICAL FACTORS

Earlier in this chapter an overview is given of individual and social factors that may have an effect on the relation between noise and the annoyance it causes. This section presents an overview of non-acoustical factors that have been found to influence the relation between wind turbine sound and annoyance.

Economic Benefit

The Dutch survey showed that little annoyance with sound from wind turbines was reported by residents if they had economical benefits from wind energy. This contrasts sharply with the result that for those who did not benefit economically, wind turbines were more annoying than other major noise sources. Also, those benefiting economically from wind turbines felt significantly less disturbed in their sleep by noise. Despite this clear difference in annoyance and disturbed sleep between the two groups, there was no difference in the ability to hear the sound between respondents that benefited economically from wind turbines and those who did not. And in fact, those benefiting were on average exposed to higher sound levels. They were relatively young, well educated, worked at home and/or were self-employed. Also, they were healthier (less often exhibiting high blood pressure, cardiovascular disease or migraine), and they had less difficulty in falling asleep. In short, they could be typified as 'healthy farmers' and entrepreneurs who have to earn their living by making use of the land. This may help to explain the different opinions on wind farms, arising from different views on landscape utility and use, between those that benefited and those that did not.

Influence of Visibility

Respondents that could see at least one wind turbine from their dwelling were more likely to be annoyed by the sound than those who did not see any wind turbines (Pedersen *et al*, 2009). When respondents did not see wind turbines, the turbines were either relatively small, *i.e.* distant, or respondents lived in built-up areas. When the sight of the wind farm was blocked, then the sound could be (partly) blocked too, leading to lower sound levels, which could possibly explain the lower levels of annoyance. Alternatively, the enhanced annoyance when the wind turbines were visible could also be due to a multimodal effect: the rotating blades of a wind turbine attracting attention could increase the awareness of the sound and hence also the possibility of noise annoyance (Pedersen *et al*, 2007).

Masullo et al (2011) investigated the interaction between rotor speed and colour, sound level and wind turbine form in a preliminary laboratory study. The 33 subjects were shown short (30 s) videos of wind turbines in realistic landscapes while recorded wind turbine sound was played at one of three levels. Visual parameters were rotor speed (0, 18, 30 and 38 rpm), rotor colour (grey, purple, red, white, yellow and azure), wind turbine form and dimensions (lattice or tubular mast, 55 or 105 m hub height, 46 or 72 m rotor diameter), and the number of wind turbines (from 1 to 5). It was found that the perceived loudness was influenced by sound level, as expected, but also by the number of wind turbines and the speed and colour of the rotor. Annoyance was influenced by sound level and by the rotor speed and wind turbine form.

Predictability and Control

Wind turbine sound varies unpredictably in level within a relatively short time span (minutes to hours). Although the sound level of modern, tall wind turbines may at night be lower for part of the time, it probably is either higher or does not change clearly at night. This was confirmed in the Dutch survey where only a small proportion of respondents reported hearing wind turbine sound less clearly at night. In this respect wind turbine sound is different from traffic noise or industrial noises that usually return to lower levels at night. As a consequence residents that are bothered by wind turbine sound are never sure whether the sound at night will allow them to be at home without being disturbed.

According to Taylor (1997) a healthy environment "provides safety, opportunities for social integration, and the ability to predict and/or control aspects of that environment". In the Dutch survey there was little annoyance amongst respondents with an economical benefit of wind turbines. One of the reasons may be that they have more control over the wind turbines: most of them have taken part in the decision to put up the turbines and they can stop them if they want. One respondent remarked spontaneously that if a turbine close by caused too much noise for him or his neighbour, the turbine could be stopped.

Attitude and Fairness

In the Dutch survey annoyance with wind turbine sound was associated with a negative attitude towards the wind turbines, especially with regard to the impact on landscape. Annoyance was also correlated with changes for the worse in the living environment as perceived by respondents (but not with changes for the better). A strong correlation between noise annoyance and a negative opinion of the impact of wind turbines on the landscape was found in earlier studies and confirmed in the Dutch survey (Pedersen *et al*, 2009). The theory of 'place attachment' may explain this correlation. In this theory, a new technical device that is perceived as not being beneficial for the living environment induces a negative reaction (Pedersen et al, 2009).

In a review of recent research literature Tyler (2000) concluded that people are more willing to accept decisions when they feel that those decisions are made through decision-making procedures they view as fair. People evaluate a procedure as fair when all parties at stake have opportunities to participate, the authorities are neutral, the motives of the authorities are trusted, and people are treated with dignity and respect in the process. This fairness can influence noise annoyance as Maris et al (2007) have shown. In a laboratory experiment 117 participants were exposed to medium or loud aircraft noise (50 and 70 dB(A)). Part of the participants could give their preference for one of three sound samples (many short passages, some long passages, or in between) to be played when doing a task, others were not asked. In fact, the same sample was presented to all participants. Results showed that annoyance ratings were significantly lower when a preference could be given compared to the situation this was not asked, though this was found only at the exposure level of 70 dB(A). Thus, at a high noise level a 'fair procedure' can reduce noise annoyance. Illustrations of this at an individual level are given by Bowdler (2011) showing how residents can develop complaints and illness driven by the "arrogance and intransigence" of the operator and local authority.

Wolsink and Breukers (2010) compared the development of wind energy in three regions (North Rhine Westphalia, the Netherlands and England) and found that "an approach that focuses on implementing as much wind power as possible, relying on technocratic reasoning and hierarchical policies is in practice the least successful, whereas collaborative perspectives with more emphasis on local issues and less on the interests of the conventional energy sector were particularly dominant in the most successful case".

SLEEP DISTURBANCE

Sleep disturbance due to noise can manifest itself in different forms such as a difficulty to fall asleep, or being awakened by a noise, or feeling tired in the morning. An overview of all effects is given in the guidelines for night time noise of the World Health Organization (WHO, 2009): sleep disturbance also leads to biological effects that are less directly noticeable, such as changes in cardiovascular activity and hypertension, movement during sleep, changes in sleep structure and fragmentation of sleep, and changes in (stress) hormone levels. As a result it can also lead to drowsiness or tiredness and increased daytime irritability, the use of sleeping drugs, impaired cognitive performance, impaired social contacts, depression, psychiatric disorders and (occupational) accidents. The effects depend on the sound level and the number of noise events in the bedroom. The WHO night time noise guidelines are based on observed effect thresholds: the level above which an effect starts to occur or shows itself to be dependent on the dose. This can be an adverse effect such as a heart infarct or a possibly adverse increase in a naturally occurring effect such as movement during sleep.

For the maximum level of a noise event the indoor threshold event levels above which various biological effects can occur varies from 32 to 42 dB(A). As a façade will usually reduce the sound level with at least 10 dB (with windows open) and often higher (windows slightly open or closed), the maximum outside noise event level must be at least 42 dB(A) but often up to and over 50 dB(A). Of course *average* noise levels over the night will be lower as the events occur for only part of the time. When related to the average outdoor night time noise level (Lnight), threshold levels are 40 to 42 dB. Although individual reactions will differ and effects have been observed, no severe general adverse effects seem to appear at outdoor night time levels below 40 dB(A) (WHO, 2009). For levels from 40 to 55 dB adverse health effects do occur in an exposed population and many people have to adapt their lives to cope with the noise at night. Vulnerable groups such as children, chronically ill and elderly people are more susceptible for night time noise and will be more severely affected at high noise levels.

When exposed at night to a certain noise level, there may be a few loud events or a larger number of less loud events, in each case yielding the same night time average. For sleep disturbance, the 'worst case' is when the noise events levels are just above (about 5 dB) the effect threshold, but not very loud. In that case there is a high number of events above threshold and the probability of being disturbed is highest (Health Council, 2004). For wind turbines distinct noise events may occur if there is sufficiently loud swishing or beating, or when other, less frequent sounds arise such as when the rotor stops or turns.

Sleep disturbance was investigated in the Swedish and Dutch studies. In the Swedish surveys respondents could answer yes or no to a question about their sleep quality. In the Dutch survey respondents were asked how often their sleep was interrupted by (any) sound and the answer categories were: (almost) never, at least once in the past year, at least once in the past month, at least once in the past week, (almost) daily. These data were analyzed by Janssen et al (2008) and the results are plotted in Figure 6 for respondents that had no economic profit from the wind turbines. The figure shows the percentage of residents being (consciously) disturbed in their sleep at least once in the past month. This percentage depends significantly on the wind turbine sound level. There is no significant relation when respondents benefiting from the turbines are included, as they are far less disturbed by the sound.

This relation has been analyzed further by Bakker et al (2011) with the Dutch wind turbine noise survey data. Two groups of respondents without economic benefits from wind turbines were

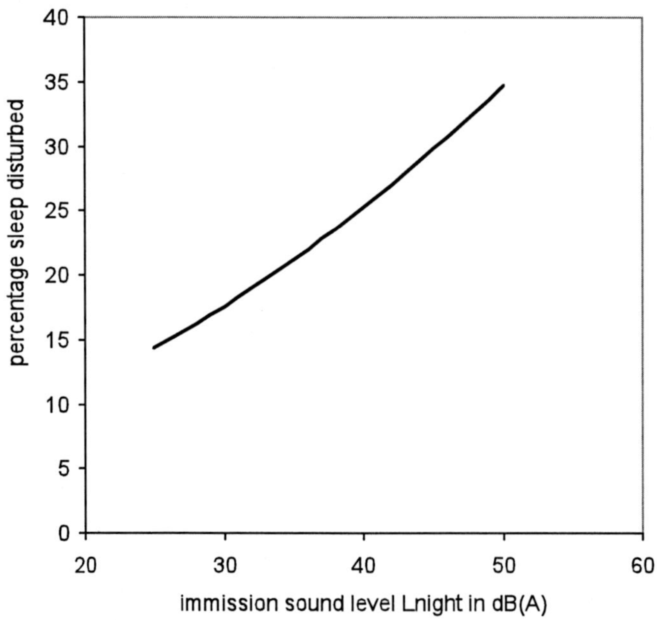

Figure 6. Percentage of residents being (consciously) disturbed in their sleep at least once in the past month

distinguished: those in quiet rural areas, with no main road nearby, and those in more noisy areas, that is: in built-up areas and rural areas with a main road less than 500 m from the closest wind turbine. When applied to those who had not heard wind turbine sound at their home (and were thus not annoyed), there was no significant influence of the sound level on sleep disturbance. This is as one would expect when the sound is not audible. For this group, sleep disturbance and age were related significantly.

A further analysis was applied to respondents who stated they could hear the sound from wind turbines at their home, no matter whether they were or were not annoyed by the sound. This analysis showed that in the quieter rural area a higher sound level from wind turbines led to a significantly higher percentage of annoyed residents and the annoyance in turn led to significant more sleep disturbance. The direct relation between sound level and sleep disturbance was not significant, indicating that it is the annoyance rather than the sound as such that leads to more sleep disturbance. These relations were weaker and less significant for the population in the noisier area types: here only annoyance and sleep disturbance were significantly related and the sound level did not influence either annoyance or sleep disturbance in a significant way. Possibly the impact of wind turbines is less dominant in these noisier areas.

In plain terms, these results show that audible noise may lead to annoyance and the annoyance aggravates sleep problems. Thus, perhaps sleep disturbance due to wind turbine noise will not occur if noise limits are low enough to prevent annoyance. To be able to set a separate night time noise limit, apart from a daytime limit for noise annoyance, more research is needed. For field studies Hanning (2011) recommends quantifying sleep disturbance using dedicated questionnaires like the Epworth Sleepiness Score (ESS) or the Pittsburgh Sleep Quality Index (PSQI).

OTHER HEALTH EFFECTS

Irritability or annoyance and sleep disturbance are common effects of noise because they occur at relatively low noise levels and thus are bound to come up first. On the other hand, hearing impairment and cardiovascular effects are adverse health effects that occur at relatively high levels only and are –at least as yet- not relevant when considering wind turbine noise. Other effects that are associated with noise (WHO, 2000) are effects on communication by speech, mental health, performance and behaviour.

Speech Communication

For vocal communication a threshold level is given above which effects may occur at home and in school. A threshold noise level of 35 dB(A) enables people to understand one another at close distance (1 meter) without having to raise their voice. Assuming a façade sound insulation of at least 15 dB, this means that effortless speech is possible at outdoor sound levels of 50 dB(A) and possibly higher.

Mental Health, Performance and Behaviour

It is clear that sound can have effects on mental well-being. The sound of pleasant music, birds or a familiar voice can raise one's spirits. But these very same sounds can also evoke negative feelings if the listener is not in the mood to appreciate them or when the sounds are not at the right time (e.g. music at night) or place (e.g. voice in the dark). The level of the sound may be less important than the information content of the sound: where does the sound come from, what does it mean?

Even sounds that are not audible to others can have a profound influence on people's well-being and in general their quality of life. Several studies have shown that the impact of tinnitus or aural hallucinations is highly influenced by feelings of fear (Van den Berg, 2009b).

Although we know from everyday life how sound can affect us mentally, not much research has been published on the adverse mental effects of noise and the results are not straightforward. According to the WHO (2000) environmental noise can have various mental effects such as effects on anxiety, emotional stress, nervous complaints, nausea, headaches, mood changes and social conflicts. It can also have an effect on psychiatric disorders such as neurosis, psychosis and hysteria. As we know, noise –for example from a neighbour- can drive a person crazy, though of course we usually do not really mean the sound causes mental insanity. But it does mean that the effect can be very serious.

There are two studies on mental health in relation to wind turbine noise. Bakker et al (2011) analyzed the Dutch wind turbine noise survey data for noise in relation to sleep disturbance (see 6.5) and non-specific psychological distress. Here, psychological distress was assessed with the 12-item version of the General Health Questionnaire which is designed to detect psychiatric disorders in community samples and non-psychiatric clinical settings. Two groups of respondents without economic benefits from wind turbines were distinguished: those in quiet rural areas, with no main road nearby, and those in more noisy areas, that is: in built-up areas and rural areas with a main road less than 500 m from the closest wind turbine. For those who had not heard wind turbine sound at their home, sleep disturbance, psychological distress and age were interrelated significantly, but there was no significant influence of the wind turbine sound level (as one would expect when the sound is not audible). For respondents who stated they could hear the sound from wind turbines at their home the analysis showed that in the quieter rural area a higher sound level from wind turbines led to a significantly higher percentage of annoyed residents and the annoyance in turn led to significantly more psychological distress (as well as sleep disturbance, which was in itself related to distress). The direct relation between sound level and psychological distress was not significant, indicating that it is the annoyance and sleep disturbance rather than the sound as such that may lead to more distress. These relations were weaker and less significant for the population in the noisier area types: here psychological distress was not significantly related to annoyance or sleep disturbance (and not to sound level).

Thus, these results can be interpreted as follows: the audible noise leads to annoyance and the annoyance as well as sleep problems, partly as a result of annoyance, aggravate the distress.

Pedersen (2011) analyzed the results of all three (Swedish and Dutch) wind turbine noise surveys and found that health variables and wellbeing were not consistently related to the sound levels for all studies. Tinnitus and diabetes were positively related to the sound levels in one survey each, but not in the other studies. However, several measures of stress were associated with annoyance due to wind turbine noise: feeling tense or stressed and feeling irritable was associated with noise annoyance in all three studies. Headache was associated with annoyance in two of the three studies and undue tiredness in one study only.

Considering the lack of publications, studies investigating other noise sources could help to understand the relation between noise and mental health.

In a laboratory study with young adults Standing and Stace (1980) investigated the effect of low, medium and high sound levels of white noise on the outcome of several psychological tests. One test measured State-anxiety that assesses the individual's current level of anxiety, whereas Trait-anxiety reflects the usual feeling of anxiousness as an individual characteristic (and was found to be correlated to neuroticism[9]). The results showed that in the very noisy condition State-anxiety was significantly higher than in the quiet or noisy condition while Trait-anxiety did not change. Also, the variability of the State-anxiety scores (and not the Trait-anxiety scores) increased significantly for both the noisy and very noisy condition, implying that the degree to which test persons were affected spread over a wider range when noise level increased. Although the persons differed in their baseline anxiety (as measured with the Trait-anxiety score), the increase in anxiety was similar for those with the lowest scores when compared to the others.

In a study following British men of 50 to 64 years of age over a period of five years (Stansfeld et al, 1996) no clear relation between the level of road traffic noise and the prevalence of minor psychiatric disorders (as determined with the General Health Questionnaire) was found, but there was some evidence for an increase in anxiety. The authors concluded that the study confirmed the results of previous studies: environmental noise does not appear to be an important cause of overall psychiatric disorder but it can contribute to feelings of anxiety.

[9] According to the Encyclopedia of Mental Disorders (www.minddisorders.com) a neurosis is a nonpsychotic condition which triggers feelings of distress and anxiety and impairs functioning

[10] The section on VVVD has been written with the help of D.M. Faber, clinical psychologist

VVVD[10]

According to Pierpont the (infra)sound of wind turbines can cause Visceral Vibration Vestibular Disease (VVVD). This is a new diagnosis characterized by the following symptoms: "a feeling of internal pulsation, quivering or jitteriness, and it is accompanied by nervousness, anxiety, fear, a compulsion to flee or check the environment for safety, nausea, chest tightness, and tachycardia" (Pierpont, 2009). However, these symptoms can also occur when persons are put under stress. If these symptoms comply with an existing diagnosis, there may not be a need for a new diagnosis. This will be explored here.

Common symptoms of excess stress include sleep disturbances, muscle tension, headache, gastrointestinal disturbances, fatigue, nervousness, anxiety, changes in eating habits including overeating, loss of enthusiasm or energy, and mood changes (Medicine Net, 2010). The degree of stress a person experiences depends upon the stressor itself, but also upon a number of individual factors such as physical health, the quality of relationships, the number of commitments and responsibilities, the degree of others' dependence and the amount of support from others, expectations, and the number of changes or traumatic events that have recently occurred in a person's live (Medicine Net, 2010).

The Diagnostic and Statistical Manual of Mental Disorders (abbreviated to DSM; the current version is the fourth: DSM-IV) lists a number of disorders that are related to stress. Here disorder is a neutral medical term referring to a disturbance of a 'normal', healthy state.[11] There are three such disorders that could apply to people who attribute their suffering to a wind turbine or wind farm: an adjustment disorder, a panic disorder and a generalized anxiety disorder. According to DSM-IV (1994) the main criteria for an adjustment disorder are:

- Emotional or behavioral symptoms develop in response to an identifiable stressor within three months of the onset of the stressor.
- The symptoms are clinically significant:
  1. there is marked distress in excess of what would be expected from exposure to the stressor; and/or
  2. there is significant impairment in social or occupational (academic) functioning
- When the stressor has terminated, the symptoms vanish within six months.

A major life stress can cause a panic attack, a sudden surge of intense fear or discomfort. The panic is probably not a result of only one cause, but the result of a combination of biological vulnerabilities, ways of thinking and social stressors. Panic attack symptoms include trembling or shaking, a fear of losing control, fear of dying, a need to escape, nausea, chest pains and a racing or pounding heartbeat (Medicine Net, 2010).[12] A panic attack as such is not necessarily a disorder. However, when panic attacks recur unexpectedly and are followed by a period of concern about additional attacks and/or the implications or consequences of an attack (such as losing control, having a heart attack, "going crazy"), and/or a significant change in behaviour related to the attacks it may be a panic disorder (DSM-IV, 1994). Finally, people with a generalized anxiety disorder (GAD) are more vulnerable to stress and hence may react more strongly than others to a stressor. Symptoms of a GAD include trembling, restlessness or a feeling of being "edgy", excessive worry and tension, an unrealistic view of problems, nausea and muscle tension and these may become worse during periods of stress (Medicine Net, 2010).[13]

[11] According to Wikipedia a mental disorder is a psychological or behavioral pattern generally associated with subjective distress or disability that occurs in an individual, and not a part of normal development or culture

[12] Other possible symptoms of panic disorder are stomach upset, dizziness, difficulty breathing (a sense of feeling smothered), tingling or numbness in the hands, hot flashes or chills, dreamlike sensations or perceptual distortions

[13] Other possible symptoms of a generalized anxiety disorder are irritability, headaches, sweating, difficulty concentrating, the need to go to the bathroom frequently, tiredness, trouble falling or staying asleep, being easily startled

Of course, the above cannot serve as a diagnosis for people reacting strongly to the presence of wind turbines, but it does show that a psychological view on the symptoms may help to understand such reactions.

Pierpont's approach of studying people severely affected by the presence of wind turbines is very different from epidemiological studies of environmental factors that are designed to find effects on a population level. Although the epidemiological studies may try to identify risk groups or vulnerable groups, individual diagnoses and symptoms are not considered. If Pierpont's selection procedure would be applied to other sources of noise one will probably find similar results: there are also people that suffer from, for example, the sound of aircraft, neighbours, barking dogs or untraceable hums. The selection procedure is bound to find people who suffer the most, for whatever reason. It is possible that selected people are by their nature above-average anxious and are thus most sensitive to what they feel is an intrusion of their home. The list of symptoms above shows that VVVD could very well be the same as a stress related disorder. As yet it is not clear whether they are the same and needs to be investigated.

VAD

According to Alves-Pereira and Castelo Branco (2007) the infrasound and low frequency sound of a wind turbine can cause Vibroacoustic Disease (VAD). They postulated that at least some of the effects of sound from a wind turbine on residents could be qualitatively explained by low frequency sound (including infrasound). Marciniak *et al* (1999) stated that VAD is caused by long term (10 years or more) exposure to high noise levels (at least 90 dB) at low frequencies (below 500 Hz), when it cannot be explained by other possible causes. They found VAD, identified by a thickening of the mitral valve (one of the valves in the heart) and the pericardium (a sac containing the heart), in aeronautical workers who were exposed to high sound levels over long periods of time. Several years later Alves-Pereira and Castelo Branco (2007a) stated it was not clear at what sound level VAD can occur and dose-response relations for infrasound and low frequency noise did not exist and thus the risk for developing VAD cannot be assessed. Nevertheless, in their investigation of a family with two wind turbines at 322 and 642 m from the dwelling (Alves-Pereira and Castelo Branco, 2007b), they concluded that VAD occurred and was caused by low frequency sound. The measured noise levels were substantially lower (20 dB or more) than levels at which VAD was thought to occur by Marciniak *et al* (1999) and the spectral levels were below the normal hearing threshold for a considerable range of frequencies in this range. If the same sound energy would be presented at the most sensitive audible frequencies, it would not cause hearing damage although the ear is the most sensitive organ. It is therefore highly unlikely that the sound as such could have effected physiological damage. Of course the sound and sight of the wind turbines could have disturbed the family.

Quality of Life

One could argue that, in whatever way a sound may influence a person (physically, psychologically, in social or personal well-being, in amenity), in the end a sound source affects a person by improving or degrading the quality of life of that person. The term quality of life (QoL) is used to describe the general well-being of an individual. There are a multitude of questionnaires to assess QoL, either in general or related to physical, psychological and social health (Health Related QoL or HRQoL). A questionnaire may address the general public or a specific target group such as patients. HRQoL Questionnaires that are widely used in health surveys are the full WHOQOL and short WHOQOL-BREF from the World Health Organization (WHO), the EQ-5D from the EuroQol group and the Short Form SF-36.

Shepherd et al (2011) used the WHOQOL-BREF for a masked survey comparing residents in a New Zealand community within 2 km from a wind farm with a control group at least 8 km to a wind farm matched for area type (rural) and for geographic, demographic and socio-economic characteristics. The overall HRQoL score was significantly lower in the 'turbine group' compared to the control groep, due to lower scores on the physical and environmental domains of HRQoL. There were no significant differences in the psychological and social domains. The low physical score was due to differences in two items: perceived sleep quality ("How satisfied are you with your sleep?") and self-reported energy-level ("Do you have enough energy for everyday life?"). The low environmental score was again due to differences in two items: environmental condition ("How healthy is your physical environment?") and the individual space ("How satisfied are you with the conditions of your living place?"). The latter is in line with a lower amenity perceived by the turbine group.

CONCLUSION

Wind turbine noise appears to be a relatively disturbing sound when compared to other major noise sources. Acoustical and non-acoustical factors may explain this. Perhaps most important is the swishing or beating character of the sound. Its unpredictability, the perceived lack of control and the perceived balance of advantages and disadvantages are other relevant factors related to the annoyance from wind turbine sound (and sight). Sleep disturbance is a second important health effect to consider, but more research is needed to quantify the relation between this effect and wind turbine noise. Psychological distress is related to sleep disturbance and annoyance, and may be a result of stress, but it is not necessarily an effect of wind turbine noise. Also the unwanted presence of wind turbines and the fairness of the licensing process, possibly in combination with other stressful factors, may contribute.

Measures to mitigate noise annoyance from wind turbines are reducing the sound level at the dwellings of residents, especially at night, and decreasing the modulation level. Changing the meaning of the sound can have a significant impact. If procedures and the distribution of pros and cons are perceived as just by residents, they will probably have a less negative evaluation of the sound.

REFERENCES

M. Alves-Pereira, N.A.A. Castelo Branco (2007a), Public health and noise exposure: the importance of low frequency noise, *proc. Internoise2007*, Istanbul

M. Alves-Pereira, N.A.A. Castelo Branco (2007b), In-Home Wind Turbine Noise Is Conducive to Vibroacoustic Disease, *proc. 2nd Int. Meeting on Wind Turbine Noise*, Lyon

R.H. Bakker, E. Peersen, W. Lok, G.P. van den Berg, R.E. Stewart, J. Bouma (2011), *Effects of wind turbine sound on health and psychological distress, submitted to Science of the Total Environment*

D. Bowdler (2011), Why turbine noise annoys, proc. *4th Int. Meeting on Wind Turbine Noise*, Rome

J.S. Bradley (1994), Annoyance caused by constant amplitude and amplitude modulated sounds containing rumble, *Noise Control Eng* 42, 203-208.

Devine-Wright P. (2005) 'Beyond NIMBYism: Towards an integrated framework for understanding public perceptions of wind energy', *Wind Energy*, vol 8, pp125–139

K. Dittrich, D. Oberfeld (2009), A comparison of the temporal weighting of annoyance and loudness, *J. Acoust. Soc. Am.* 126(6), pp. 3168-3178

DSM-IV (1994), *Diagnostic and Statistical Manual of Mental Disorders*, Fourth Edition, American Psychiatric Association (a slightly updated version is the Text Revision (DSMIV-TR, 2000)

I.H. Flindell, P.J.M. Stallen (1999), Non-acoustical factors in environmental noise, *Noise & Health*, vol. 1 (3), pp. 11-16

B.J. Frey, P.J. Hadden (2007), Noise radiation from wind turbines installed near homes: effects on health, www.windturbinenoisehealthhumanrights.com

C. Hanning, M. Nissenbaum (2011), Selection of outcome measures in assessing sleep disturbance from wind turbine noise, *proc. 4th Int. Meeting on Wind Turbine Noise*, Rome

M. Hayes (2006), "The measurement of low frequency noise at three UK wind farms", Hayes Mckenzie Partnership Ltd, report to the Department of Trade and Industry, UK

Health Council (2004), Effects of noise on sleep and health, Health Council of the Netherlands, publication no. 2004/14, The Hague

S. von Hünerbein, A. King, J, Hargreaves, A. Moorhouse, C. Plack, T.H. Pedersen (2011), Perception of noise from large wind turbines, *proc. 4th Int. Meeting on Wind Turbine Noise*, Rome

J. Jakobsen (2005), Infrasound Emission from Wind Turbines, *Journal of Low Frequency Noise, Vibration and Active Control*, vol. 24 (3), pp. 145 – 155

Janssen S.A., Vos H., Eisses A.R, and Pedersen E (2009), Exposure-response relationships for annoyance by wind turbine noise: a comparison with other stationary sources, *proc. Internoise2009*, Edinburgh, October 26-28

Janssen S.A., Vos H. and Eisses A.R. (2008), Hinder door geluid van windturbines – dosis-effectrelaties op basis van Nederlandse en Zweedse gegevens (Annoyance from wind turbine sound – dose-effect relations base don Dutch and Swedish data – in Dutch), TNO-report 2008-D-R1051/B, Delft

Johansson M. and Laike, T. (2007) 'Intention to respond to local wind turbines: The role of attitudes and visual perception', *Wind Energy*, vol 10, no 5, pp435–451

S.V. Legarth (2007), Auralization and assessments of annoyance from wind turbines, *proc. 2nd Int. Meeting on Wind Turbine Noise*, Lyon

S. Lee, K. Kim, S. Lee, H. Kim, S. Lee (2009), An estimation method of the amplitude modulation in wind turbine noise for community response assessment, *proc 3rd Int. Meeting on Wind Turbine Noise*, Ålborg

G. Leventhall (2003), A Review of Published Research on Low Frequency Noise and its Effects. Defra Report, Department for Environment, Food and Rural Affairs, London

Marciniak W, Rodriguez E, Olszowska K, Atkov O, Botvin I, Araujo A, Pais F, Soares Ribeiro C, Bordalo A, Loureiro J, Prazeres De Sá E, Ferreira D, Castelo Branco MS, Castelo Branco NA. (1999), Echocardiographic evaluation in 485 aeronautical workers exposed to different noise environments, *Aviat Space Environ Med.*vol. 70(3 Pt 2), pp. 46-53.

E. Maris, P.J. Stallen, R. Vermunt, H. Steensma (2007), Noise within the social context: Annoyance reduction through fair procedures, *Journal Acoustical Society of America*, vol. 121 (4), pp. 2000–2010, pp. 117-125

M. Masullo, G. Iannace, S. Basturk, L. Maffei, V.P. Senese, F. Ruotolo, The influence of vision on noise annoyance evaluation of wind farms, *proc. 4th Int. Meeting on Wind Turbine Noise*, Rome

H.M.E. Miedema, C. G.M. Oudshoorn (2001), Annoyance from Transportation Noise: Relationships with Exposure Metrics DNL and DENL and Their Confidence Intervals, *Environmental Health Perspectives*, Vol. 109, No. 4, pp. 409-416

H.M.E. Miedema, H. Vos (1999), Demographic and attitudinal factors that modify annoyance from transportation noise, *Journal Acoustical Society of America*, vol. 105 (6), 3336-3344

H.M.E. Miedema, H. Vos (2004), Noise annoyance from stationary sources: Relationships with exposure metric day–evening–night level (DENL) and their confidence intervals, *Journal Acoustical Society of America*, vol. 116 (1), pp. 334-343

A. Moorhouse, M. Hayes, S. von Hünerbein, B. Piper, M. Adams (2007), *Research into Aerodynamic Modulation of Wind Turbine Noise*, University of Salford en Department for Business, Enterprise & Regulatory Reform, UK

A.T. Moorhouse, D.C. Waddington, M.D. Adams (2007), The effect of fluctuations on the perception of low frequency sound, *Journal of low frequency noise, vibration and active control* 26(2), pp. 81 – 89

A.T. Moorhouse, D.C. Waddington, M.D. Adams (2009), A procedure for the assessment of low frequency noise complaints, *J. Acoust. Soc. Am.* 126(3), pp. 1131-1141

C. S. Pedersen, H. Møller (2010), An analysis of low frequency noise from large wind turbines, *proc. of 14th Int. Meeting on Low Frequency Noise and Vibration and its Control*, Aalborg

E. Pedersen and K. Persson Waye (2004), Perception and annoyance due to wind turbine noise – a dose-response relationship, *Journal of the Acoustical Society of America* 116(6), pp. 3460-3470

E. Pedersen (2007), Human response to wind turbine noise: Perception, annoyance and moderating factors. Thesis, Göteborg University.

E. Pedersen, LR-M. Hallberg, K. P. Waye (2007), Living in the Vicinity of Wind Turbines — A Grounded Theory Study, *Qualitative Research in Psychology*, Vol. 4 (1), pp 49 – 63

E. Pedersen, F. van den Berg, R. Bakker, J. Bouma (2009), Response to noise from modern wind farms in The Netherlands, *Journal of the Acoustical Society of America* vol. 126 (2), pp. 634-643

T.H.Pedersen, K.S. Nielsen (1994), Genvirkning af støj fra vindmøller (Annoyance by noise from wind turbines), Report 150, DELTA Acoustic & Vibration, Lydtekniske Institut, Copenhagen, Denmark

T.H. Pedersen (2008), A procedure for evaluation of the audibility for low frequency sound and a literature study, report AV1098/08 of EFP-06 project "Low frequency noise from large wind turbines", performed for Danish Energy Authority, Delta, Denmark

K. Persson Waye, E. Öhström, Psycho-acoustic characters of relevance for annoyance of wind turbine noise, *J Sound Vib* 250 (1), pp. 65-73 (2002)

N. Pierpont (2009), Wind Turbine Syndrome – A report on a natural experiment, K-Selected Books, Santa Fe, USA

A.N. Salt, T.E. Hullar (2010), Responses of the ear to low frequency sounds, infrasound and wind turbines, *Hearing Research* 268, pp. 12-21

A.N. Salt, J. A. Kaltenbach (2011), Infrasound from Wind Turbines Could Affect Humans, accepted by *Bulletin of Science, Technology & Society*

D. Shepherd, D. McBride, D. Welch, K.N. Dirks, E.M. Hill (2011), Wind turbine noise and health-related quality of life of nearby residents: a cross-sectional study in New Zealand, *proc. 4th Int. Meeting on Wind Turbine Noise*, Rome

B. Søndergaard, D. Hoffmeyer, B. Plovsing (2007), Low frequency noise from large wind turbines, *proc. 2nd Int. Meeting on Wind Turbine Noise*, Lyon

L. Standing, G. Stace (1980), The effects of environmental noise on anxiety level, *Journal of General Psychology*, vol. 103, p.263-272

S. Stansfeld, J. Gallacher, W. Babisch, M. Shipley (1996), Road traffic noise and psychiatric disorder: prospective findings from the Caerphilly study, *British Medical Journal* vol. 313 (7052), pp. 266-

Y. Takahashi (2009), Vibratory Sensation Induced by Low-Frequency Noise: A Pilot Study on the Threshold Level, *Journal of Low Frequency Noise, Vibration and Active Control*, vol. 28 (4), pp. 245 – 253

S.E. Taylor, R.L. Repetti (1997), Health psychology: what is an unhealthy environment and how does it get under the skin?, *Annual Review of Psychology*, vol. 48, pp. 411-447

T.R. Tyler (2000), Social Justice: Outcome and Procedure, *International Journal of Psychology*, vol. 35 (2),

G.P. van den Berg (2004), Do wind turbines produce significant low frequency sound levels?, *proc. 11th Int. Meeting on Low Frequency Noise and Vibration and its Control*, Maastricht

G.P. van den Berg (2005), The beat is getting stronger: the effect of atmospheric stability on low frequency modulated sound of wind turbines, *Journal of Low frequency noise, vibration, and active control*, Vol. 24(1)

G.P. van den Berg (2006a), The sounds of high winds (doctoral thesis), University of Groningen, Groningen

G. P. van den Berg (2006b), Wind-induced noise in a screened microphone, *Journal of the Acoustical Society of America*, vol. 119 (2), pp. 824-833

F. van den Berg (2008), Criteria for wind farm noise: L_{max} and L_{den}, *proc. Acoustics08*, Paris

F. van den Berg, E. Pedersen, J. Bouma, R. Bakker (2008), WINDFARMperception: Visual and acoustic impact of wind turbine of wind turbine farms on residents. Report FP6-2005-Science-and-Society-20. University of Groningen and Göteborg University

F. van den Berg (2009), Why is wind turbine noise noisier than other noise?, *proc. Euronoise 2009*, Edinburgh

F. van den Berg (2009b), Low Frequency Noise and phantom sounds, *Journal of Low Frequency Noise, Vibration and Active Control*, vol. 28 (2), pp. 105-116

S. Wagner, R. Bareiss, G. Guidati (1996), *Wind turbine noise*, Springer, Berlin

WHO (2000), Guidelines for Community Noise, World Health Organization (Geneva) and Institute of Environmental Epidemiology (Singapore)

WHO (2009), Night noise guidelines for Europe, World Health Organization Regional Office for Europe, Copenhagen

M. Wolsink, S. Breukers (2010), Contrasting the core beliefs regarding the effective implementation of wind power -an international study of stakeholder perspectives, *Journal of Environmental Planning and Management*, vol. 53 (5), pp. 535–558

M. Wolsink, M. Sprengers, A. Keuper, T.H. Pedersen, C.A, Westra (1993), Annoyance from wind turbine noise on sixteen sites in three countries, *proc. European Community Wind Energy Conference*, 8–12 March, Lubeck, Travemunde, Germany

S. Yamada, M. Ikuji, S. Fujikata, T. Watanabe, T. Kosaka (1983), Body sensations of low frequency noise of ordinary persons and profoundly deaf persons, *Journal of Low Frequency Noise and Vibration*, vol. 2, pp. 32-36.

Chapter 7. Measuring and Analyzing Wind Turbine Noise

David Hessler (USA), Hessler Associates, Inc.
David@HesslerAssociates.com

INTRODUCTION

The sound or noise, depending on one's perspective, produced by wind turbines differs fundamentally from the noise emitted by other power generation facilities. This is primarily because the magnitude and nature of wind turbine noise is almost entirely dependent on the local environment, specifically the time-varying wind and atmospheric conditions, whereas a conventional fossil-fueled power station operates, often continuously and steadily, in a manner that is completely independent of the local environment. Consequently, a combustion turbine plant, for example, is most apt to be perceptible and a potential noise problem during calm and still weather conditions while a wind turbine project would, in all likelihood, be idle and utterly silent under those same conditions. During windy conditions, on the other hand, the wind turbine project would be generating noise, usually in competition with natural wind-induced sounds, while the perceptibility of the gas-fired plant would be reduced by elevated levels of background sound. In addition, the testing of conventional power stations, or any kind of industrial noise source for that matter, is generally prohibited in wind speeds above the 3 to 6 m/s range by various ISO [1, 2] and ANSI [3, 4] standards whereas wind turbine noise, by definition, can only be measured during moderately high wind conditions.

What all this suggests is that a completely different field measurement methodology is required for wind turbines that is tailored to, and takes into account, the unique circumstances and technical challenges surrounding their noise emissions. At the present time, a lone standard, IEC 61400-11 Ed. 2.1 [5] exists for the specific purpose of measuring the sound power level of a single wind turbine operating in isolation but there are no other standards in any kind of general use that are directly applicable to measuring and evaluating noise from operational wind farms.

This chapter seeks to fill in, at least to a certain extent, this blank spot by outlining and explaining some practical procedures for measuring wind turbine noise that have been developed essentially through trial and error field experience and that have been found to produce reasonably good results - in the sense that the project-only sound level exclusive of background contamination, the key quantity sought in such tests, can be surmised fairly well and compares favorably to model predictions. Following some general comments on the characteristics of wind turbine noise, the bulk of the discussion centers around a suggested methodology for carrying out a field test in which the noise emissions from a newly operational project are evaluated on a site-wide basis for compliance with planning conditions or other regulatory noise limits. Examples are provided of typical results at on-site measurement locations along with a comparison of the measured levels to model predictions.

This general, site-wide test methodology is subsequently adapted to the purpose of investigating the project sound level in much greater detail at a specific location in response to complaints, with an emphasis on determining what the project-only sound level is, where it is coming from, if several turbines are involved, and why, or under what conditions, it becomes undesirably loud.

Finally, the IEC 61400-11 test standard is briefly discussed in terms of some of the practical considerations involved in the test.

CHARACTERISTICS OF WIND TURBINE NOISE

As discussed in more detail in Chapter 2, aerodynamic noise associated with wind turbine rotors is not usually the continuous bland hum associated with other types of power generation facilities, but rather it can be highly variable with time, time of day and weather conditions. The fundamental noise generation mechanism, the turbulent interaction of air flow over the moving blades, is dependent on the characteristics of the air mass flowing into the rotor plane. For example, when the airflow is fairly constant and steady in velocity over the swept area noise is generally at a minimum. While such ideal, laminar flow conditions may exist much of the time, particularly during the day, they do not occur all of the time, and the reality is that the wind often blows in the form of intermittent gusts separated by periods of relative calm rather than as a smooth, laminar stream of constant velocity. In addition, the flow may contain turbulent eddies, may be unstable in direction and the mean velocity may vary considerably over the vertical diameter of the rotor, which is usually quite large; normally in the 77 to 112 m range on the utility scale turbines now in common use. These uneven and unstable air flow conditions generally cause more noise to be generated and it is generated sporadically as each gust sweeps past and as the wind varies amorphously in speed or direction over the rotor plane. Such unstable conditions can lead to sound levels that change very noticeably in the short-term not only in general volume but also in character.

Qualitatively, under average or normal daytime circumstances rotor noise, as perceived at a typical set back distance of around 400 m, might be described as a churning, mildly periodic sound, particularly when there are several units at comparable distances from the point of observation. The normally non-synchronized and incoherent sounds from multiple units tend to blur the sound and minimize the perception of periodicity, although this quality is frequently weak during "normal" circumstances even if only one unit is present. Another common description is that the noise is reminiscent of a plane flying over at fairly high altitude. This comparison is probably partly due to the basic similarity of the two sounds but also to the phenomenon where the sound can fade in and out randomly. In the case of an actual plane it is the intervening non-homogeneous atmosphere that alternately enhances or hinders sound propagation from the distant (constant) source producing this effect while, in the case of the wind turbine, it may be path effects due the atmosphere, short-term variations in noise generation at the source itself, or a combination of both.

At night atmospheric conditions, in particular the vertical wind speed and thermal gradients, can often lead to an increase in the perceptibility of turbine noise primarily due to an increase in noise generation at the rotor and only secondarily because the background sound level may have decreased. It is generally under conditions such as these, and also during periods when the wind is simply erratic or unstable, that pronounced amplitude modulation, or a periodic swishing or thumping sound, can occur.

What all this implies is that the sound level of an operational wind project at a particular point of interest cannot be adequately determined by holding up a sound level meter and observing it for a few minutes. This approach, which amounts to a snapshot, will not only be likely to miss the intermittent times of peak noise but, among other things, will be prone to errors from the wind blowing over the microphone and from background noise that will tend to artificially inflate the result. In short, it is extremely easy when measuring wind turbine noise to get a wrong answer that is too high and very difficult to get an answer that accurately quantifies the noise due exclusively to the project. In fact, quite a bit of equipment, time and know-how are required to get a meaningful result.

PROJECT-WIDE COMPLIANCE TESTING

Historical Approaches

For the reasons and challenges alluded to above, it has been difficult, historically, to devise or settle on a completely satisfactory methodology for testing newly completed wind projects for the purpose of determining whether or not they are in compliance with permit or planning conditions. One of the principal stumbling blocks has generally been accounting in some meaningful way for background noise, since the total measured sound level at the typically substantial distances to residences and, therefore, the point of measurement, commonly contains a very prominent background component that cannot be disregarded without causing the result to be erroneously high. It is, of course, the project-only sound level and not the total sound level that is limited by regulations. Consequently, it is the project-only sound level that is sought in such surveys.

Existing guidelines and standards that mention the topic of compliance testing at all do not lay out or detail test procedures that are entirely satisfactory in this and other respects. For example, the often beleaguered [6] ETSU-R-97 report *The Assessment and Rating of Noise from Wind Farms* [7] addresses the issue of background noise in one sentence, quoted below, by suggesting simply that one might want to measure operational turbine noise at night.

> To minimize the effects of extraneous noise sources it may be necessary to perform these measurements during night-time periods when other human and animal activity noise sources are likely to be at a minimum [7, p. 87].

This approach, which involves measuring only for a relatively short period of time (20 to 30 $L_{A90, 10\,min}$ samples), is connected with the idea of taking measurements only at, or close to, a specific critical wind speed identified from "monitoring", carried out in an unspecified manner, and correlated to logged observations by complainants as to when the "noise is most intrusive" [ibid]. In short, the idea is for the test engineer to be physically at the location and ready to take measurements when the wind conditions that result in maximum noise are occurring so long as those conditions are happening at night on a night when the background sound level is, hopefully, negligible (i.e. roughly 10 dBA or more lower in magnitude than the turbine sound level). As might be imagined, the unfortunate reality is that the probability of all these things coming together at the same time is miniscule. In particular, it is typically difficult, for a number of reasons, for a test engineer to schedule a site visit to coincide with a particular wind speed or direction and the phrase "you should have been here yesterday" is frequently heard.

In general, the notion of being on hand to observe and measure wind turbine noise when it is at its loudest may sound reasonable on paper but it is seldom practical to actually do it.

Another approach to the issue of background noise that has been used, for example in the New Zealand Standard NZS 6808:1998 *Acoustics – The assessment and measurement of sound from wind turbine generators* [8], is to measure the background level at one time, say, prior to construction or start-up, and the operational noise from the project at another time - and then subtract the two to derive the project-only sound level. While this is often thought of or suggested as a reasonable approach, the problem is that both the background and wind turbine sound levels are extremely dependent on circumstances that vary significantly with time in both the short and long-term. The two sounds are highly specific not only to the prevailing wind speed at a particular time but also to factors such as the stability of the wind (whether it's gusty or constant in nature, for instance), wind direction, shear gradient, thermal gradient, time of day and time of year. Moreover, the background level is also exclusively influenced by foliage (bare trees vs. leafed out trees, for example), insects, frogs, distant or nearby traffic, farm equipment and a myriad of other human activities that occur sporadically and unpredictably. Consequently, a background sound level measured days or months before can't be used with a tremendous amount of confidence to correct a later measurement of operational noise, even if both have been normalized to similar wind speed conditions, because so many other unquantifiable factors may have had a hand in shaping the final results. What is needed, of course, is the background sound level that would have existed at that particular time and at that place if the project had not been operating.

This latter objective can sometimes be essentially realized by using the technique of temporarily shutting down, or parking, the nearest turbines to a measurement position, if not the entire project. While this technique has its applications, which will be discussed later, it is not usually a practical method that can be used for a general site-wide compliance test. Widespread or complete shutdowns would be required repeatedly over a variety of wind speed conditions and times of day to get even a minimally complete set of usable background levels.

Thus, there are certain impracticalities associated with the few existing guidelines, standards or common practices that deal with the testing of operational noise from wind turbine projects.

Test Methodology

The methodology outlined below, which has been developed over time through field experience on a variety of wind projects, does not purport to completely solve the problems of background noise and capturing the periods of maximum noise, among other things, but it has been found to work very well in numerous field applications.

Survey Duration and Scheduling

In order to overcome the problem of being on hand to take short-duration measurements when conditions might favor noise generation at the source and/or sound propagation from the turbines to typical receptor points, a long-term, continuous monitoring approach is needed in which multiple instruments are set up at key locations and programmed to run day and night for a period of about two weeks or more. In essence, it is necessary to cast a wide net in order to capture sound levels during a variety of wind and atmospheric conditions; something that is extremely difficult to achieve by taking intermittent manned samples, which amount to static snapshots of a dynamic situation.

Field experience suggests that an adequate range of wind speeds, from 0 to 10 m/s at 10 m above ground level, will usually be observed over any given 14 day period at most wind energy project sites, except perhaps during the low wind season at sites that might have very pronounced seasonal wind characteristics. Probably the principal reason for this observation is that this length of time is large relative to the time normally taken for weather patterns, wind directions and general atmospheric conditions to change, which essentially ensures that the data are statistically independent, as discussed in great detail in ANSI S12.9-1992/Part 2 [9]. Data independence implies that test results can be taken to represent the longer-term acoustic situation for that area, at least for the general time of year of the test. However, if a review of the weather conditions that occurred during the survey period shows that the winds were unusually calm or if an insufficient number of data points were collected at the higher wind speeds, the survey can normally be extended for another two weeks. Low wind conditions are most commonly captured and the vast majority of the measurements will be for conditions below or just above the cut-in wind speed. High winds normally occur intermittently over a few hours or a few days separated by sometimes lengthy periods of relatively calm conditions. It may sound counterintuitive, but it is not critical to capture extremely high wind conditions, say higher than about 12 m/s at 10 m, since most complaints and issues with wind turbine noise occur during moderate or even light wind conditions, while background noise tends to predominate under very windy conditions.

As a practical matter, the instruments for such a survey are set up, started and left to run unattended for the nominal two-week test period following which they can be retrieved and downloaded by a technician. Of course, one could stay on site through the test making additional intermittent manned measurements and observations but the very high, and probably prohibitive, cost of such an effort would be difficult to justify in the sense that it would not necessarily guarantee a better or more definitive result than could be derived from the monitor data alone.

In terms of scheduling, it is highly preferable to conduct this type of survey during cool season, or wintertime, conditions to eliminate or at least minimize possible contaminating noise from summertime insects, frogs and birds. In addition, it is best for deciduous trees to be leafless at sites where they are present in quantity to decrease this source of wind-driven background noise and maximize the signal to noise ratio. Human activity, such as from farm machinery or lawn care, is also normally lower during the winter. While summertime surveys have been successful they should, as a general rule, be avoided wherever possible because nocturnal insect noise, for instance, can easily render the project sound level indeterminate at some or all of the measurement positions. If measurements are required during the summer, and they often are for reasons of project scheduling, high frequency contamination can be analytically factored out by taking the measurements in octave or 1/3 octave bands and correcting the spectra, as will be discussed later in greater detail.

In addition to seasonal concerns, it is desirable; when practical, to attempt to schedule the survey set up to just precede a predicted period of moderate or high winds. This not only ensures that the survey period will capture these winds but also creates an opportunity for manned observations and measurements to be made for a day or two to augment to the longer term monitoring survey. There is generally nothing to observe or measure at a wind turbine site when the winds are calm, so if one can be on site with the proper equipment just before a windy period useful short-term measurements can probably be made that can later be viewed within the context of the long-term monitor results for that time period.

As an alternative or supplemental approach, another opportunity for these supplemental manned observations can sometimes be arranged by coordinating the instrument retrieval visit with a predicted windy period. The specific end date for the survey is usually flexible, although instrument battery life is normally the limiting factor. The principal danger in carrying out manned measurements just before the end of a survey, however, is that all of the long term monitors may not still be recording due to power supply issues or any number of other lamentable and sometimes comical things, such as tampering, weather damage or the removal of the windscreen by livestock. On one occasion a meter was even trampled by a herd of antelope.

Test Positions

The test positions should be selected to capture data at a number of potentially sensitive receptors (usually non-participating residences within or near the site area) or other possibly relevant points of interest, such as property lines, where maximum project sound levels might be expected either from modeling or a simple inspection of the site plan. In just about every case, it is not practical or even possible to establish a monitoring station at every house in the vicinity of a project so it is necessary to carefully select a limited but adequate number of sites that are representative of the worst-case exposures at potentially sensitive receptors in all relevant settings. Examples of specific settings would be: homes in sheltered valleys below ridge top turbines; homes on high, open ground with exposure to the wind and nearby project turbines; homes in generally flat open country with turbines in multiple directions; homes in wooded area; homes on the outer edge of a project area, etc. Because every site is unique the number of monitoring stations required to adequately evaluate project noise will vary but the general concepts are to reasonably account for different settings, to cover a number of points were maximum project sound levels are likely to occur at residences and to cover the entire project area with a generally even but somewhat random distribution. Adding one or two deliberately random positions can help increase the statistical independence of the data and avoid inadvertent bias. For sparsely populated sites in open and uniform farm country only about 4 or 5 on-site monitors might be needed while at more densely populated sites with more complex topography the number of monitoring stations would only be limited by the quantity of equipment reasonably available to the test engineer either from in-house stock or outside rental. Realistically, it is seldom possible to gather enough equipment for more than about 10 to 14 on-site monitoring points, but that is normally enough. A typical survey at a fairly large project site with numerous residences intermixed with the turbines might call for about 10 positions at receptors within the project area.

As mentioned above, the general objective is to capture sound levels throughout the site area at key receptors in all distinct settings within the project area. In addition, it is commonly necessary and desirable to establish a measurement position at all homes where complaints or concerns about noise have been expressed to the operations staff. In these instances, it is sometimes possible to enlist the help of residents by having them try to keep a date and time log of when the noise becomes particularly noticeable or unusually loud or when other non-project sounds are present; for example, from lawn moving, farm activity, etc. When this is actually done (experience, unfortunately, suggests that the good intentions to keep a log usually only last for a day or two into the survey period) the comments can provide some valuable insights that help explain and identify peaks in the recorded sound levels.

It is often assumed that project noise is of no concern to project participants who were, and presumably still are, favorably disposed to the project and are receiving lease royalties for units on their land; however, experience at a number of sites suggests that this is not always the case largely due to the confluence of two factors: (1) these residences are typically the closest ones to turbines (sometimes only a few hundred meters away) and (2) the actual sound levels from these nearby units can turn out to be substantially louder than they expected them to be or they were led to believe. Consequently, monitoring at the homes of project participants in response to complaints is fairly common – even though participants are often, but not always, technically exempt from ordinance or permit noise limits. Formal permit compliance is normally only examined at non-participating residences.

It is usually best to start the site selection process a week or two in advance of the actual survey by circling proposed measurement areas on a site map or sound contour plot and submitting this to operations personnel at the site for their input on who, within or near each designated area, might be willing to host a sound monitor at their house and where else, outside of these proposed areas, it might be also be desirable to measure (at complaint locations, for instance). The objective of this preparatory review is to obtain approval and permission from homeowners to set up equipment on their property prior to arrival. Although it is desirable to inspect the proposed locations and make a judgment as to their suitability in person, attempts to arrange for permission on the day of the survey are often unsuccessful due to the simple fact that people are not at home and cannot be reached. Calling ahead usually settles the issue before the equipment is shipped to the site. Setting up the equipment in the rear yard of a house where permission has been obtained generally ensures that the equipment will still be there upon returning at the end of the survey, that the equipment won't be interfered with and that it can be minimally attended to, if necessary (replacing the windscreen after the family dog has run off with it, for example). Positions that are not at anyone's house, such as on utility poles along the public right-of-way, are sometimes necessary to collect data at strategic locations without a suitable host, but they do not have any of these advantages and, in fact, the risk of theft or tampering is uncomfortably high.

In terms of the specific placement of the monitor at each position, it should be located in an area representative of but away from the house, or any other building with large reflective surfaces, and that is not prone to frequent activity or contaminating local noises, such as from air conditioning units, milking machines at dairy farms or flowing streams or rivers. Some contaminating local noise sources are not apparent at the time the equipment is set up so it is usually beneficial to have a word with the homeowner about the subject and ask about any plans for noise making activities in the next two weeks; as from construction, mowing and plowing, for example. Those things can certainly proceed, of course, it is just helpful to know what caused the noise spikes and when so that those particular measurements can be disregarded.

As a final note on placement, it is best to avoid using fences or posts to mount the monitor or microphone in areas where livestock or other domestic animals may be able to get at the equipment during the survey. Microphone windscreens are evidently of keen interest to cows, horses and dogs, among others.

Background Noise

On the important issue of background noise, an approach that has worked well in a number of field applications is to set up a number of monitoring stations outside of the project area in settings similar to those at the on-site monitor positions. Of course, considerable judgment is involved in selecting these positions but in an ideal situation of, say, an isolated project in open farm country that is largely uniform in character both within and beyond the project area one would want monitors at least 3 or 4 km from the perimeter of the project (nearest turbines) in the four cardinal directions. The locations should be far enough away that project noise is negligible and yet close enough that they are reasonably representative of the site area. At the end of the survey the off-site positions can then be evaluated for consistency. If the levels are generally similar, and somewhat surprisingly this is usually the result, the average can be taken as a time history record of the background sound level that probably would have existed within the site area and then used to correct the on-site measurements taken, importantly, at the same time under identical environmental conditions.

Figure 1 below is an example from a site in the Eastern United States where the landscape is rural and generally homogenous in nature within the project area and for some distance beyond it in terms of topography (rolling hills), vegetation (a mix of farm fields and wooded areas) and population density (farms and residences scattered more or less uniformly over the site area). The 80 or so 1.5 MW turbines are spread throughout a roughly 20 sq. mi (5200 ha) project area on numerous parcels of private land and thoroughly intermixed with the residences in the area. Proxy background measurement positions were set up about 3 km beyond the perimeter of the turbine array to the northwest, east and south of the project (a neighboring wind project to the west prevented measurements in that direction) at locations that were similar in character to the various settings near on-site residences: one was on an open and exposed hilltop, another was at the edge of a field with nearby trees and a third was essentially in a forested area. The expectation was that there might be a consistent difference between these different positions – with the sheltered forest location being quieter than the windy hilltop, for instance – in which case background corrections for a particular setting would be applied to on-site measurements at positions with comparable settings. However, as can be seen from the figure, the levels at all three locations, each many miles from the others, were largely the same at any given time and, perhaps more significantly, no one position is consistently higher or lower than the others. Consequently, the arithmetic average of all three, with the site area physically lying

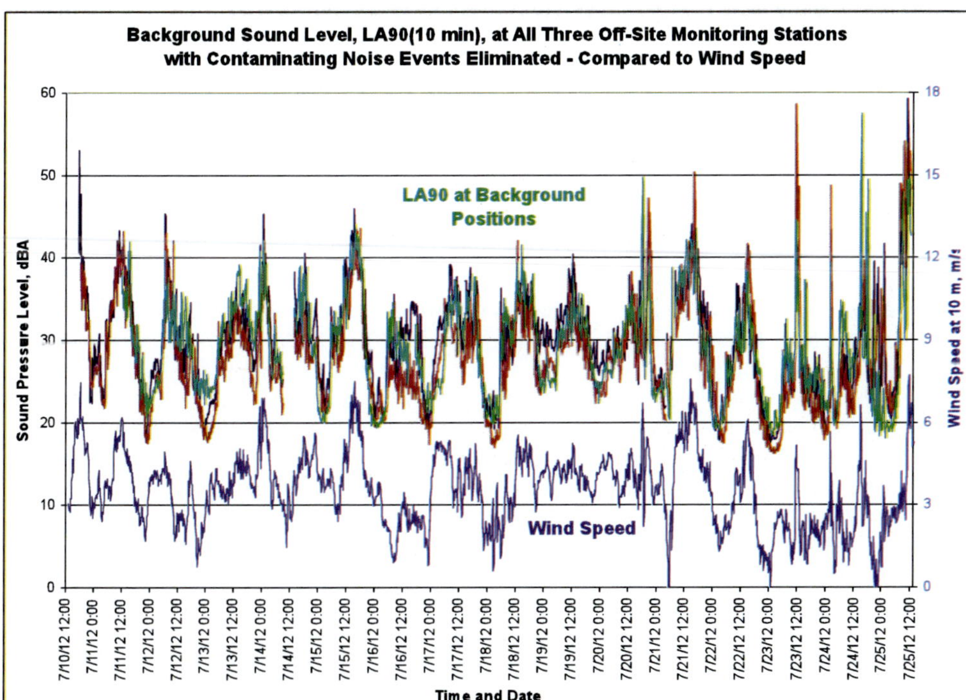

Figure 1. Measured background sound levels at three off-site proxy positions

between them, can be taken as a reasonably reliable estimate of the on-site background level at any particular time that accounts for the specific wind speed, direction, time of day and atmospheric conditions prevailing during that 10 minute period.

The data in Figure 1 have been edited to remove noise spikes that were observed only at one position and not at any others, indicating a contaminating local noise event that is not representative of the area as a whole. Spikes were also deleted (from both the on-site and background data) if there were no concurrent spike in wind speed, even if they may have occurred at multiple locations, on the premise that the noise was not associated with the turbines and may have been due to thunder, rain, a helicopter flyover or some other area-wide noise event.

The results shown in the Figure 1 example are not unique to that site and a similar consistency between the off-site proxy location sound levels has been observed at a number of other projects in rural areas even though the background monitors are deliberately set up in diverse settings. Fortunately, for the purpose of estimating simultaneous background sound levels, most wind projects are located in rural areas but, of course, not all of them are and other situations exist. In urban settings or near major highways the background sound is no less important, in fact more so, but its dependence on wind and atmospheric conditions is greatly diminished, if not relegated into complete insignificance. In such cases, the proxy background technique is still theoretically viable although the selection of background positions that are representative of receptors potentially affected by project noise becomes highly specific to the circumstances at each receptor. In the case of a highway, for instance, one might try to find a background position that is the same distance from the roadway as the actual point of interest and similar in all other ways but far enough from any turbines that they are undetectable. In this kind of a complicated situation where the background level is more dependent on man made noise than natural, wind-induced sounds it may be necessary to perform a pre-construction survey at the key receptors near turbines and at a number of candidate background positions to evaluate the validity of the proxy locations before the project turbines become operational.

Sound Test Equipment and Set up

As with any field sound survey, what equipment is used and how it is deployed must adhere to certain minimum technical standards. These requirements are generally described in numerous standards, such as ANSI S12.9-1992/Part 2 [9]; however, the focus of this section is not to repeat and belabor those details but rather to point up what adaptations need to be made for the specific application of performing general site-wide testing of wind turbine noise. As mentioned earlier, no standard exists that can be directly used for this purpose, if only because they limit data collection to low wind conditions.

In terms of instrumentation, most environmental sound measurement standards recommend the use of Type 1 precision equipment per IEC 61672-1 Ed. 1 [10] or ANSI S1.43-1997 [11] while also allowing, somewhat grudgingly, for the use of Type 2 equipment. There is certainly no reason on technical grounds to oppose this recommendation but, from a practical perspective, it is often necessary to use Type 2 equipment for surveys of this type because of the large number of instruments needed. The utterly intangible difference in technical performance between these two instrument classes is totally inconsequential within the inherently and unavoidably imprecise nature of this type of survey. It is much more important that the equipment is durable, reliable and specifically designed for extended use in the outdoors. Delicate and expensive Type 1 precision grade equipment can be unreliable in such applications or even unable to be programmed as a lowly data logger.

Although high cost and extreme precision are not essential, the functional capabilities to statistically integrate sound levels over a user defined time period and automatically store the results are necessary. Because the on-site wind and weather monitoring towers, or met towers, normally integrate and store measurements in 10 minute increments it is convenient, if not necessary, to measure and store sound data in synchronization with the wind data collected by these towers for later correlation. It is evidently universal practice for met towers to store data 6 times an hour in 10 minute intervals that begin at the top of the hour; as in 9:00, 9:10, 9:20, etc. Consequently, sound data logging should be started using a trigger function to begin at the top of an hour and not randomly by the manual push of the start button. The timers on all instruments should be exactly synchronized to local time or to the project's SCADA control system clock, if it is different from the actual time, which it often is.

Of course, all of the instruments must be field calibrated at the beginning of the survey and checked again for drift at the end of the survey.

Because this long-term survey approach involves unattended monitoring, the instrument and the microphone must be capable of withstanding damage, interference or outright destruction from rain and snow, which, among other things, means that the ground plate technique specified in IEC 61400-11 [5] – where the microphone is laid flat in the center of a board on the ground and covered with one or more hemispherical windscreens – is not a viable option despite its otherwise highly desirable advantage of minimizing wind-induced pseudo noise. Consequently, the microphone must be mounted above ground level and protected from wind-induced distortion by a spherical weather-treated windscreen, which normally entails a higher density foam that is hydrophobically treated to shed water (windscreens and wind-induced noise are discussed in detail later). As a general rule, a slightly lower than normal microphone height of about 1 m above ground level is preferred for this application on the premise that wind speed diminishes exponentially with decreasing elevation theoretically going to zero at the surface, or boundary layer. To illustrate this, the nominal wind speed profile, or shear gradient, per Eqn. (7) in IEC 61400-11 is illustrated below in Figure 2 for a common turbine operating condition where the wind speed is 6 m/s at the standard elevation of 10 m above ground level.

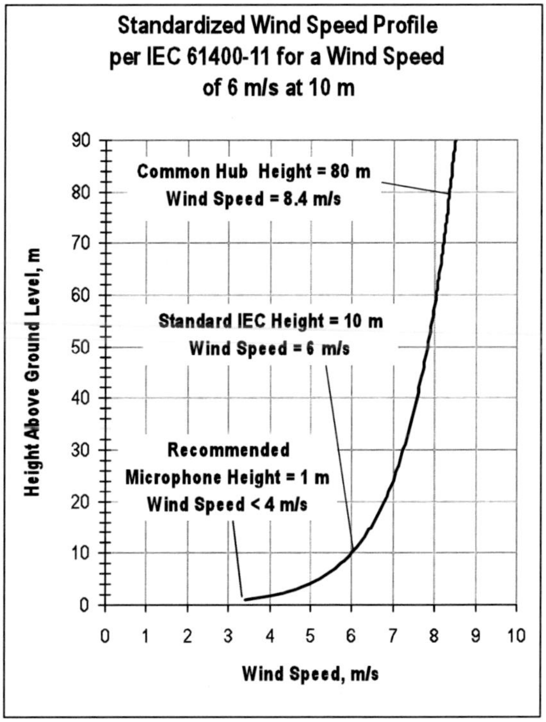

Figure 2. Standardised wind speed profile

For these moderate wind conditions, which are often when turbine noise tends to be most prominent relative to the background level, the wind speed at a 1 m microphone height would be less than about

3 or 4 m/s, which as shall be seen later, means that distortion from wind blowing through the windscreen is of little or no consequence with respect to the A-weighted sound level. Measurements of the actual wind speed at 1 m over a number of multi-week surveys, for typical projects in the Eastern United States at least, show that wind speeds often tend to follow the empirical IEC curve in Figure 2 and remain fairly low, below 4 m/s, most of the time as shown in Figure 3, for example.

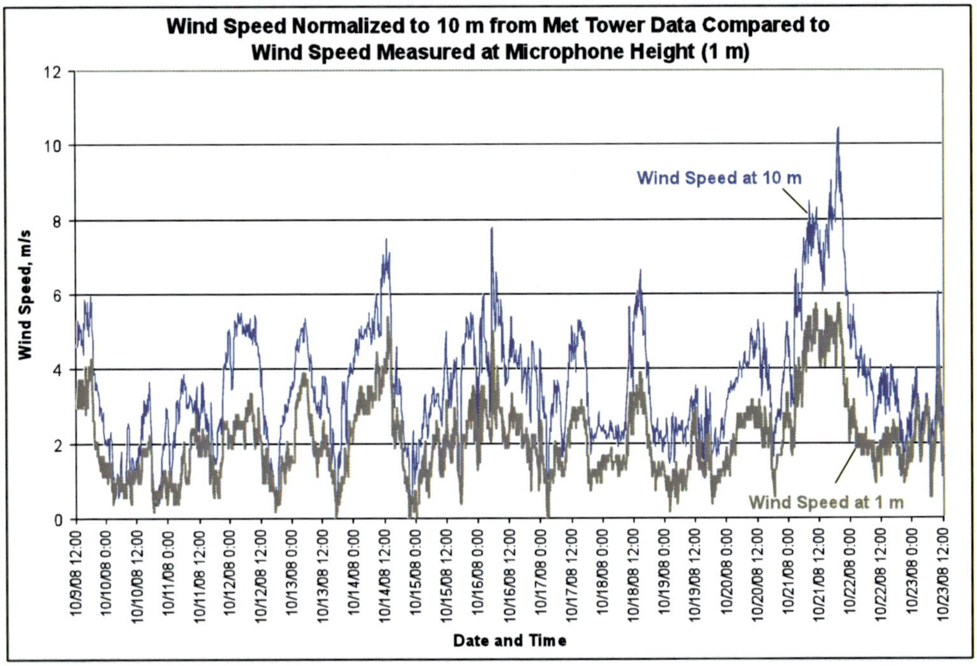

Figure 3. Typical 10m wind speed compared with wind speed at microphone height

However, in some instances, mostly associated with flat, wide open sites, there can be an apparent disconnect between the derived 10 m wind speed and the wind speed measured close to the ground where they do not parallel each other, as illustrated in Figure 4. In this instance, the site was evidently affected by an exaggerated summertime wind shear at night. The tendency in these cases is for the microphone height wind speed to remain even lower than might otherwise be expected.

In addition to arranging for the microphone to be about 1 m off the ground so that it is not adversely affected by precipitation, it is also necessary to keep the instrument itself dry and secure in a waterproof case, which is best mounted above the ground on a fencepost, utility pole or other support. Wrapping the instrument itself in a plastic bag along with several desiccant packs is also highly recommended as a defense against humidity and condensation.

While the microphone can be remotely connected to the instrument with a cable and independently supported, another practical option is to use a self-contained system where the microphone is attached to the instrument case with a rigid boom to hold the microphone away from the box and the entire assembly is mounted 1 m above ground level with a strap as shown, for example, in Figure 5. While there is nothing wrong with supporting the microphone separately on a tripod there is a tendency, unique to wind turbine survey work, for tripods to blow over, even after being weighted down and/or firmly staked to the ground. The use of temporary metal fence posts to support either the microphone alone or the entire system is a more reliable option and is sometimes the only option in places where there are no existing supports, such as in open fields.

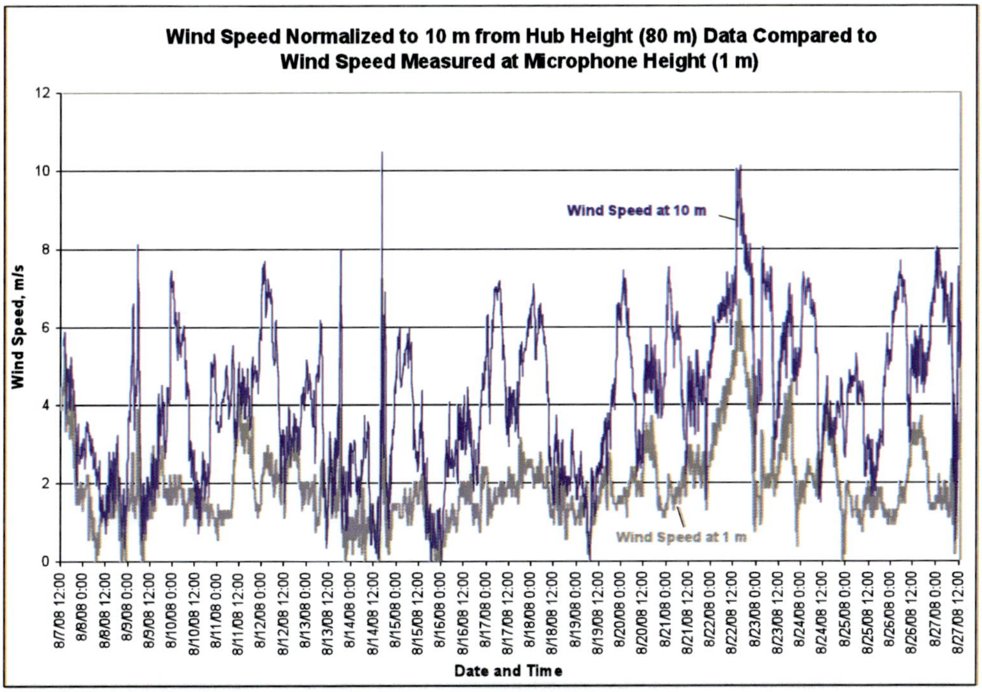

Figure 4. 10m wind speed compared with wind speed at microphone height on flat site

Figure 5. Typical Integrating Sound Monitor

Weather Stations and Wind Speed Monitoring

In addition to the sound monitors it is also advisable to establish at least one temporary weather station at the sound monitoring position with the most exposure to wind. The primary reason for this station is to measure the maximum wind speed at microphone height (about 1 m) for use in correcting the measured sound data for wind-induced distortion as described in a later section. Wind speed at 1 m, direction and rainfall are the primary parameters to be recorded by this station, or others set up in other settings as appropriate, such as at a sound monitoring position sheltered from the wind by the local terrain (to demonstrate, for instance, that wind-induced distortion is negligible at such locations). This data should be integrated and stored in 10 minute blocks in synchronization with the sound monitors.

This temporary anemometer at 1 m above ground is solely there to evaluate microphone wind exposure and it is the on-site met tower anemometers, usually at 50 to 80 m above ground level, that should be used to correlate the measured sound levels at ground level to the wind speed essentially experienced by the turbine rotors. Turbine nacelle anemometers scattered throughout the site may also be used to determine wind speed, but this is somewhat less desirable because a free field correction usually needs to be applied to this data to account for the energy extracted from the wind by the rotor just upstream of the wind speed sensor.

It is customary to normalize mast top or nacelle wind speeds to a standard elevation of 10 m above grade per IEC 61400-11. It is this result that is compared to the measured sound levels.

Measurement Quantities and Parameters

The objective of a compliance survey is to extract the project-only sound level from the total soundscape and compare that result to the permissible limit. As such, the principal challenge is identifying and eliminating contaminating noises that are unrelated to the project over many days and thousands of measurements. If it were practical to take a manned sample for 20 minutes, removing spurious noises by pausing the instrument or discarding contaminated subsamples, and declare the result as the performance of the project it would be a trivial matter; however, over a relatively long time period of unattended monitoring it is necessary to use the L_{A90} statistical measure to generally perform this function in an automated manner, since it captures the consistently present sound level during relatively quiet periods between common interfering and identifiable noise events like cars passing by or planes flying over. A 10 minute sampling duration has been found to work very well since it allows direct correlation with met mast wind speed data and is generally short enough that fairly rapid changes in project noise are captured.

Of course, the use of this parameter as a measure of wind turbine noise immediately calls into question its ability to capture very short-term fluctuations in turbine noise; i.e. if the turbine sound level as manifested at a test location is variable within a particular 10 minute measurement period, the L_{A90} statistical will tend to capture its baseline or near-minimum sound level and blindly exclude any short-lived spikes or excursions in turbine noise in the same way, for instance, that 30 seconds of barking dog would be eliminated. While the possibility of missing brief noise spikes may intermittently exist during periods of unstable wind, extensive field observations suggest that under most normal conditions turbine noise does not usually fluctuate dramatically within a fairly short 10 minute period - meaning that in most cases *any* statistical level could theoretically be used to quantify the sound, setting aside for the moment the possibility of sporadic contamination from background sources, since all measures collapse to a single value for a constant source. Although turbine noise may not be steady in a perfectly ideal sense, the advantage of being able to tease out the fundamental, wind-driven project sound level with the L_{A90} generally outweighs the ability to capture every blip and nuance that occurs over the two week survey period.

It is sometimes suggested that the true average sound level (L_{Aeq}) of a wind turbine, which can be variable in the short-term, can be approximated by adding an explicit factor, generally on the order of 2 dBA, to the measured L_{A90} sound level; however, the quantitative basis for this, if there is one, is unclear and the value of the factor would appear to be dependent on the ability to accurately measure the L_{Aeq} and know what the differential is, which is often not practical.

The use of the average, or $L_{Aeq,\ 10\ min}$, sound level or a finer time resolution of, say, 1 minute come to mind as alternatives to the L_{A90}, but these approaches have their own serious drawbacks. If the L_{Aeq} is used to measure at on-site positions with the idea of better quantifying turbine sound levels, then the L_{Aeq} measured at the proxy background positions must be used as an apples-to-apples correction factor.

But the L_{Aeq} is often completely unusable for this application. As an example, multiple statistical measures were recorded at the off-site background measurement positions previously mentioned in connection with Figure 1, including the L_{Aeq}. Figure 6 below shows the average L_{A90} and L_{Aeq} levels measured at all three locations compared to wind speed.

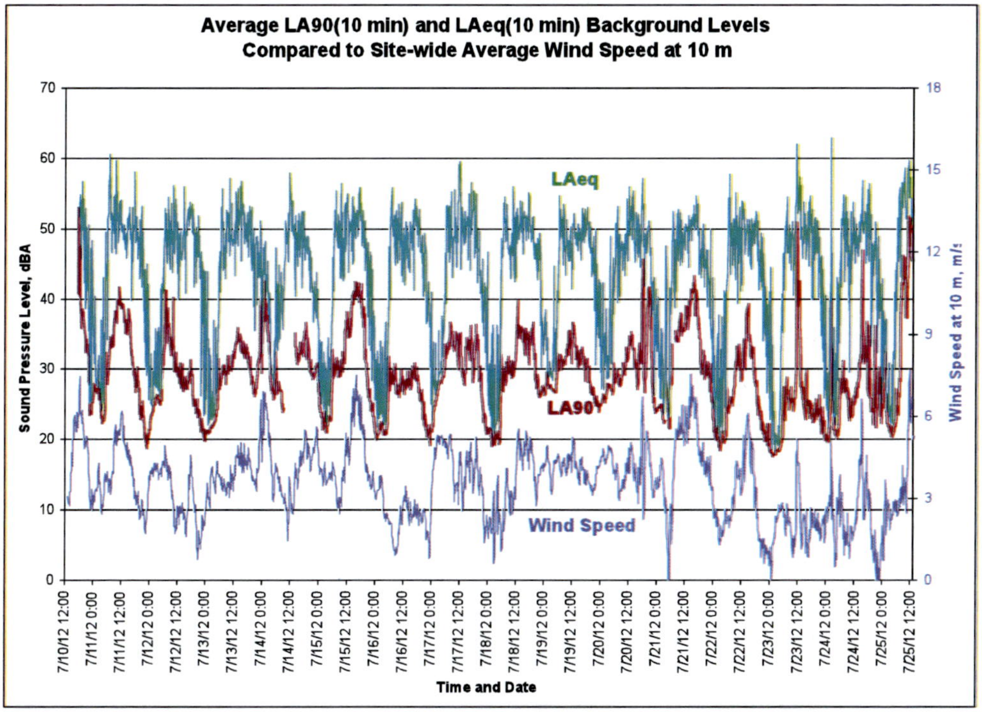

Figure 6. Sound measurements with L_{Aeq} and L_{A90} compared

What is immediately obvious from this plot is that the $L_{Aeq,\ 10\ min}$ level is clearly driven by daily human activity; primarily intermittent vehicular noise on nearby sparsely traveled roads (noise that is filtered out by the L_{A90}). The L_{Aeq} levels rise to about 53 dBA every morning, stay there all day irrespective of the wind conditions and then gradually fall off in the evening hours bottoming out briefly somewhere around 23 dBA every night. The L_{A90} level, on the other hand, is clearly more attuned to the natural environmental sound level, which in rural areas like this one is normally a function of wind speed. The unsuitability of the $L_{Aeq,\ 10\ min}$ as a measure that might quantify project noise can be seen in Figure 7 where the average background L_{Aeq} level from Figure 6 is compared to the L_{Aeq} level measured at a typical, randomly selected on-site receptor.

The $L_{Aeq,\ 10\ min}$ sound levels at both positions are virtually indistinguishable meaning that the project-only sound level simply cannot be deduced. Furthermore, it could even be reasoned that project noise is utterly inconsequential at this location because the on-site level is about the same or even lower than the off-site level, which is entirely free of any turbine noise, but, as we shall see later, that is not at all the case at this particular test position.

A second alternative approach to using the $L_{A90,\ 10\ min}$ might be to use a much finer time resolution on the order of 1 minute. Apart from the fact that a two week survey would go from producing about 2000 data points per position to an unwieldy 20,000 data points, such a short time duration would cause

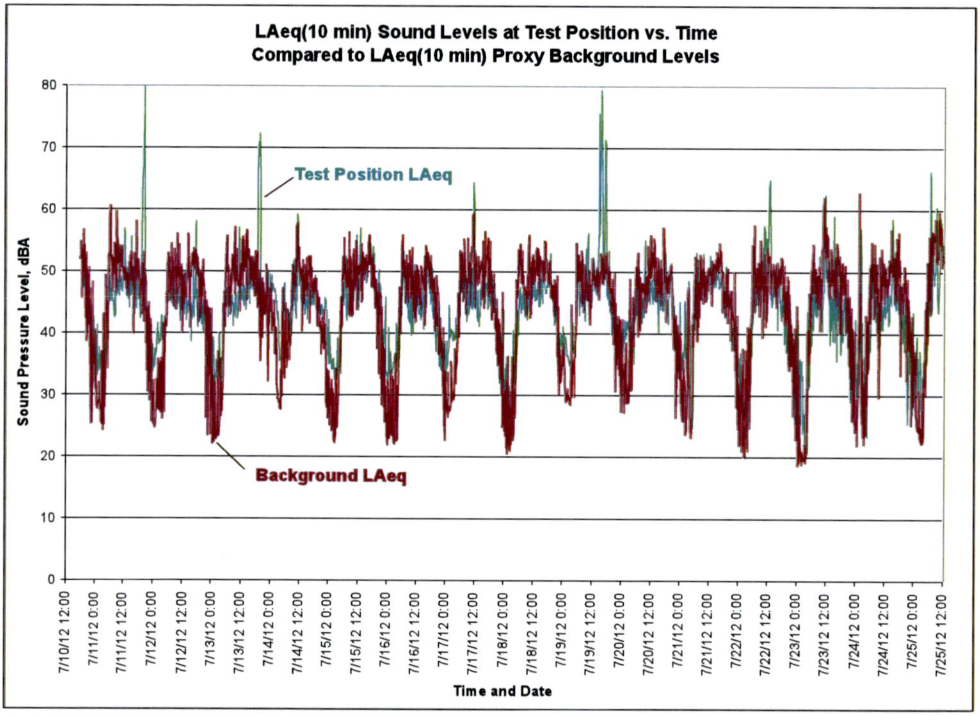

Figure 7. Sound measurements as L_{Aeq} at two locations

the time history plot to be extremely choppy irrespective of which statistical parameter is used because the measurement period would be comparable in duration to, and statistically dependent on, many contaminating noise events that one is seeking to eliminate. For example, a single car passing by on a rural road might significantly elevate the sound level at a monitoring position for 20 seconds or more as its sound increases gradually, quickly peaks and then fades slowly. Thus even the $L_{A90,\ 1\ min}$ would be significantly skewed by this event. In addition, met tower wind speed data is not normally stored in 1 minute increments making it very difficult to reconcile the sound data with concurrent wind conditions.

Even though the $L_{A90,\ 10\ min}$ is normally the only measure that will likely yield a meaningful result from a long-term survey in terms of quantifying the project sound level, it is not the only measure that should be recorded during the survey. At least the L_{Aeq}, L_{A10}, L_{A50}, L_{Amin} and L_{Amax} should also be stored by the instrument for every 10 minute period. The reason for this is that these other measures can be used to examine specific periods of interest - such as unusually loud, average or below average levels - in further detail, if desired. For example, a noise spike occurring over one or two measurement 10 minute periods and suspected of being project noise can be broken down by looking at the complete set of statistical measures for those periods to get a sense of how high the sound went on a short-term or even instantaneous basis and how long it might have actually lasted.

As a further tool for evaluating suspicious noise events many integrating sound level meters now have an audio record function that can be programmed to start at a user defined trigger level. This is typically set at, or just below the permitted noise limit, assuming it is an absolute limit, to investigate any possible overages. Thus any period where relatively high sound levels were recorded can theoretically be played back to help identify the source of the noise and, in particular, determine whether it was due to the turbines. While this function is often useful, informative and usually worth doing, if one's instruments are so equipped, it has its practical drawbacks. First of all, there is an art to setting the trigger level. If set too low the function is activated almost all the time possibly resulting in a single recording that is two weeks in duration; if set too high it may record only a few noise events. With a low trigger point not only is there an overload of data but it is likely that the memory capacity of the instrument will be quickly exceeded, which may well prevent the instrument from continuing to record numerical sound levels throughout the entire survey period. Secondly, it is often surprisingly difficult to identify just what it is that is being heard during playback – for example, leaf rustle sounds a lot like rain, which sounds a lot like water flow noise, which sounds a lot like turbine noise. In any event, it is certainly better to have this data for possible use in interpreting the survey results than not have it.

Finally, it is highly desirable to use instruments capable of measuring the frequency spectrum in 1/3 octave bands at one or two key locations with, usually Type 2, monitors measuring overall A-weighted levels at the majority of positions. The large number of instruments required for this kind of survey usually precludes the possibility of using frequency analyzers at all positions. Nevertheless, the use of one or more at key positions allows for some frequency analysis, although great caution must be exercised with the lower frequency bands, as discussed later, since wind-induced false signal noise is largely inevitable and the low frequency results cannot be taken at face value. Fortunately, this phenomenon does not significantly affect the measurement of A-weighted sound levels, however.

The use of 1/3 octave band analyzers is largely essential for surveys that, for one reason or another, must be conducted during summertime conditions when insect, frog or cicada noise is present. Measurements taken under these unfavorable conditions can be "corrected" to a certain extent by smoothing the high end of the frequency spectrum, where this kind of noise is usually obvious, and then recalculating the overall A-weighted sound level as shown in the (generic) example below.

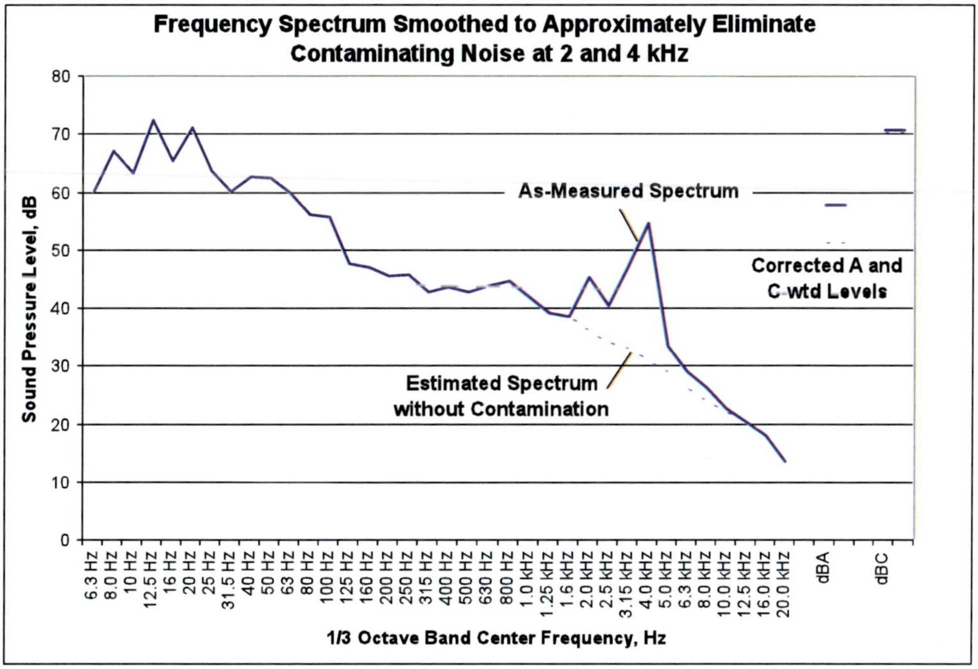

Figure 8. Principle of Ai-weighting

Of course, this correction would be laborious to perform for thousands or even just dozens of measurements so it is usually necessary to determine a typical correction, such as the -7 dBA adjustment that resulted in the example above, and apply that to all periods when this noise was apparently present. This is, of course, an imperfect remedy and the best policy is to avoid, if possible, measuring under these circumstances in the first place.

A solution to this common problem is currently being proposed by Hessler [15] and Schomer [16] in the form of a modified A-weighted network, termed "Ai-weighting", where all of the measured sound above 1000 Hz, or the 1250 Hz 1/3 octave band, is disregarded in situations where insect noise is present and an adjusted A-weighted sound level is calculated from the truncated spectrum. As detailed in Reference 16, this apparently radical approach has almost no detrimental effect on the accuracy of the measurement in the sense that this correction was applied to 100 random samples of environmental sound without insect noise and the recalculated sound level changed on average by less than 1 dBA. It is hoped that Ai-weighting can be incorporated into sound level meters and data loggers to automatically record this level, along with the unadjusted A-weighted sound level, of course, in situations where high frequency insect or frog noise is present and not the subject of the survey.

Wind-induced Microphone Distortion

One of the principal errors in measuring wind turbine noise is false signal noise from wind blowing through the windscreen and over the microphone tip, which is manifested in the form of artificially elevated sound levels in the lower frequency bands. Taken at face value any measurement made in moderately windy conditions will ostensibly indicate relatively high levels of low frequency noise, irrespective of whether a wind turbine is present or not. This measurement error is probably one of the principal reasons wind turbines are mistakenly believed to produce high, if not harmful, levels of low frequency and infrasonic noise.

Some degree of distortion is essentially inevitable in any measurement taken above ground level when the wind is blowing, even when using an extra-large windscreen. It is in an effort to minimize this error that the IEC 61400-11 test procedure prescribes measuring on a reflective plate at ground level, where the wind speed is theoretically, although often not actually, zero. As previously mentioned, this ground plate technique is fine for short-term, attended measurements but is impractical for long-term surveys due to the potential for rain or melted snow to damage the microphone. Consequently, for lengthy compliance and evaluation surveys it is necessary to measure above ground level using a large, weather-treated windscreen - perhaps augmented with a very large secondary windscreen, although the practicality of such devices is questionable in harsh winter conditions.

Because environmental sound measurements of most other sources apart from wind turbines are not generally conducted in windy conditions as mandated by applicable standards, the significance and even existence of this measurement error has long gone unnoticed. Although this phenomenon and its physical basis were theorized decades ago by Strasberg [17, 18] it is only fairly recently that its relevance to wind turbine sound measurements has been examined in detail and quantified. In particular, the subject of wind generated self-noise was thoroughly reviewed in 2006 by van den Berg [19] where he showed that the magnitude of the distortion depends not only on the mean incident wind speed but also on the amount of atmospheric turbulence present at the microphone position (largely a function of the local surface roughness) and on atmospheric stability. Measurements taken at 1 or 2 m above a smooth surface during stable, nighttime atmospheric conditions, when the surface winds are usually light, generally contain the least amount of self-generated noise ultimately replicating the case where the principal noise generation mechanism is wake turbulence trailing off the windscreen. In other less ideal circumstances self-noise levels can be developed by estimating the local surface roughness and atmospheric turbulence factor, Ψ, from wind speed measurements at two heights and/or from observations of cloud cover, time of day, general wind conditions, or meteorological data, if available.

The minimum level of false-signal noise due to wind, excluding the effect of atmospheric turbulence, can be estimated based on an empirical wind tunnel study carried out by Hessler and Brandstätt in 2008

[12] in which conventional ½" microphones fitted with an array of common windscreens and were subjected to known wind velocities in a massively silenced wind tunnel. This is similar to the approach used by Strasberg [18]. The measured sound levels during each test were essentially a direct measure of the false-signal noise – although for more or less laminar flow conditions corresponding to an outdoor setting with a very low surface roughness in neutral atmospheric conditions. Nevertheless, for the specific windscreens examined it is possible to generally estimate both the overall A-weighted or un-weighted sound level and frequency spectrum of the distortion from the microphone height wind speed and then subtract it from the total measured level to *largely* reverse the error.

An example is shown in Figure 9 where the overall A-weighted level of self-noise is calculated as a function of wind speed and subtracted from the as-measured sound level. The plot is a three day detail of a wind turbine survey where oversized 175 mm (7") diameter treated windscreens (ACO Model WS7-80T) were used. This particular windscreen was found to be the best performer, in terms of minimizing wind-induced self-noise, in the wind tunnel study.

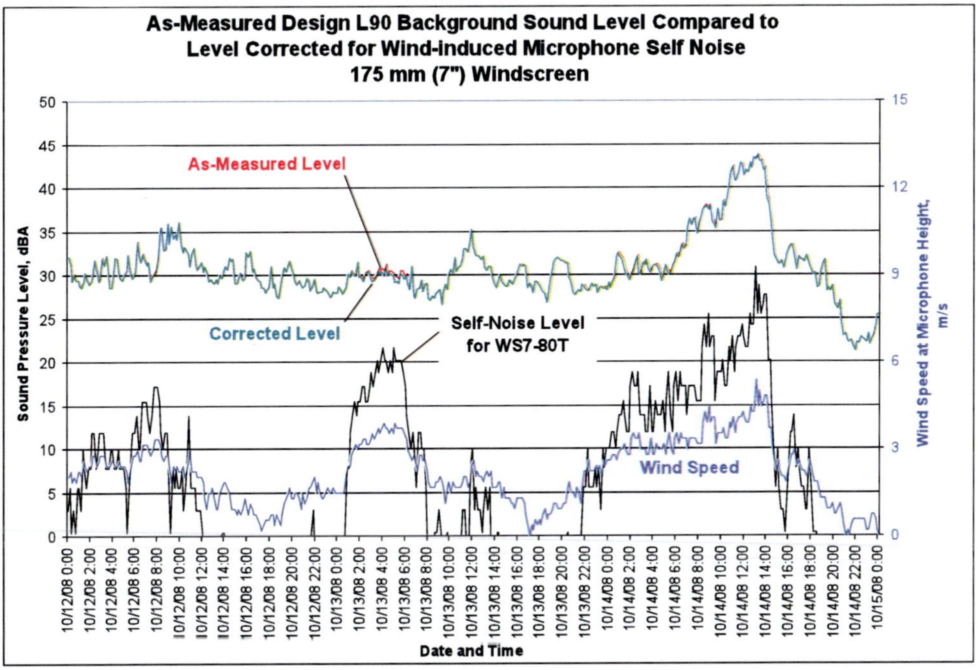

Figure 9. Correction for wind induced microphone noise – 175mm windscreen

Figure 9 shows the very typical result, at least where extra-large windscreens are used, that the correction is insignificant and can be essentially neglected when it comes to A-weighted sound levels. This is because with a large windscreen the distortion is confined to the very lowest frequencies where it has almost no impact on the A-weighted sound level. With a conventional 75 mm (3") windscreen, on the other hand, wind-induced noise begins to become significant in the mid-frequency region, between about 63 and 400 Hz, where it has much more influence on the A-weighted sound level. If a 75 mm windscreen had been used in the example above the effect of self-generated noise would have been considerably more prominent, as illustrated hypothetically in Figure 10 where the false signal noise is at times comparable in magnitude to the actual environmental sound level. If a 75 mm windscreen had actually been used to make these measurements the as-measured level would have been significantly higher in the regions where self noise peaks.

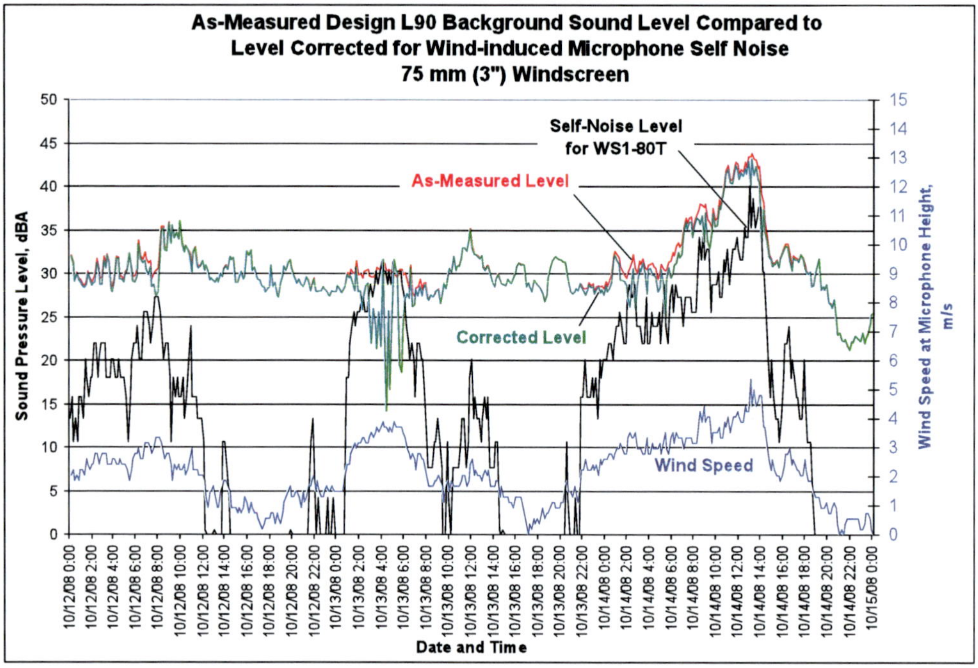

Figure 10. Correction for wind induced microphone noise – 75mm windscreen

Because a 175 mm (7") windscreen is capable of substantially suppressing interfering wind-induced noise, while others are not, this size windscreen is recommended for wind turbine field work. The weather treated variety is also suitable for use in long-term surveys where rain or snow are a possibility although its attenuation or acoustical non-transparency is rather pronounced in the high frequencies, which usually results in the measurement of A-weighted sound levels that are slightly lower than they actually are. Of course, the overall attenuation is a function of the frequency content of the sound but for a typical wind turbine which has very little high frequency content when measured at a few hundred meters away the effect of this is usually small, as will be seen.

The empirical study [12] results for 175 and 75 mm treated windscreens are shown in shown in Figure 11 and 12.

The overall level of self-generated noise for these windscreens may be estimated from the general expression below with the understanding that local atmospheric turbulence is not accounted for and a neutral atmosphere is assumed.

$$L_{p,self} = A \ln(v) + C, \text{ dB} \quad \text{for } v>1.5 \text{ m/s} \qquad [1]$$

Where A and C are constants given in the table below and v is the normally incident wind speed at the microphone in m/s.

The 1/3 octave band frequency performance can also be approximated for the larger 175 mm treated windscreen using Eqn. (2) and the constants in Table 2.

$$L_{p,self,\,f} = A \ln(v) + C, \text{ dB} \quad \text{for } L_p > 10 \text{ dB} \qquad (2)$$

An example spectrum for a 10 m/s wind speed is plotted is plotted in Figure 13. Note that the C-weighted sound level at 87 dBC is ostensibly well above the typical threshold for the onset of

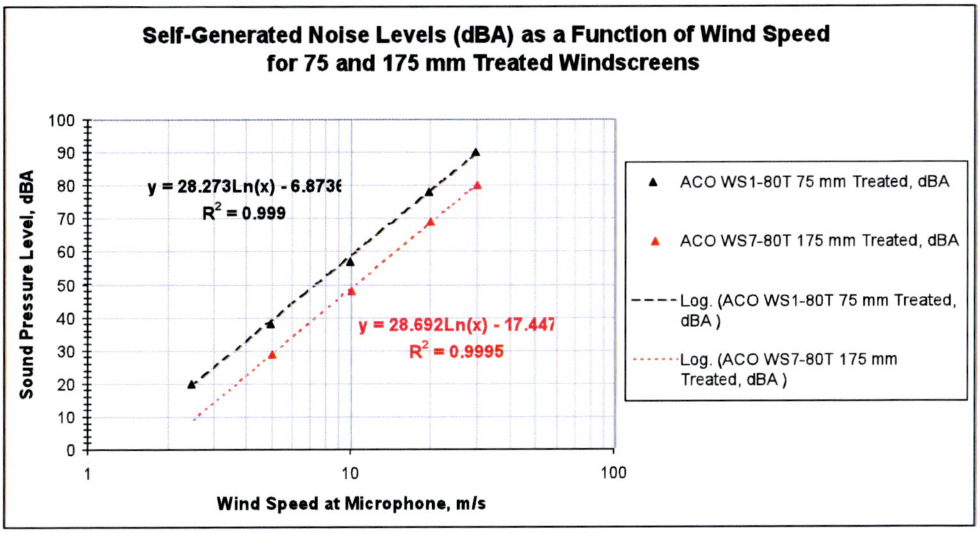

Figure 11. Comparison of wind induced noise – A-weighted

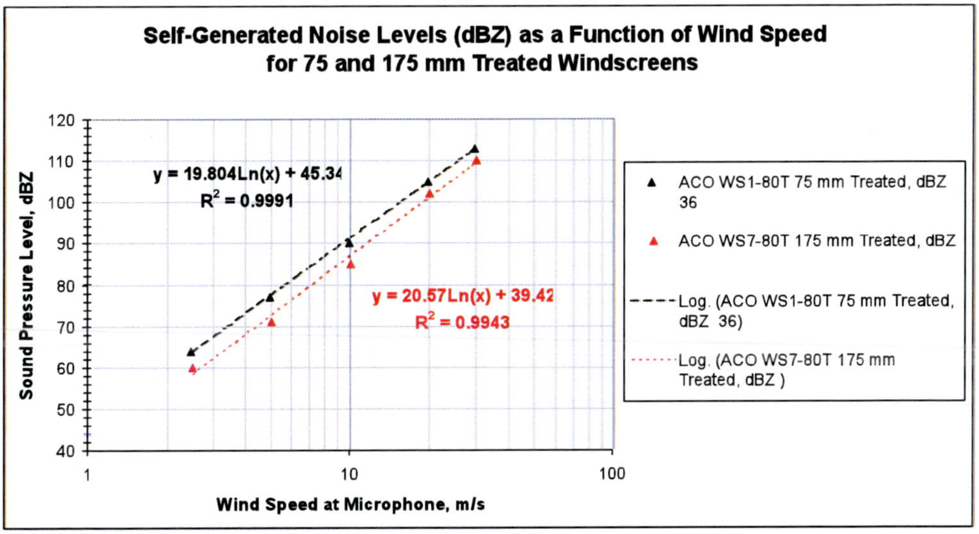

Figure 12. Comparison of wind induced noise Z-weighted

perceptible vibrations and other problems – but all of this noise is internal to the measuring system and no actual low frequency noise exists.

In a real atmosphere the sound level may be higher or lower than given in Table 1, depending on the turbulent energy present, which again depends on the stability of the atmosphere. In a neutral atmosphere, which occurs at higher wind speeds (> 6 m/s at 10 m height) or in very clouded conditions, the wind-induced level might be anywhere from 5 to 9 dB higher than the levels shown above. After sunset, when the atmosphere is more prone to be stable, the wind-induced noise levels will be more similar to the values given above.

Table 1. Constants for A and Z-wtd self-noise calculation algorithm (Neglecting Atmospheric Turbulence)

Windscreen Type	A-weighted Sound Level, dBA		Un-weighted Sound Level, dBZ	
	A	C	A	C
75 mm (3") Treated	28.273	-6.8736	19.804	45.34
175 mm (7") Treated[1]	28.692	-17.447	20.57	39.42

Table 2. Constants for calculating 1/3 octave band frequency levels of self-noise for a 175 mm treated windscreen and attenuation factors (Neglecting Atmospheric Turbulence)

1/3 OBCF, Hz	Un-weighted Sound Level, dB		Windscreen Attenuation, dB
	A	C	
6.3	20.160	33.541	
8	19.855	33.406	
10	20.694	32.222	
12.5	20.834	30.827	
16	22.683	21.755	
20	24.002	15.974	
25	22.975	16.900	
32	21.767	19.473	
40	23.970	11.667	
50	25.427	5.516	
63	28.098	-3.581	
80	28.846	-8.120	
100	26.948	-5.961	
125	27.139	-9.587	
160	26.466	-11.478	
200	32.274	-31.352	
250	31.555	-32.200	
315	30.444	-30.036	
400	27.385	-23.327	
500	25.287	-19.871	
630	25.904	-20.312	
800	27.593	-24.213	0.4
1000	27.222	-24.633	0.7
1250	27.762	-27.198	1.8
1600	27.693	-28.173	2.2
2000	27.906	-30.456	3.7
2500	29.701	-36.029	3.8
3150	30.151	-39.032	4.5
4000	30.391	-41.695	5.3
5000	30.662	-44.181	5.8
6300	31.289	-48.000	7.2
8000	32.257	-53.039	8.0
10000	32.925	-57.412	9.4

[1] It should be noted that the constants for the 175 mm A-weighted case in Table 1 vary slightly from those given as an example in Eqn. (2) of Reference 12, which was based on a spurious data point for the 2.5 m/s test and gives a somewhat higher result in the lower wind speeds than it really should. The constants in Table 1 are essentially an improved interpretation of the raw data.

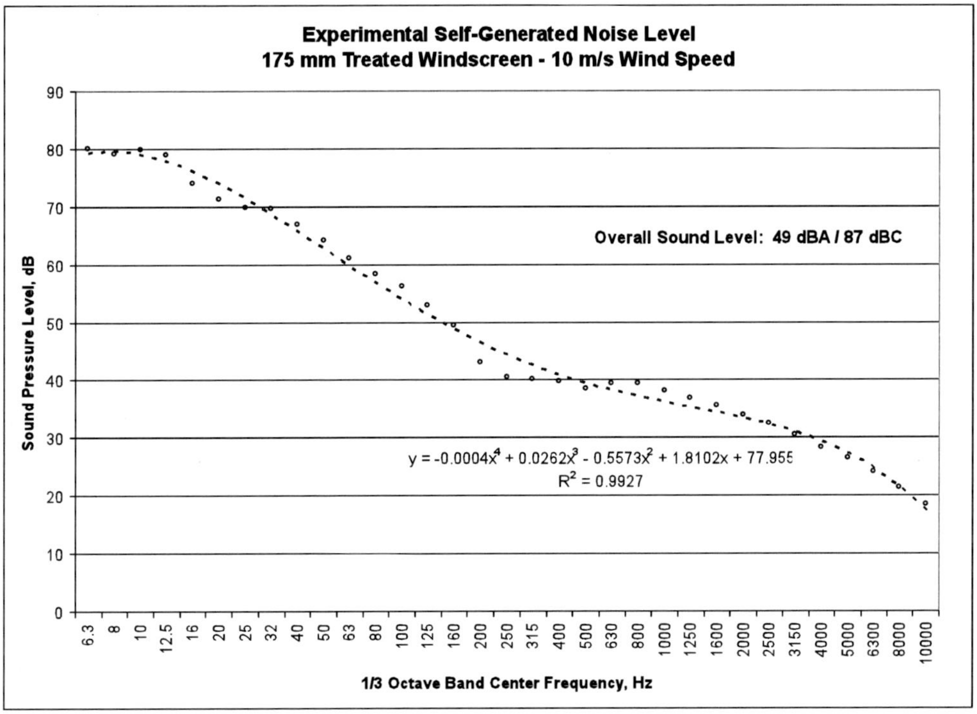

Figure 13. Frequency spectrum of wind induced noise 175mm windscreen

The attenuation or noise reduction inherent in nearly all common windscreens is very small and can normally be neglected; however, the 175 mm treated windscreen that is desirable for use in wind turbine measurement work is unique in that it has an unusually high attenuation in the higher frequencies as found from measurements in an anechoic chamber and reported in Reference 12. This not the case for all other windscreens tested in the study, including the 75 mm treated or 175 mm untreated windscreens, which have negligible attenuation. The attenuation factors for the 175 mm treated windscreen are given in Table 2 above. These factors would be added in each 1/3 octave band to the measured value to correct the spectrum from which a new A-weighted sound level can be calculated by re-summing the modified spectrum. An example is shown below where the sound level measured with a 175 mm treated windscreen at 300 m from a typical wind turbine is corrected in the higher frequencies.

For this example relatively close to the unit (usually the minimum set back distance ever encountered) the actual change in the A-weighted sound level is fairly small at about 1 dBA. Further from the unit one would expect the higher frequencies to diminish reducing the effect and significance of windscreen attenuation. As a side note, the level of wind-induced distortion for the conditions under which this particular measurement was made is plotted for reference in Figure 14. Since both the as-measured and self-noise lines converge in the lower frequencies it is evident that all the measured data below about 40 Hz is false-signal noise that is not attributable to the turbine. This also means that the measured level of 67 dBC is artificially high and cannot be taken as a quantification of the low frequency noise emissions of the turbine.

Correction for Background Noise

Once a design L_{90} background sound level has been developed from averaging the data collected at the off-site proxy positions it can then be subtracted in the usual logarithmic manner from the levels measured

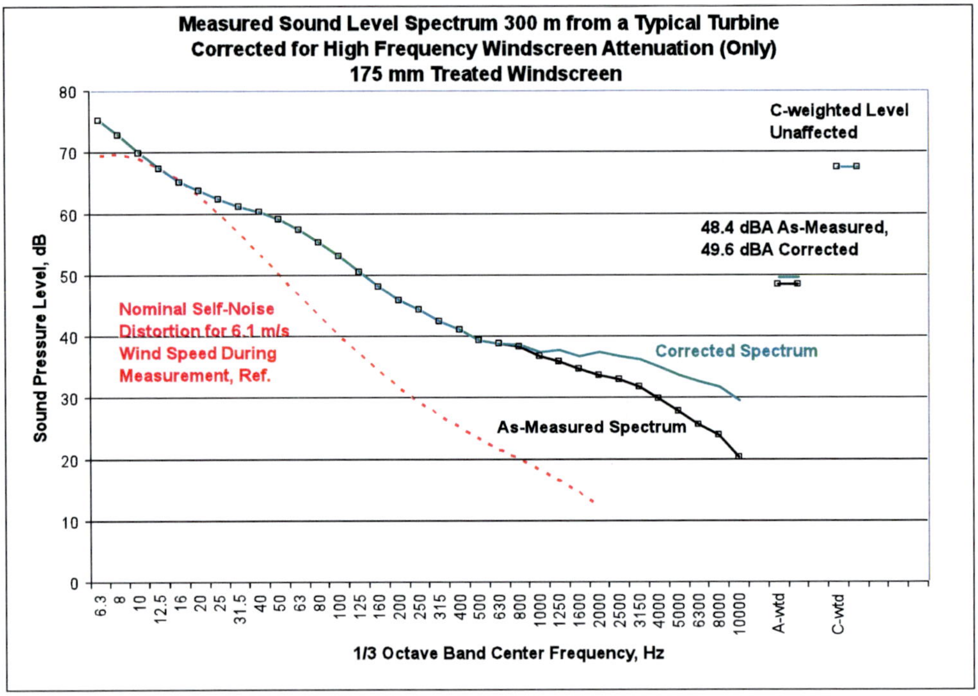

Figure 14. Correction for attenuation of windscreen

at each of the on-site positions to deduce the project-only sound level. However, this correction process is only relevant to samples recorded while the turbines were actually in operation and not necessarily to all samples; consequently, the data must be sifted to ignore all periods of calm winds. This can be accomplished by dealing only with data sets collected above the effective cut-in wind speed for the turbine model in question (bearing in mind whether that wind speed is measured at 10 m or hub height) or, more preferably, by comparing the measured data to a time history of project electrical output obtained from the SCADA, or project control system. For this latter option it is best to compare the operational output of the 2 or 3 units closest to each on-site measurement position rather than the total project output because this not only accurately defines the on and off times at each monitoring station but also may reveal, the fairly common occurrence, that certain units were temporarily down for maintenance or due to some unexpected malfunction. The relevance of this, of course, is that the measurements of project noise during this period would not have captured the maximum possible sound level.

Because the proxy background level is, for practical reasons, an inexact estimation of the site-wide background level, there will usually be instances when the background level exceeds the total measured level at certain on-site positions. Under this circumstance, and when the background level is below but within 3 dB of the total level, the project-only sound level would normally be considered indeterminate. In such cases, a decision has to be taken whether to simply discard these samples or take a slightly less cautious tack and assume the background level is dominant and that turbine noise must be at least 10 dB below the measured level – and therefore subtract 10 dB from the total measured level to estimate the project-only sound level. Both approaches have their merits with the second option leading, probably not unfairly, to a slightly lower result. A third possible path, that of taking no correction under these circumstances and assuming the total level is due entirely to turbine noise, probably is unfair and is not recommended.

While the calculation of the project-only sound level is mathematically possible when the background level is below but within 3 dB of the total level, doing so tends to create spurious mathematical artifacts where the project level can be estimated at unrealistically low and obviously incorrect sound levels. Since most standards, such as ISO 3746 [2], essentially disallow this calculation it is best to follow that policy here as well. IEC 61400-11 [5] takes the somewhat unusual hybrid approach of allowing the calculation only where the difference between the total and background levels is 6 dB or more and suggests using a constant of 1.3 dB as a universal correction for differences of between 6 and 3 dB.

Typical Test Results and Comparison to Model Predictions

Representative examples from typical test positions within two different wind projects using two different turbine models and located in two different states are discussed below as a way of illustrating the outcome of the test methodology outlined above.

Example 1

The first example is from a test position at a residence within a project in a rural area in the Eastern United States where the turbines and homes are thoroughly mixed together – a common situation in this region and the Midwest. This location is surrounded in nearly all directions by a number of turbines at various distances, the closest being about 490 m (1600 ft.) away from the home with another 11 lying within a 1500 m (4900 ft.) radius. The terrain is gently rolling hills with a mixture of open fields and wooded areas. Mild complaints about noise had been received by the project from the residents of this home, which is the primary reason it was selected as a monitoring position.

The overall test results from a two week measurement survey in terms of the total measured level at the test point, the design background level derived from proxy positions and the normalized 10 m wind speed, are shown in Figure 15. This is same test position that was previously discussed in conjunction with Figure 7 and L_{Aeq} sound levels.

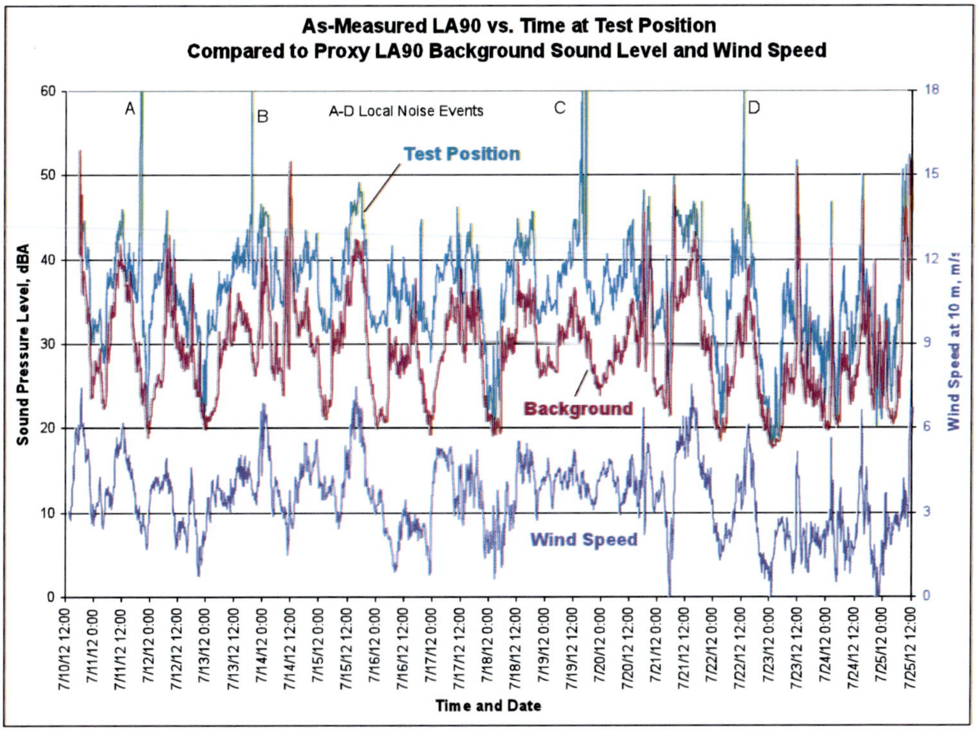

Figure 15. Overall measurements compared with proxy background noise

Although the raw results may appear unintelligible at first glance, a closer look reveals that the design background level (developed from an average of three off-site measurement positions as shown in Figure 1) and the sound level at the test position both generally parallel the wind speed indicating that the measured levels are due to wind-induced sounds associated with the natural environment in the first case and to both natural and wind turbine sound in the second. As expected, the on-site level at the position surrounded by a dozen turbines is usually substantially higher than the background whenever a moderate wind is blowing and, also as expected, the on-site level is similar to the background during calm conditions when the project is not operating. It is the difference between these two levels during windy conditions that essentially constitutes and quantifies the noise impact of the project. As is evident from the plot, it is an ever-changing dynamic situation where the project sound level variously exceeds the background by anywhere from 0 to 10 dBA. This figure graphically points up the inadequacy of attempting to determine the project's noise emissions from a few short-term manned samples. The greatest differentials between the on- and off-site level tend to occur at night but it is important to note that while the project level may be quite a bit higher than the background, the sound level at the receptor point often remains very low in absolute terms with unadjusted raw levels commonly in low to mid 30's dBA.

Taking these test results through the next steps of correcting the on-site level for background noise and parsing out the low wind periods when the project was idle produces the following plot where the nominal project-only sound level is shown as a function of time over the survey period.

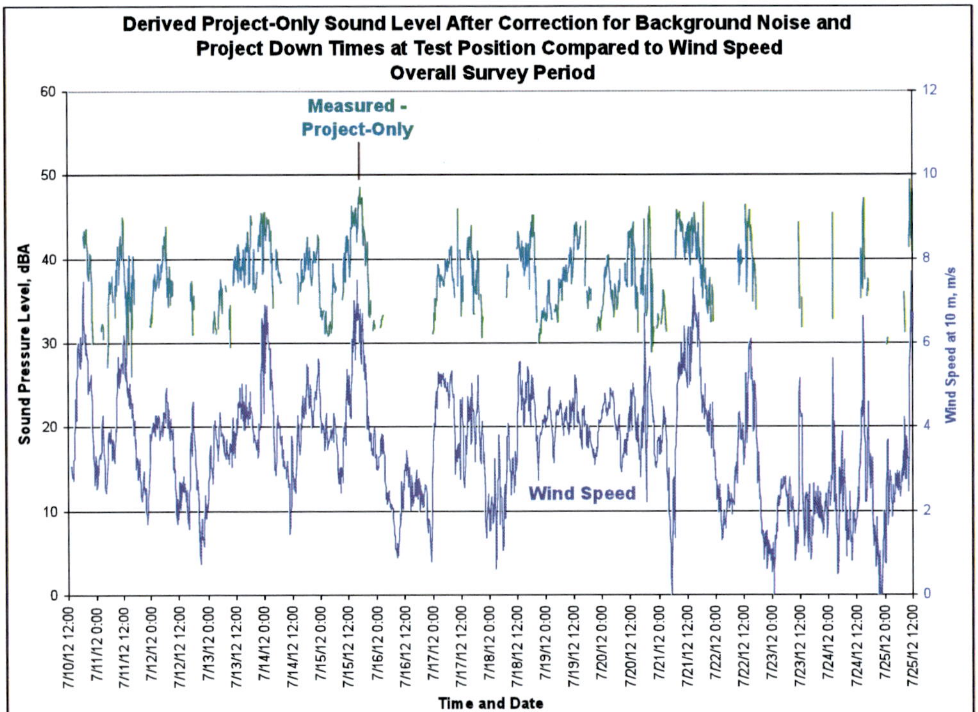

Figure 16. Derived turbine noise levels

In terms of magnitude the project apparently generates sound levels ranging from 30 to 49 dBA at this location, depending largely but not only on wind speed. The fact that the project sound level does not exactly parallel the wind speed (which was derived from high elevation, rotor height anemometers) shows that other atmospheric factors play a significant role in determining exactly how loud the project is at this location at any given moment.

What Figure 16 is theoretically showing is the baseline - L_{A90} - project sound level that is consistently present during each 10 minute measurement period. This means that somewhat higher sound level excursions lasting a few seconds to a few minutes are possible, if not probable, but, as discussed earlier, it is not practical to capture the moment to moment variation over the lengthy survey period needed to adequate evaluate long-term project sound levels. However, comparing these results to model predictions based on the L_{Aeq} turbine sound power level indicates that the L_{A90} approach does not inadvertently underestimate project levels, as might be suspected. Figure 17 plots the modeled project sound level at this test position against the measured project-only sound level. For clarity a detail of a representative three day period from the third to the sixth day of the survey is shown.

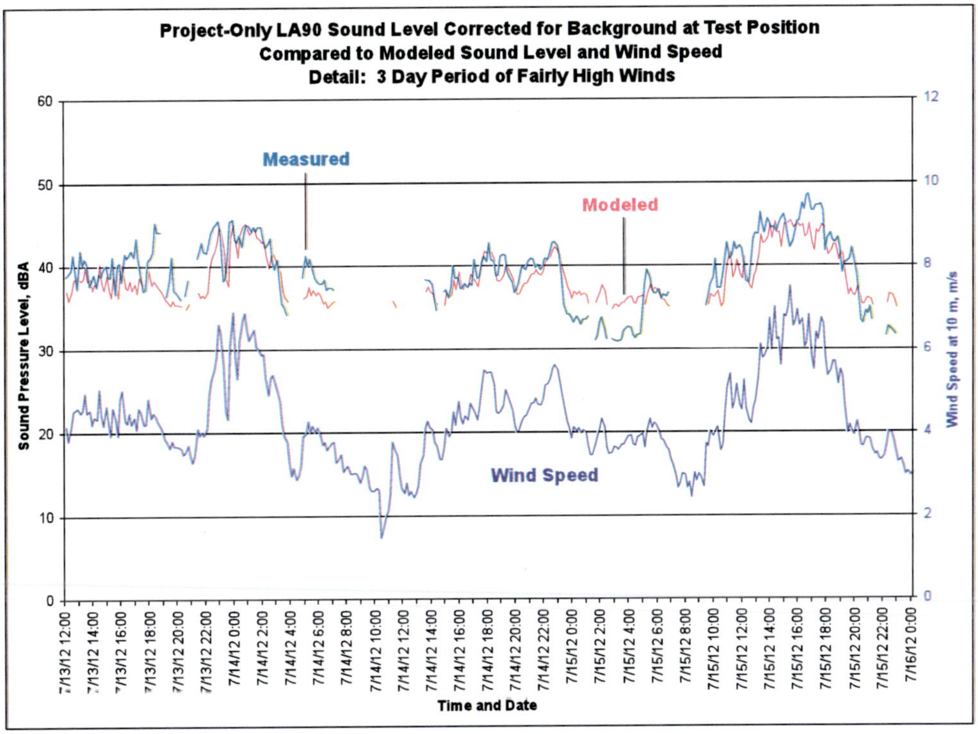

Figure 17. Measured turbine noise levels compared with modelled

The modeled level is derived using a curve-fit polynomial function based on the predicted project sound level at integer wind speeds, which in turn is based on the turbine sound power level at those wind speeds taken directly from an IEC 61400-11 field test report. The modeling software used is simply an automated version of ISO 9613-2 *Attenuation of sound during propagation outdoors* [13] where a mid-range ground absorption coefficient of 0.5 was assumed along with uniform omni-directional noise radiation from each turbine, which is modeled as a point source 80 m above ground level. In general, Figure 17 shows that the model prediction, based solely on the turbine's L_{Aeq} sound level at specific wind speeds, provides a reasonably good approximation of the actual observed sound level. In fact, the model prediction captures the *mean* project sound level remarkably well as can be seen in Figure 18 if the measured project-only sound level is plotted as a function of wind speed rather than time.

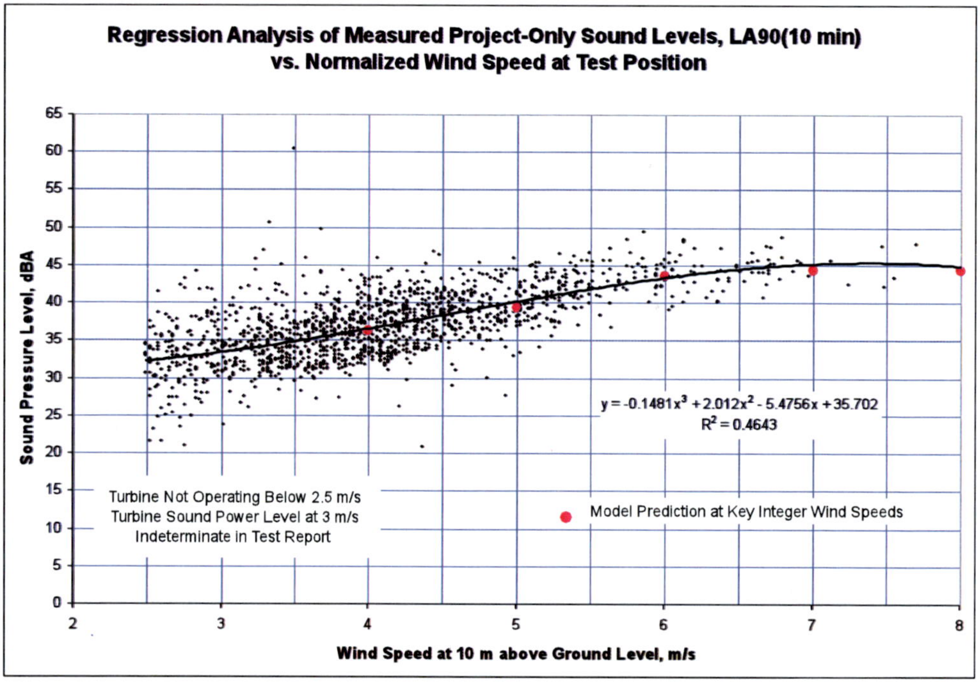

Figure 18. Measured turbine noise vs wind speed compared with modelled

The predicted levels at integer wind speeds fall very close to the mean trend line through the measured data. The scatter above and below the mean is essentially due to the variable nature of turbine noise which is affected in an ongoing but largely unpredictable way by unstable winds and ever-changing atmospheric conditions.

Example 2

The second example is from a site in the Midwestern United States where the turbines are again intermixed with scattered homes and farms in a rural setting. This particular test location was adopted in response to, what turned out to be understandable, complaints about noise from a participant's "own" turbine that had been sited at the unfortunate distance of only 180 m (600 ft.) from the house. The raw test results are summarized in Figure 19.

In this instance, the total sound level at the house is consistently and not surprisingly well above the background level developed from four off-site monitoring stations, meaning that much of the time background noise was largely insignificant, if not inaudible. The corrected project-only sound level for a three day windy period near the beginning of the survey is shown below compared to model predictions.

In this instance, as with Example 1, the predicted level intertwines with the measured level, sometimes over-estimating, sometimes underestimating but generally capturing the mean project sound level. The variation above and below the predicted level is largely a measurement of how all other factors beyond the simple wind speed are affecting the total sound level perceived at this location. One of these factors may be unique to the turbine model used at this site, which, based on other surveys and observations, appears to have a tendency to produce sound levels in excess of the manufacturer's stated performance in high wind conditions, which may be part of the reason the actual level significantly exceeds the expected levels in the second half of this sample period. This same departure

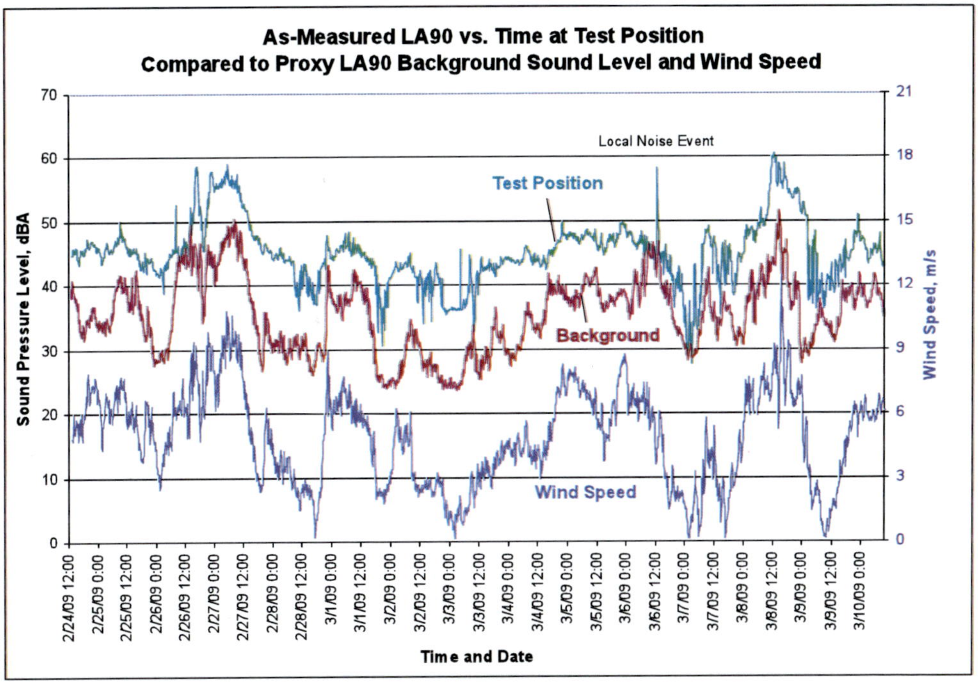

Figure 19. Turbine noise and background noise with wind speed

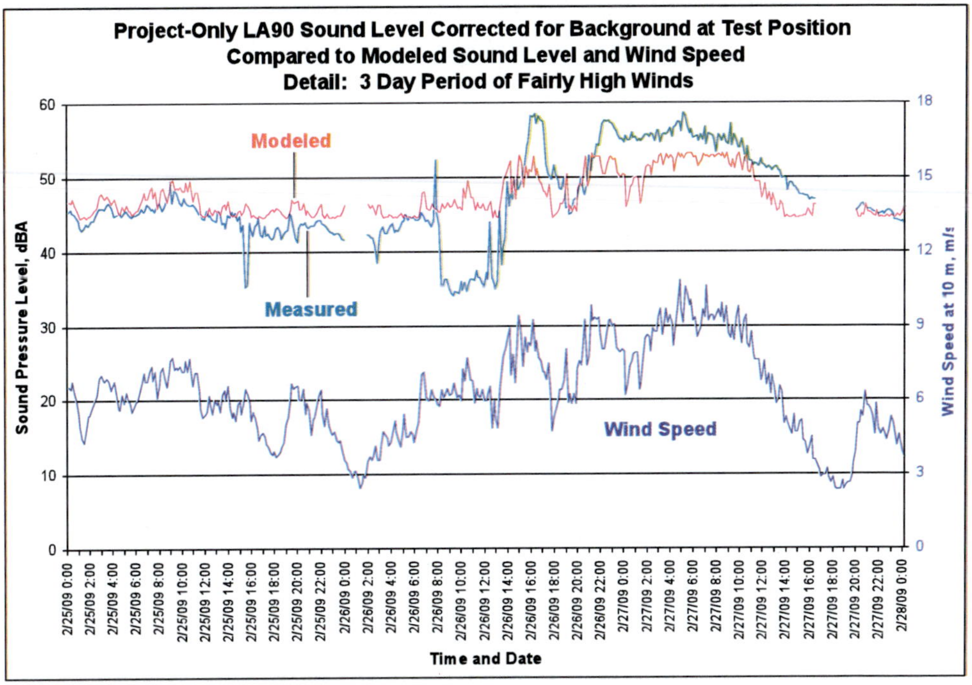

Figure 20. Turbine noise and modelled noise wind speed

between the predicted and measured levels also appears in the regression analysis, Figure 21, where the project-only sound levels are plotted as a function of wind speed.

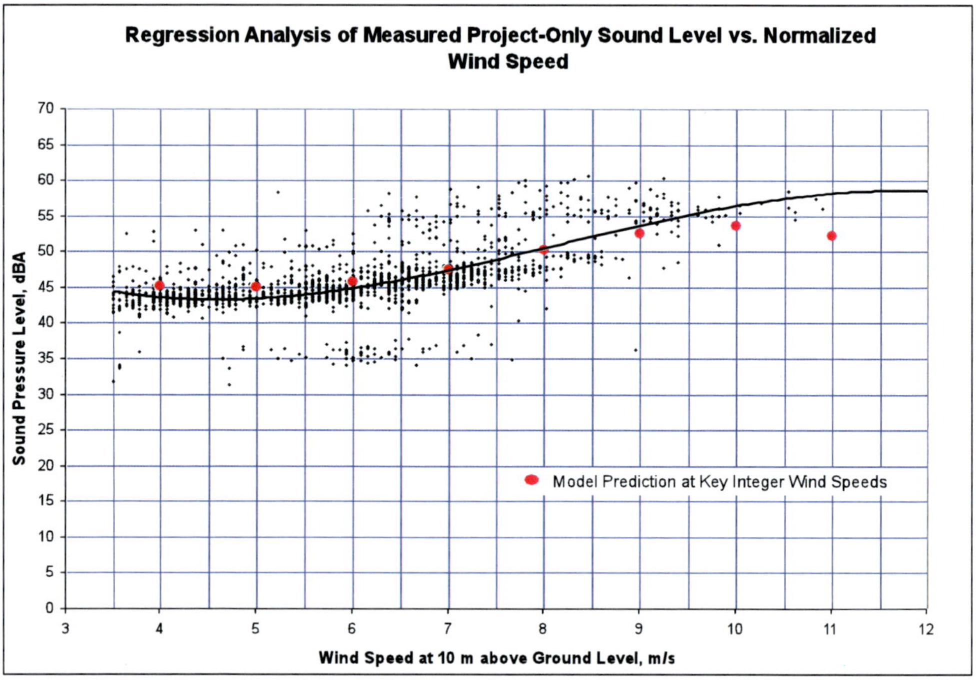

Figure 21. Turbine noise vs wind speed

Good agreement with the mean trend is evident up to about 9 m/s but not beyond it.

These two examples are presented to illustrate the outcome of the test methodology and are generally representative of the typical results obtained at a number of test positions over a number of such surveys. That is not to say, however, that the method is infallible and that mismatches between measured and predicted levels will never be found. Testing wind turbine noise is challenging and inherently imprecise because the sound sources themselves and the propagation of sound from them to a given point of interest is dependent on the environment in general and amorphous wind and atmospheric conditions in particular.

Interpretation of Test Results Relative to Permit Limits

The examples above, and particularly the regression plots (Figures 18 and 21), exhibit the typical behavior where there is a scatter to the test results and the project sound level is not a perfectly fixed quantity at a given wind speed. This is an unavoidable consequence of the nebulous atmospheric conditions mentioned above. The question that this raises, however, is how to interpret the results of the survey relative to the absolute, or in some cases relative, noise limits contained in planning consent or permit conditions. Excursions, sometimes very substantial excursions, above the mean project sound level are inevitable and under all normal circumstances it would be a complete impossibility to design and lay out a project so that the sound level never exceeded a specific value at a particular point or, more realistically, at a large number of residences within the vicinity of the project. Only projects in obviously remote locations could ever be comfortably designed to such a limit. Consequently, the possibility, even likelihood, that project noise will occasionally spike for short periods should be factored in to regulatory limits. That this issue is not addressed in current laws or limits pertaining to wind turbines is simply a result of the understandable fact that few are aware that it is even an issue.

As a suggestion, it seems reasonable to conclude that a project is in compliance with an absolute regulatory limit if the measurements indicate that the project-only sound level is lower than the stated limit 95% of the time, taking that number from the commonly used statistical confidence interval.

Single Site Investigations

In addition to evaluating operational sound levels on a project-wide basis with regard to regulatory compliance, it is sometimes necessary to carry out dedicated field surveys, usually in response to complaints, that are focused only on a specific point. Although each of these situations is certainly unique, the general test approach outlined above can generally be applied with the exception that more resources can be brought to bear on understanding the project sound level at that particular location.

General Test Design

The general test set up for a diagnostic or investigative sound survey at a single point would follow the procedures described for a site-wide test in terms of survey length, equipment and measurement technique with the following enhancements.

The primary measurement position will be outside the residence or point of interest where it is usually prudent to use multiple instruments for redundancy and/or increased functional capability. For example, it is highly desirable to measure the overall A-weighted sound level, the frequency content in 1/3 octave bands and to store audio recordings whenever an appropriate trigger level is reached. While all three of these things can be achieved by some instruments, it would be safer to use the 1/3 octave band analyzer to store numerical data and use a second instrument to store both back-up A-weighted data and the audio files. As mentioned earlier, the trigger setting for audio recording can dramatically alter how much or how little audio data is stored, with the general tendency being for many recordings to be made, which may fill up the memory of the instrument and effectively shut it down before the survey period has run its course. A potential alternative method for this particular application may be to put the audio record function under the control of the resident with a manual trigger. In any case, having multiple instruments can also allow for additional time resolutions (beside the standard 10 minute periods) to be recorded at the same time; 1 minute or 1 hour data, for instance. In addition to the sound recording equipment a weather station recording wind speed at microphone height, wind direction and rainfall, among other common parameters, should be set up nearby.

The specific measurement position should be at a location with exposure to all of the nearest turbines or at a place that replicates the exposure of the residence to the project but is removed from any sources of local contaminating noise (HVAC equipment, farm machinery, human activities, etc.).

As with a more general survey, the background level is still of just as much concern so 2 to 3 proxy background measurement positions should be found in opposite directions that are remote from any turbines and, in this particular case, replicate as closely as possible the setting of the principal test location in terms of terrain, exposure to wind and exposure to other noise, such as from a road.

The principal and proxy background positions above will theoretically determine what the project sound level is at the residence but may not indicate why it is. To this end several additional monitoring stations close to the 3 or 4 nearest turbines are recommended that are ideally located in line with the principal position at the standard IEC 61400-11 test distance of the hub height plus half the rotor diameter (typically around 125 m). A hypothetical test set up involving four nearby turbines is shown in Figure 22.

Note that several of the intermediate positions are slightly off the direct sight line to keep them in open and reasonably accessible areas. Although this hypothetical example was conveniently conducive to this test set up, additional complications are likely to arise; in particular access to private property, which may call for some creativity in designing the test layout. Nevertheless, the idea is to gauge the individual contribution from all of the nearest units over a variety of wind directions and weather conditions to determine if the problematic noise levels are principally associated with perhaps

Figure 22.

one unit or a particular set of wind conditions. Moreover, the principal purpose for measuring the noise emissions of all the nearest units is to be able to estimate the actual sound power level of each unit and analytically calculate, by means of a simple spreadsheet model, or modeling software, the total sound level at the house for comparison to the measured level there. This approach allows the individual contribution from each unit to be quantified for different conditions and also helps confirm, in a manner independent from the proxy monitoring approach, how much of the received signal at the principal measurement location is due to the project and how much is background noise. In addition, the sound power level of each unit can be informally checked against the manufacturer's warranty value.

While the ground board technique specified in IEC 61400-11 is not practical for long-term, unattended measurements - mainly because of concern about rain - a comparable, if somewhat less rigorous, result can be obtained from measuring at 1 m above grade by placing the microphone or monitor on a tripod or temporary post at the appropriate distance. In Figure 23, for example, measurements were made simultaneously at 1 second resolution with a microphone on a ground plate and with two additional microphones at 1 and 2 m above it. The average and consistent differential between both above ground positions and the microphone on the reflective plate was 2.7 dB, which is close to the ideal 3 dB differential that one would expect.

This example illustrates that it is possible under certain circumstances to reasonably measure the apparent A-weighted turbine sound power level above ground level without serious degradation due to wind distortion. Of course, this may not be true when it is particularly windy at 1 m above ground

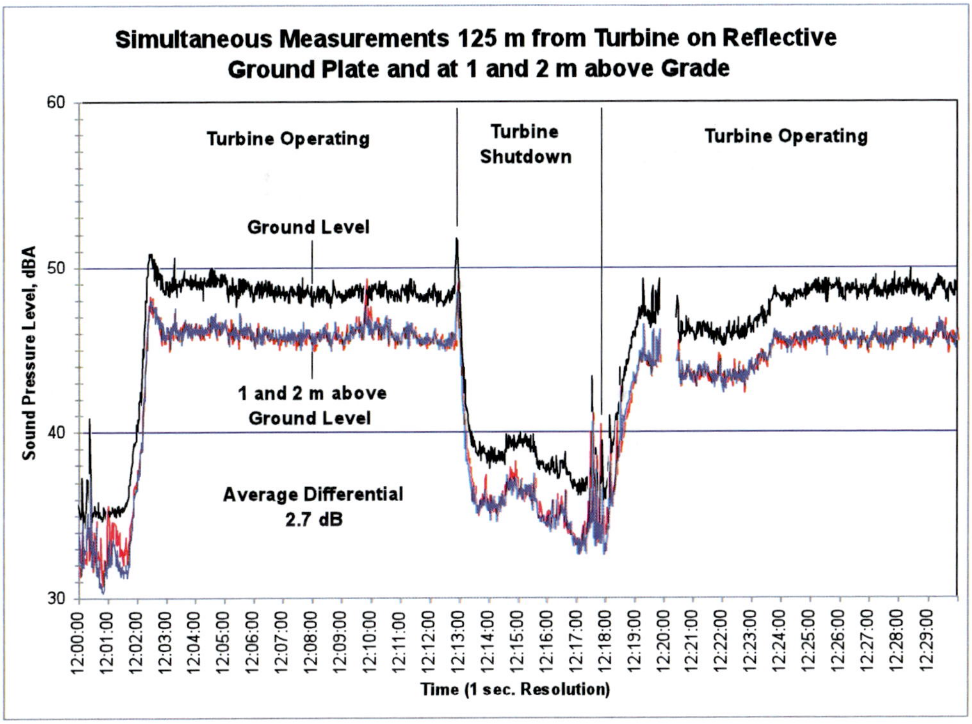

Figure 23. Effect of a ground board

level. Another potential complication arises when multiple turbines are in unusually close proximity to each other, as they are in Figure 22, and background noise or cross-contamination from one unit to another must be taken into account in such cases. In general, however, the only substantive modification to the IEC 61400-11 process for calculating sound power level would be to change the constant "6" to "3" in Eqn. (9) of the standard [5] since above ground measurements are being used.

As suggested by Figure 23 an additional tool that is normally useful and practical for single site investigations is to temporarily shutdown, for 10 to 20 minutes, the nearest turbines to the point of interest, if not all those that could conceivably be affecting the sound level there, in order to obtain direct measurements of the background level so the project-only level can be derived with some confidence from the operational sound levels occurring just before or after the shutdown. A short-duration shutdown helps ensure that the wind and weather conditions are essentially identical for both the on and off measurements. This technique also offers a way of verifying the validity of the levels measured at the off-site background positions. It is usually during the times of peak noise that it is most desirable to have an exact measurement of project's sound level, since these are the noise levels that most likely engendered the complaint in the first place. Consequently, it becomes a matter of either being there when these conditions occur, which is frequently at night, to organize the shutdown - or putting control over the shutdown in the hands of the resident who can call in by pre-arrangement to the control room if and when the noise becomes objectionable in terms of its overall magnitude and/or begins to exhibit some adverse character, such as from amplitude modulation. Although this latter approach of allowing the resident identify the time of maximum noise has been used successfully to quantify the overall magnitude of project noise and its frequency content in 1/3 octave bands, one must really be on hand to manually measure amplitude modulation, since it calls for the use of an extremely fine time resolution, on the order of milliseconds, to capture the sound oscillations that normally have a period of roughly 1 second. Such manual measurements can be taken indoors, where this kind of noise is most often observed to be objectionable, as well as outdoors.

Only with attended measurements it is possible, and then only occasionally, to measure indoor sound levels in any kind of meaningful way because contaminating noises can be observed and, hopefully, factored out. Long-term monitoring is effectively limited to the outdoors for the fundamental reason that there is no way to ascertain the background sound level inside of a dwelling at a particular time with the project operating. This is because the background sound level indoors is driven by a unique set of seemingly minor but significant sound sources that cannot be replicated by a proxy measurement position. Indoor background sound levels are partially a function of the outdoor conditions, particularly when it is windy or raining, but are also driven by such things as air flow from the heating and air conditioning system, appliances, computers and, of course, human activity even when it is in a distant part of the house. These usually very minor sounds are significant because the intruding noise level from the project is often very low or extremely low in terms of the A-weighted sound level. For example, it would not be unusual for a project sound level to be in the vicinity of 30 dBA inside of the house (perhaps being in the 40 to 45 dBA range outdoors). The successful measurement of the project-only sound level would then require the indoor background level to be 20 dBA or less, which is usually not the case. Sound levels in a bedroom at night are commonly at least 30 dBA even when no wind project is present.

In any event, it is sound level outside of dwellings that is normally (but not always) restricted by regulations or permit conditions and this level can typically be measured with the long-term monitoring methodology described above.

Sound Power Level Testing

The only existing standard for testing wind turbine sound levels is IEC 61400-11 *Wind turbine generator systems – Part 11: Acoustic noise measurement techniques* [5], which is only intended to quantify the sound power level of an isolated unit measured under rigorously controlled conditions so that the sound emissions of diverse turbines can be compared on a consistent basis and to provide input data for modeling analyses. This type of field testing is utterly different from the evaluation of site-wide sound levels for compliance with permit conditions or the investigation of complaints. This standard, contrary to the widespread misconception, cannot be used as a test protocol for compliance testing and, in fact, has no real application to the measurement of entire operational projects except where some of its concepts might be co-opted and adapted for use in diagnostic testing as discussed above in connection with Figure 22.

The general test procedure, described in full detail in the standard, is to measure the sound pressure level from a specific unit with a microphone laid flat on a ground plate at a position downwind that is one tower height and one-half of the rotor diameter away while the turbine is operating in wind speeds ranging from 6 to 10 m/s, as measured at 10 m above ground level. These sound pressure levels are then converted to sound power levels based on the hemispherical surface area of the wave front at the measurement location using the diagonal distance from the hub to the microphone as the radius. At least three one minute L_{Aeq} measurements must be obtained in each of the 5 wind speed bins (for example from 5.5 to 6.5 m/s for the 6 m/s integer wind speed) both with the unit operating and with it off – to get the background, which is then subtracted from the total measured level with the unit running to get the turbine-only sound level.

The results are expressed as the overall A-weighted sound power level at each of the 5 mandatory wind speeds along with the 1/3 octave band frequency content, ideally at each speed. Sound power levels at wind speeds outside of the mandatory range can be voluntarily reported and often are, particularly for the lower wind speeds.

The test also includes a rather complicated process for evaluating the potential tonal content of the sound spectrum based on narrowband FFT measurements. In general, most turbine models do not produce any significant tones but occasional tonality is sometimes indicated by this test, although when it does appear it is usually only found at one or two particular wind speeds.

This is the procedure in broad outlines for the current edition (Ed. 2.1) of the standard. A revision (Ed. 3.0) is currently being contemplated that would, among other things, expand the mandatory wind speed range to 3 to 14 m/s and require reporting in 0.5 m/s increments rather than at integer speeds [14].

The primary practical challenge inherent in the IEC 61400-11 procedure, at least in its current state, lies in being on hand and set up to measure in 6 to 10 m/s wind conditions, at least within a reasonable period of time. It is usually possible to immediately obtain data in the lower wind speeds but wind speeds in the 9 and 10 m/s region are often elusive. Of course, every wind project site is different and what may be commonplace at one is rare at another, but it is normally necessary to carefully plan the test for a time period when high winds are generally predicted.

Another challenge is maintaining a test set-up - both in terms of the sound measurement equipment and the required 10 m met mast - that is appropriate to the current wind direction. The sound measurement point must be kept within a fairly narrow angular range (+/- 15 deg.) of directly downwind and the 10 m anemometer must be kept within a certain range upwind of the turbine to avoid any possible downwind wake effects. The sound equipment can usually be easily moved but the met mast is quite difficult to re-position on short notice.

Although the 1 minute sound samples can certainly be taken manually, it is generally advantageous to set the instrument(s) to automatically record and store 1/3 octave spectra every minute. In this way, and provided no precipitation is expected, it is not necessary to be on hand continuously awaiting a change in wind conditions from low to high or vice versa that may take hours or days to occur. To further automate the process excellent results have been obtained by arranging with the control room to regularly shut down the test turbine for about 10 to 20 minutes every hour. This approach during a 10 or 12 hour period when high winds are expected, for instance, allows both on and off measurements

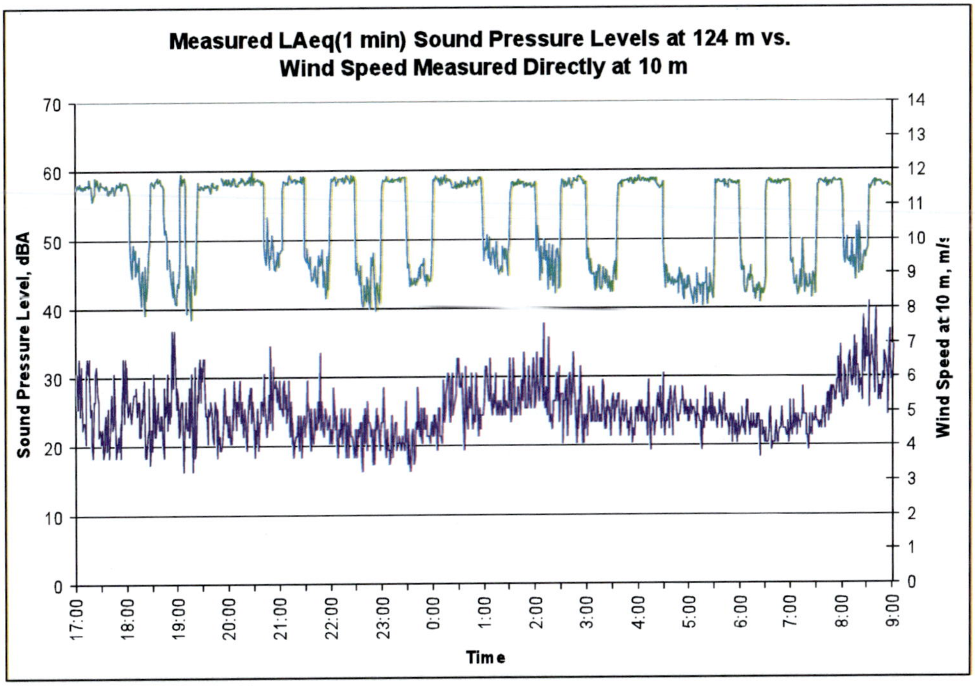

Figure 24. Monitoring using regular turbine shut-downs

to be made continuously as the wind speed changes over time. At least three 1 minute unit-on and three background sound level measurements need to be taken in all wind speed bins. With this continuous measurement approach a large number of samples are obtained from which representative samples can be selected or developed from regression trends. An example of the results from such a test is shown below for an overnight measurement period.

In this instance the turbine being studied had the unusual characteristic of producing a largely constant sound level not only during higher wind conditions, as would be expected, but also during low wind conditions when lower aerodynamic sound levels are normally observed. Background sound levels during this test generally remained 10 dB or more below the turbine sound level resulting in negligible corrections.

Continuous monitoring at the IEC test position can also capture other unusual behavior or unexpected trends, such as in the example below where the same turbine was measured during an overnight period on another occasion when a cold front with accompanying high winds was expected. The high winds arrived as anticipated between 10 and 11 p.m. at which point two neighboring turbines were shut down to minimize background noise and the official test was started. The turbine was put into a mode where it operated for 40 minutes and then automatically shutdown for 20 minutes to allow for the collection of background sound levels. At the end of the first shutdown period the temperature dropped to freezing and a very light mist began to build up as ice on the rotor blades. As can be seen in the figure, the turbine sound level steadily increased with the ice build-up until it became about 11 dBA higher than normal – at which point the unit tripped due to icing.

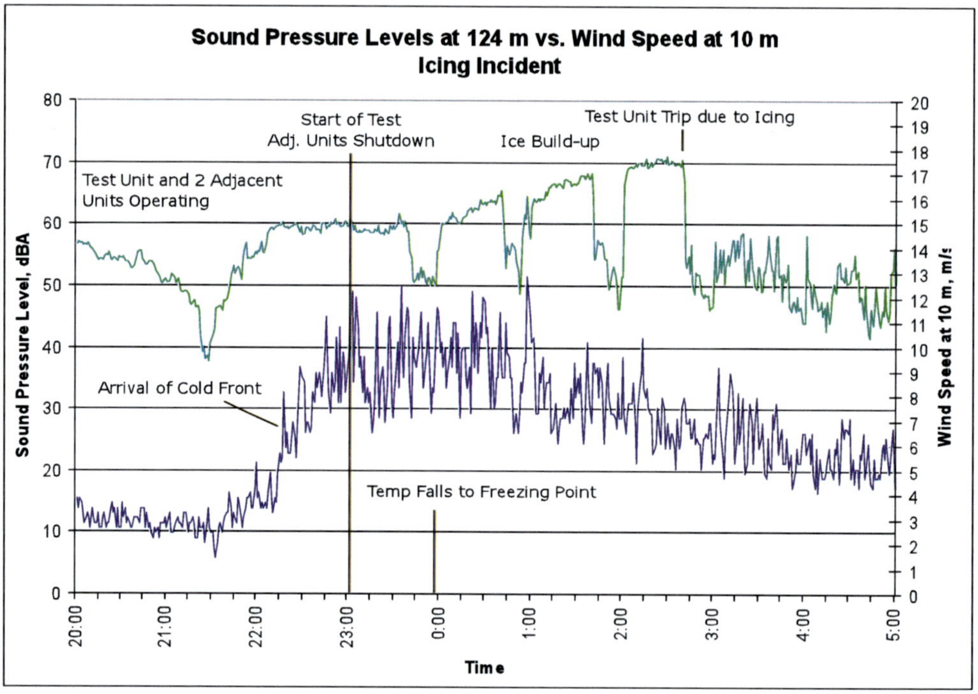

Figure 25. Example of an unusual event

REFERENCES

[1] *Gas turbines and gas turbine sets – Measurement of emitted airborne noise – Engineering/survey method,* ISO 10494:1993, International Standards Organization, Geneva, Switzerland, 1993.

[2] *Acoustics – Determination of sound power levels of noise sources using sound pressure – Survey method using an enveloping measurement surface over a reflecting plane,* ISO 3746:1995, International Standards Organization, Geneva, Switzerland, 1993.

[3] *American National Standard Quantities and Procedures for Description and Measurement of Environmental Sound. Part 3: Short-term measurements with an observer present,* ANSI S12.9-1992/Part 3 (R2008), Acoustical Society of America, New York, NY, 2008.

[4] *American National Standard Gas Turbine Installation Sound Emissions,* ANSI B133.9-1977 (R1989), America Society of Mechanical Engineers, New York, NY, 1989.

[5] *Wind turbine generator systems – Part 11: Acoustic noise measurement techniques*, IEC 61400-11 Ed. 2.1, International Electrotechnical Commission, Geneva, Switzerland, 2006.

[6. Bowdler, D., *ETSU-R-97 Why it is Wrong,* Internet White Paper, New Acoustics, Dunbartonshire, Scotland, July 2005.

[7. Department of Trade and Industry (UK), *The Assessment & Rating of Noise from Wind Farms,* ETSU-R-97, September 1996.

[8] *Acoustics – The Assessment and measurement of sound from wind turbine generators*, Standards New Zealand, NZS 6808:1998, Wellington, New Zealand, 1998.

[9] *American National Standard Quantities and Procedures for Description and Measurement of Environmental Sound. Part 2: Measurement of Long-Term, Wide-Area Sound,* ANSI S12.9-1992/Part 2 (R2008), Acoustical Society of America, New York, NY, 2008.

[10] *Electroacoustics – Sound Level Meters – Part 1: Specifications,* IEC 61672-1 Ed. 1.0, International Electrotechnical Commission, Geneva, Switzerland, 2002.

[11] *American National Standard Specifications for Integrating-Averaging Sound Level Meters, ANSI S1.43-1997 (R2007), Acoustical Society of America, New York, NY, 2007.*

[12] Hessler, G. F., Hessler, D. M., Brandstätt, P., Bay, K., "Experimental study to determine wind-induced noise and windscreen attenuation effects on microphone response for environmental wind turbine and other applications", *Noise Control Engineering Journal*, J.56, July-August 2008.

[13] *Acoustics – Attenuation of sound during propagation outdoors – Part 2: General method of calculation,* ISO 9613-2:1996, International Standards Organization, Geneva, Switzerland, 1996.

[14] Søndergaard, B., "The next version of the IEC 61400-11 measurement method", Third International Meeting on Wind Turbine Noise, Aalborg, Denmark, June 2009.

[15] Hessler, G. F., "Measuring ambient sound levels in quiet environments", Inter-Noise 2009, Ottawa, Canada, Aug. 2009.

[16] Schomer, P. D., Slauch, I. M., Hessler, G. F., "Proposed 'Ai'-weighting; a weighting to remove insect noise from A-weighted field measurements", Inter-Noise 2010, Lisbon, Portugal, Jun. 2010.

[17] Strasberg, M., "Non-acoustic noise interference in measurements of infrasonic ambient noise", *Journal of the Acoustical Society of America* **66**, pp. 1487-1483, 1979.

[18] Strasberg, M., "Dimensional analysis and windscreen noise", *Journal of the Acoustical Society of America* **83** (2), pp. 544-548, 1988.

[19] Van den Berg, G. P., "Wind induced noise is a screened microphone, *Journal of the Acoustical Society of America* **119** (2), pp. 824-833, 2006.

Chapter 8. Criteria

Mark Bastasch (USA), CH2M HILL.
Mark.Bastasch@CH2M.com

As with other sources of environmental sound, the standards established for wind energy facilities by planning and permitting authorities vary in both their approach and quantitatively. The goal of this chapter is to present an overview and examples of varying regulatory approaches, rather than a critical assessment of noise regulations. Because regulations may change over time, the reader is encouraged to confirm a given standard with the relevant authority. As noted below, several standards were recently updated in 2010 and it is important to ensure one understands their practical implementation which may evolve over time. Criteria may also dictate compliance assessment methods, including modeling and perhaps measurement methods. When such methods are stated, it is important to review them in tandem with the criteria to obtain a clearer understanding of each jurisdiction's requirement. It may not be appropriate to directly compare criteria across different jurisdictions as their associated assessment methods vary. Where clearly stated, these methods are highlighted in the discussion that follows, but the reader is encourage to conduct a more detailed, focused review of the jurisdictions of interest. The regulatory limits and approach established for wind energy facilities may or may not be consistent with how other sources of environmental noise are evaluated. When comparing regulations between jurisdictions, the consistency with how that jurisdiction regulates other environmental noise sources is an important aspect to consider, as is population density and renewable energy policy. For example, the impact imposed by a 2 km residential setback requirement in Australia would have a less dramatic affect on the potential wind energy generating capacity than in smaller more densely populated areas such as the Netherlands or Japan. Overall, there are two distinct approaches: (1) establishment of absolute limits, or (2) relative limits. These approaches may also be combined as described below.

ABSOLUTE LIMITS

Absolute limits establish a fixed numeric threshold that may not be exceeded. The numeric threshold is commonly presented in terms of overall A-weighted level, although other criteria (for example, octave band or one-third octave band) may also be established. Criteria may take the form of a project-specific limit that is, limiting the noise attributable to the specific project under consideration or establishing an overall "noise budget" that requires consideration of other sources of noise in the vicinity.

RELATIVE LIMITS

The fact that sound emissions from a wind energy facility depends on wind speed is a defining characteristic. In a rural setting, it is normal for the level of existing noise to increase with increasing wind speed. Relative limits, otherwise known as a background plus approach, attempt to take this into account and establish a threshold that is relative to the existing sound level, thereby regulating the change or increase resulting from a new noise source. When winds at a residence or receptor are calm and a lower level of background noise is documented, a lower overall project level would be required compared to times when a higher level of existing noise under more substantial wind conditions. As described elsewhere, the wind speed near ground level, at the receptor, may not always correlate with the wind speed at hub height. This differential is increasingly important to understand and consider when addressing relative limits given that turbine hub heights have increased from 40 to 100 meters in the past decade. This matter is discussed briefly in Appendix 2. For relative limits, the methods and metrics for determining existing or baseline noise levels must also be established. As with most natural systems, some degree of temporal and spatial variability is to be expected. When authorities rely on relative limits, establishing guidance on how such variability or scatter is to be addressed provides the necessary certainty for both the project and its neighbors.

COMBINATION OF ABSOLUTE AND RELATIVE LIMITS

A combination or hybrid of both approaches may also be established. Such approaches may establish an effective floor (lower bound or absolute minimum level) or ceiling (higher bound or absolute maximum level), regardless of the existing level of noise documented. For example, if an 8 dB relative limit is established and the existing level is documented to be 48 dB, a regulatory ceiling of 50 dBA would be the controlling limit rather than 56 dB (48 + 8) potentially allowed by a relative limit. Similarly, a regulatory floor may be established such that if 30 dBA was established as the existing level and an 8 dB relative threshold would require a 38 dB limit, a regulatory floor of 40 dBA would allow a higher level. When working with relative limits, the establishment of regulatory floors or ceilings assists in regulating the potential extremes.

While many regulations establish higher limits for daytime hours (or in one case nighttime hours), a project may choose to design to the most restrictive limit rather than restrict nighttime operations. Other topics mentioned in criteria documents, but not always regulated, include infrasound, low-frequency noise, tonality, and amplitude modulation.

EXAMPLES OF REGULATORY APPROACHES

What follows are brief examples of the varying regulatory approaches. As noted above, the reader is encouraged to confirm interpretation by reviewing permit conditions or consultation with relevant authorities.

Australia

The Australian Environment Protection and Heritage Council (EPHC) prepared the draft *National Wind Farm Development Guidelines* (2010), which provide complementary guidance to Australian Standard AS 4959 (2010), "Acoustics—Measurement, prediction and assessment of noise from wind turbine generators." The guidelines are intended to assist regulatory authorities in developing criteria suitable for their communities. The guidelines suggest the establishment of A-weighted relative limits based on the existing L_{90} levels with provisions for a floor or lower bound. The existing lower bounds vary among the jurisdictions summarized and vary between 35 dBA to 40 dBA. The metrics also vary and include L_{eq}, L_{10}, L_{90}, and L_{95}. The guidelines suggest that project participants or stakeholders may agree to higher noise levels though a noise floor of greater than 45 dBA is not advisable unless additional acoustical mitigation of living areas is implemented. A 5 dB penalty for tonality is suggested.

The guidelines do not address maximum noise levels. However, consideration of cumulative levels for all operating or proposed projects is suggested. It is acknowledged that doing so may render additional wind farm development in some areas impracticable. The guidelines also note the following: "Low frequency noise and infrasound levels generated by wind farms are normally at levels that are well below the uppermost levels required to cause any health effects. As a result, these Guidelines do not require specific assessment of low frequency noise and infrasound, but do present guidance on their assessment in response to complaints." In addition, "Ground-borne vibration from wind turbines is not likely to be perceptible to humans outside the wind farm site boundary and so has not been specifically addressed in these Guidelines."

In July, 2011 the Australian Senate Community Affairs Reference Committee concluded an inquiry and issued a report titled "The Social and Economic Impact of Rural Wind Farms." The committee recommendations included:

- that the noise standards adopted by the states and territories for the planning and operation of

rural wind farms should include appropriate measures to calculate the impact of low frequency noise and vibrations indoors at impacted dwellings
- that further consideration be given to the development of policy on separation criteria between residences and wind farm facilities

The Australian State of Victoria updated their policy and planning guidelines for wind energy facilities in August 2011. The updated guidelines require written consent from residences within 2 km of a proposed turbine and reference the 2010 update to New Zealand Standard 6808, Acoustics – Wind farm noise which is discussed below.

Canada

The Ontario Ministry of the Environment (MOE) guidelines for the assessment of sound from wind turbines establish relative limits with both a floor and ceiling (MOE, 2010). Rather than require monitoring to establish existing background levels, an assumed wind-induced background sound level is established. The standard also dictates the various modeling parameters to be used in ISO 9613. The standard essentially results in 40 dBA at residences in rural areas under moderate wind speeds (Figure 1).

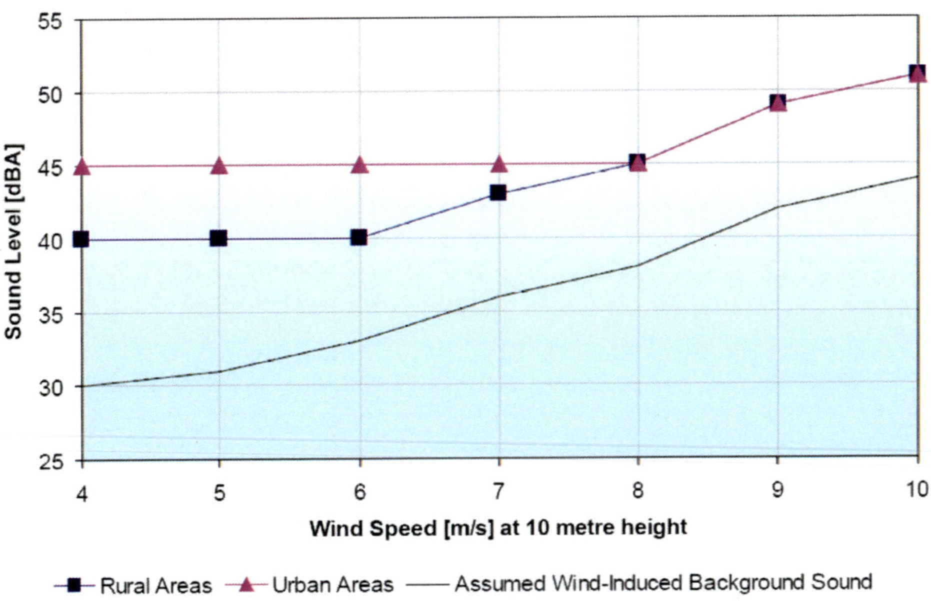

Figure 1. Summary of Sound Level Limits for Wind Turbines in Ontario, Canada
Source: MOE, 2010.

Czech Republic

The industrial guidelines established in the Czech Republic are applied to wind turbines and are reported to be 50 dB during the day (LA_{eq}, 8 hour) and 40 dB during the night (LA_{eq}, 1 hour). A 5 dBA penalty is assessed for tonality, which is determined by evaluating the Z-weighted 1/3 octave band levels between 10 and 160 Hz. The band with the potential tone must exceed the two adjacent bands by more than 5 dB, as well as the hearing threshold. Compliance is determined through a combination of IEC 61400-11 downwind measurements and project-level modeling based on ISO 9613 at the receptors. (Jiráska, 2010, 2011)

Denmark

Denmark has established absolute limits for wind speeds of 6 and 8 m/s referenced to a height of 10 meters per IEC61400-11. For outdoor living areas within 15 meters of a residence which is not owned by the owner of the turbine, the limit is 42 dBA at 6 m/s and 44 dBA at 8 m/s. For noise sensitive uses, the limits are 37 dBA at 6 m/s and 39 dBA at 8 m/s. Noise sensitive areas are defined in local planning documents and may include residential, summer homes or recreation areas. The regulations establish a prediction method based on geometric dispersion and octave band air absorption as well as measurement methods. A penalty of 5 dBA is imposed when pure tones are realized at the receptor. In November 2011 the Danish Minister of Environment announced that in response to request from industry, municipalities and citizens the limit for low frequency noise from wind turbines will be clarified. It is expected that the limit will be similar to those established for other industries in Denmark. The Ministry noted that "No evidence suggests that low frequency noise is more dangerous than other forms of noise" and that "The wind turbines we know in Denmark today emit very weak infrasound, which is below hearing threshold, even when in close proximity. Infrasound does therefore not pose a problem in regard to modern wind turbines." A final rule is expected by the end of 2011.

Germany

German wind facilities must comply with Technische Anleitung Lärm (TA Lärm, Technical Instruction Noise) which establishes A-weighted limits based on the receiving land use. The potential cumulative effect of multiple facilities is to be considered to ensure the limits in Table 1 are not exceeded. Additional penalties for tonal or impulsive noise may also apply.

Table 1. German noise limits

Receiving Land Use	**Daytime (dBA)**	**Nighttime (dBA)**
Villages and mixed Uses	60	45
Urban residential areas	55	40
Purely residential areas	50	35
Health care facilities	45	35

France

The French Neighborhood Noise Regulation of 2006 was not established to specifically regulate noise from wind turbines but is nonetheless the basis of obtaining a permit and determining compliance. These regulations establish a floor of 30 dBA outdoors and relative threshold, evaluated on an L_{eq} basis, of 5 dBA during the day and 3 dBA at night. In addition, interior levels are evaluated with a floor of 25 dBA (evaluated with the windows open or closed) and a relative octave band limit of 7 dB for octave band center frequencies between 125 to 250 Hz and 5 dB for octave band center frequencies between 500 and 4000 Hz (Bigot and Farotto, 2011; Gamba and Garrigues, 2011). Given the restrictive nature of the relative limits and challenges associated with conclusively assessing compliance, consideration is being given to establishing an alternative limit, closer to the wind turbine where the signal to noise level will allow more accurate measurements (Gamba and Garrigues, 2011). The limit would be established such that the resulting level at the receptor would comply with applicable limits (Gamba and Garrigues, 2011). When implemented, this may reflect the Czech approach discussed in Jiráska (2011).

Italy

Specific guidance for assessing wind projects compliance with applicable noise limits is underdevelopment (Artom et al, 2011). Currently, noise limits for wind turbines are established in a manner similar to other industrial sources. The absolute limits are based on the zoning or land use classification assigned by the municipal government. These classifications vary from protected areas to exclusively industrial areas (Table 2). When municipalities have not yet established an acoustical classification, nationwide classifications may be applied (Table 3) (Bartolazzi et al, 2011) though municipalities are expected to assign acoustical classifications rather than rely on the national system. In addition to the absolute limits, the regulations also establish a relative threshold of 5 dBA during the daytime and 3 dBA (Leq) during the nighttime inside the residence. Regulatory floors have been established for evaluating the relative interior criteria under both windows open and closed conditions. With the windows open, the interior level must exceed 50 dBA during the day or 40 dBA during the night for the relative test to be applied. Similarly with the windows closed, the interior level must exceed 35 dBA during the day or 25 dBA during the night for the relative test to be applied. For one project summarized by Ziliani (2009), the applicable absolute outdoor limit at residences was 70 dBA during the day and 60 dBA during the night. Bartolazzi et al (2009) report similar exterior limits at residences.

Table 2. Municipal land use compatibility guidelines

Land Use Classification	**Daytime**	**Nighttime**
I – Protected Areas	50	40
II – Residential Areas	55	45
III – Mixed Areas	60	50
IV – Intense Human Activity	65	55
V – Predominately Industrial Areas	70	60
VI- Industrial Areas	70	70

Table 3. Italy's national land use compatibility guidleines

Land Use Classification	**Daytime**	**Nighttime**
All National Territory	70	60
Zone A – Historical Urban Districts	65	55
Zone B – High Residential Areas	60	50
Industrial Areas	70	70

Netherlands

The regulations of wind turbine noise in the Netherlands were revised in 2010 (van den Berg, 2010). The previous regulations were similar to those discussed above for Ontario, Canada in that the residential noise limit was correlated with a 10-meter wind speed. The new Dutch limits are based on long term average sound power levels which are determined for each diurnal period (day, evening, night) from annual hub height wind speed statistics. The limits are expressed in terms of the preferred European Union metric, L_{den} (47 dBA) as well as L_{night} (41 dBA). As noted in van den Berg (2010) and references therein, the Dutch propagation model is similar to ISO 9613-2. In this method the propagation is calculated for downwind conditions, which are then adjusted to account for reductions afforded by periodic upwind conditions. The new requirements take into consideration the prevailing wind direction introducing a correction factor dependent on the angle under which the receiver sees the wind turbine. In addition, correction may be made for directivity effects of the wind turbine itself (if such data is available). These meteorological adjustments range from 2.5 (residence predominately downwind) to 7.5 dBA (residence predominately upwind). As compliance measurements of a wind facilities long term average L_{den} are acknowledged to be problematic, the proposed method is to evaluate operational records and calculate the actual annual sound power distribution from which the receiver sound pressure level is predicted. Measurements may be used to evaluate if the sound power of the turbine complies with the levels stated during permitting (i.e., verification of the turbine sound power levels).

The exterior L_{den} limit of 47 dBA was based on limiting the percentage of residents who are potentially highly annoyed indoors from wind turbine noise to 8%. The separate L_{night} limit of 41 dBA was established to address sleep disturbance. As discussed in van den Berg (2010) and references therein, a well developed dose-response relationship between wind turbine noise and sleep disturbance has not been established. The 41 dBA level was associated with 25% of respondents to a mail in survey noting that their sleep was disturbed at least once a month by sound, though the source of that sound was not identified. In setting the L_{night} limit of 41 dBA the World Health Organization's 2009 Night Noise Guidelines for Europe were taken into consideration. The residents evaluated in determining these thresholds were those not benefiting financially from the wind facility; those benefiting from the wind facility were found to be much less or not annoyed or disturbed during sleep.

New Zealand

In 2010, New Zealand updated its standard for wind farm noise, New Zealand Standard 6808, Acoustics – Wind farm noise. The standard states that the wind farm sound level should not exceed the greater of 40 dB $L_{A90(10min)}$ or the existing background ($L_{A90(10min)}$) by more than 5 dB at noise sensitive locations (including residences). It is noted that a relative limit is appropriate as "it would not provided an acoustic benefit to restrict the wind farm sound levels to a fixed value on occasions when background sound levels (caused by wind) are higher than this value". Additional restrictions may be imposed on high amenity areas during the evening and nighttime hours, but these limits should not be less than 35 dB $L_{A90(10min)}$. The noise limits should apply to the cumulative sound level of all wind farms affecting the noise sensitive location. If the predicted levels of proposed new wind farm are at least 10 dB less than the existing wind farm, no cumulative effect is considered. Where predictions are employed, predicted Leq sound levels are taken as the L90 sound level for the purposes of assessing compliance with the above criteria. Potential penalties for special audible characteristics (for example, tonality, impulsiveness and amplitude modulation) are provided for and the sum of all penalties shall not exceed 6 dB. It is noted that additional restrictions addressing infrasound, low frequency noise or ground-borne vibrations are not justified.

Wind farm noise levels are predicted with ISO 9613 utilizing turbine sound power levels determined in accordance with IEC 61400-11. The standard establishes procedures for measuring and evaluating background and project sound levels. Hub height wind speeds are to be used in the evaluation of both existing background and project noise levels.

South Korea

South Korea has established the guidelines shown in Table 4 for industrial sources, which may also be used with wind turbines (Lee, 2010)

Table 4. South Korea guidelines for industrial sources

Area	Daytime (dBA)	Nighttime (dBA)
Industrial area	70	60
Mixed residential and industrial area	65	55
Purely residential areas	55	45
Areas with hospital, health resorts, etc.	50	40

Sweden

The Swedish Environmental Protection Agency - Naturvårdsverket (SEPA) provides guidelines on noise from wind turbines. On the SEPA website it is stated that the noise level from a wind power plant should not exceed an L_{eq} of 40 dBA at a wind speed of 8 m/s at 10 meters height. A lower level of 35 dBA may be established in areas where quiet exists and is recognized as an important value in the applicable master planning documents[1,2]. A tonal penalty of 5 dBA is established if tonal noise is determined to be present at the receiver. Two A-weighted propagation models are established for on-shore facilities, one for when the distance to nearest turbine less than 1000 m another for distances greater than 1000 m[3].

United Kingdom

The commonly referenced noise requirements for wind turbines in the United Kingdom is the Department of Trade and Industry's, *The Assessment and Rating of Noise from Wind Farms* and generally referred to by its publication identification: ETSU-R-97. The methodology described in this document was developed in 1996 by a working group comprised of a cross section of interested persons including, environmental health officers, wind farm operators and independent acoustic experts. The stated goal of ETSU-R-97, in arriving at the recommendations was to provide:

> *"Indicative noise levels thought to offer a reasonable degree of protection to wind farm neighbours, without placing unreasonable restrictions on wind farm development or adding unduly to the costs and administrative burdens on wind farm developers or local authorities."*

ETSU-R97 states that different limits should be applied during quiet waking and night-time hours. The quiet waking hour's limits are intended to preserve outdoor amenity, while the night-time limits are intended to prevent sleep disturbance. The suggested limits are summarized in Table 5 below and are a function of wind speed. During quiet waking hours where low background noise levels are present, a noise floor of 35 - 40 dBA is established. The exact value is dependent upon a number of factors including the number of nearby dwellings, the effect of the noise limits on energy produced and the duration and level of the noise exposure.

Table 5. Permissible noise level criteria

Time of day	permissible noise level
Quiet waking hours	• 35-40 dB(A) for $L_{A90,10min}$ less than 30-35 dB(A) • $L_{A90,10min}$ + 5 dB(A), for $L_{A90,10min}$ greater than 30-35 dB(A)
Night-time hours	• 43 dB(A) for $L_{A90,10min}$ less than 38 dB(A) • $L_{A90,10min}$ + 5 dB(A), for $L_{A90,10min}$ greater than 38 dB(A)

It is interesting to note that a higher noise level is permissible during night-time hours than during the daytime or "quiet waking hours". The night-time criterion is derived from the 35 dBA sleep disturbance criterion referred to in ETSU-R-97, with an allowance of 10 dBA for attenuation through an open window and a correction of 2 dBA to allow for the use of L_{A90}, rather than L_{Aeq}.

As indicated above, the requirements rely heavily on background noise measurements to evaluate the 5 dBA relative limit over the existing L_{90}. The guidelines also note that increased levels should be considered where the affected landowner has some financial involvement in the wind farm. Bowdler (2007) presents examples from regulatory filings and discusses both the challenges and variability in interpretations of the monitoring results.

---

[1] Villkor om ljud från vindkraft - http://www.naturvardsverket.se/sv/Start/Verksamheter-med-miljopaverkan/Buller/Vindkraft/Villkor-om-ljud-fran-vindkraft/

[2] Riktvärden för ljud från vindkraft - http://www.naturvardsverket.se/sv/Start/Verksamheter-med-miljopaverkan/Buller/Vindkraft/Riktvarden-for-ljud-fran-vindkraft/

[3] Mätning och beräkning av ljud från vindkraft - http://www.naturvardsverket.se/sv/Start/Verksamheter-med-miljopaverkan/Buller/Vindkraft/Matning-och-berakning-av-ljud-fran-vindkraft/

United States

The United States does not have a federal noise requirement for wind turbines and only a few states have industrial noise regulations. Therefore, the noise requirements of commercial-scale energy projects, including wind farms, may vary from one jurisdiction to another. The permitting process imposes noise requirements that vary from state to state. For example, large renewable and fossil-fueled projects in Oregon require a permit from the State and undergo a detailed environmental review by the State. Other states, like California and Washington, do not require a state permit for wind energy projects; the local jurisdiction (city or county) is the lead permitting agency. At least one state, Oregon, revised its noise requirements specifically to address noise from wind farms.

Bureau of Land Management Programmatic EIS

The U.S. Department of the Interior (DOI), Bureau of Land Management (BLM) is the federal agency charged with managing 264 million acres of public lands, located primarily in the 12 Western States. The BLM is responsible for the development of wind energy resources on BLM-administered lands. The BLM prepared a *Final Programmatic Environmental Impact Statement on Wind Energy Development on BLM-Administered Lands in the Western United States* (BLM, 2005). Key findings relevant to assessing noise impacts of a wind project are as follows:

- At many wind energy project sites on BLM-administered lands, large fluctuations in broadband noise are common, and even a 10-dB increase would be unlikely to cause an adverse community response.
- For a typical rural environment, background noise is expected to be approximately 40 dB(A) during the day and 30 dB(A) at night, or about 35 dB(A) as DNL.
- The EPA guideline recommends a day-night sound level (L_{dn}) of 55 dB(A) to protect the public from the effect of broadband environmental noise in typically quiet outdoor and residential areas (EPA, 1974). This level is not a regulatory goal but is "intentionally conservative to protect the most sensitive portion of the American population" with "an additional margin of safety."

While the above findings are not regulations, they indicate how BLM will assess the "significance" of noise impacts on individual projects and provide guidelines on how individual projects will need to address noise.

State of Washington

The State of Washington's general noise rules are applied to wind turbines. While local jurisdictions are allowed to develop independent, state-approved noise standards, it is primarily the urban communities such as Seattle that have done so. The State's noise limits are based on the environmental designation for noise abatement (EDNA), which is defined as "an area or zone (environment) within which maximum permissible noise levels are established." There are three EDNA designations that roughly correspond to residential, commercial/recreational, and industrial/agricultural uses.

Table 6. State of Washington Noise Regulations

Statistical Descriptor	Maximum Permissible Noise Levels (dBA) from a Class C EDNA		
	Class A EDNA Receiver		Class C EDNA Receiver
	Daytime (7 a.m. – 10 p.m.)	Nighttime (10 p.m. – 7 a.m.)	Anytime
Leq	60	50	70
L25	65	55	75
L8.3	70	60	80
L2.5	75	65	85

Source: Washington Administrative Code, Chapter 173-60.

Table 6 summarizes the maximum permissible levels applicable to noise received at noise-sensitive areas such as residences (Class A EDNA) and at industrial/agricultural areas (Class C EDNA) from a Class C EDNA.

State of Oregon

The Oregon noise regulations (Oregon Administrative Rule [OAR] Chapter 340 Division 35) contain two noise standards that are generally referred to as the "Table 8 test" and the "ambient degradation test" (other portions of the rules address octave, third-octave band and tonal limits). The "Table 8 test" refers to Table 8 of the rule (reproduced here as Table 7), which limits the maximum permissible statistical noise levels generated by a project. The "ambient degradation test" specifically limits the increase in the existing L_{10} or L_{50} to a maximum of 10 dBA.

Table 7. Oregon's Table 8 limits": Maximum permissible levels for new industrial and Commercial Noise Sources

Statistical Descriptor	Daytime (7 a.m. – 10 p.m.) (dBA)	Nighttime (10 p.m. – 7 a.m.) (dBA)
L_{50}	55	50
L_{10}	60	55
L_1	75	60

Source: Oregon Administrative Rule 340-35-035.

The rule allows an applicant to address the ambient degradation standard by accepting a minimum background L50 of 26 dBA, resulting in a noise floor of 36 dBA. The evaluation of these criteria is generally conducted with all turbines operating at their maximum warranted sound level consistent with the request by the State agency overseeing the permitting of large wind energy projects. The rule also exempts properties that have entered into a noise easement from the ambient degradation standard. For residences that have entered into a noise easement, the 50 dBA "Table 8" limit becomes the controlling limit.

State of Colorado

The Colorado noise regulations do not specifically account for wind turbines. The regulations note that noise radiating from a property line in excess of 50 dBA at night in a residential area constitutes a public nuisance. In addition, activities must be conducted to ensure noise is "not objectionable due to intermittence, beat frequency or shrillness" and the regulations contain a 5 dBA penalty for "periodic, impulsive or shrill" noises. None of these qualitative terms are defined. Compliance measurements are required to be made when the wind speed does not exceed 5 miles per hour (2.2 meters per second). More than 340 local jurisdictions within Colorado may have additional noise requirements.

State of Illinois

The Illinois Pollution Control Board (IPCB) noise regulations for industrial sources have been applied to wind turbines, although local jurisdictions may also develop their own standards. The IPCB regulations are specified in terms of receiving and source land use categories. The land use classifications generally correlate to residential (Class A), commercial/retail (Class B), and agricultural/industrial (Class C) land uses. All residences, even rural residences, are considered Class A. The most restrictive noise standard is generally determined by evaluating predicted levels at the residence against the criteria established for a Class C emitting source and a Class A receiving source as summarized in Table 8. Note that limits are established solely in octave bands and for wind turbine projects, the 500- or 1,000-Hz octave band tends to be the controlling limit.

Table 8. Illinois noise regulation – sound pressure levels emitted to Class A (Residential) from Class C (Industrial) (dB unweighted)

Octave Band Center Frequency (OBCF - Hz)	31.5	63	125	250	500	1000	2000	4000	8000
Daytime Limit (Leq) (7:00 am to 10:00 pm)	75	74	69	64	58	52	47	43	40
Nighttime Limit (Leq) (10:00 pm to 7:00 am)	69	67	62	54	47	41	36	32	32

State of Maine

The State of Maine Department of Environmental Protection noise regulations for "routine operation of developments" have been applied to wind turbines, although local jurisdictions may also develop their own standards provided they do not exceed the State regulations by more than 5 dBA. When projects are proposed in areas where the pre-development hourly L_{eq} is less than 45 dBA during the day or 35 dBA during the night, the resulting limit at protected residential locations is an hourly L_{eq} of 55 dBA during the day and 45 dBA during the night. Where existing levels are found to exceed the 45 dBA/35 dBA (day/night) the resulting limit is 60 dBA during the day and 50 dBA during the night. Penalties are assessed for both tonal and short duration repetitive sounds.

In September 2011, the Maine Board of Environmental Protection voted to impose more stringent noise regulations on commercial wind turbines operating near homes or businesses. This proposal included reducing the nighttime sound limit from 45 dBA to 42 dBA, provided additional detail on how compliance would be assessed and clarified how "short duration repetitive sounds" would be evaluated. As of October 2011, the State legislature had not adopted the Board's proposal.

ACKNOWLEDGEMENTS

The author would like to thank Philip King, Bo Søndergaard, Frits van den Berg, Geoff Leventhall, Ales Jiraska, Seunghoon Lee, Rene Gamba, Roberto Ziliani, Andrea Bartolazzi and Carlo Dinapoli for providing information on regulatory approaches.

REFERENCES

Artom, F., Betti, R., Carmagnini, A., Lenzuni, P., Lodi Rizzini, L., Marchisio, A., Marisco, G., Zilani, R. (2011), *Towards a national standard to support the assessment of the noise impact of wind farms*. Proceedings of the Fourth International Meeting on Wind Turbine Noise. Rome, Italy. April 12-14.

Australian Environment Protection and Heritage Council (EPHC). (2010), *Draft National Wind Farm Development Guidelines*. July 2010.

Australian Senate. Community Affairs References Committee. *The Social and Economic Impact of Rural Wind Farms*. ISBN 978-1-74229-462-9. June 2011

Bartolazzi, A. and Mariani, M. (2009), *Optimization of energy production of a large windfarm with noise constraints: a numerical toolkit*. Proceedings of the Third International Meeting on Wind Turbine Noise. Aalborg, Denmark. June 17-19.

Bartolazzi, A. and Mariani, M. (2011), *Italian regulations on wind farm noise: A state of the art*. Proceedings of the Fourth International Meeting on Wind Turbine Noise. Rome, Italy. April 12-14.

Bowdler. 2007. *What is the real background noise*? Proceedings of the Second International Meeting on Wind Turbine Noise. Lyon, France. September 20-21.

Bureau of Land Management. (2005), *Final Programmatic Environmental Impact Statement on Wind Energy Development on BLM-Administered Lands in the Western United States*. U.S. Department of Interior. Available online at http://windeis.anl.gov/eis/index.cfm. June 2005.

ETSU. ETSU-R-97. (1996), T*he Assessment and Rating of Noise from Wind Farms*. September.

Jiráska, Ales. (2010), Personal Communication.

Jiráska, Ales. (2011), *Sound power level assessment. Is a reference position really reference?* Proceedings of the Fourth International Meeting on Wind Turbine Noise. Rome, Italy. Apirl 12-14.

Lee, Seunghoon. (2010), Personal Communication.

New Zealand Standard. (2010), *Acoustics – Wind farm noise*. NZS 6808:2010. Available from www.standards.co.nz.

Ontario Ministry of the Environment (MOE). (2008), *Noise Guidelines for Wind Farms – Interpretation for Applying MOE NPC Publications To Wind Power Generation Facilities*. October 2008. http://www.ene.gov.on.ca/environment/en/resources/STD01_078286.html.

U.S. Environmental Protection Agency (EPA). (1974), *Information on Levels of Environmental Noise Requisite to Protect Public Health and Welfare with an Adequate Margin of Safety*. EPA Report 550/9-74-004, Washington, DC. March.

van den Berg, G.P., (2010), *Rating of wind turbine noise using Lden*. Proceedings of Internoise 2010. Libson, Portugal. June 13-16.

Ziliani, R. (2009), *Wind farm noise measurements and residual noise estimation by modeling*. Proceedings of the Third International Meeting on Wind Turbine Noise. Aalborg, Denmark. June 17-19.

Appendix 1. Detection of Sound Sources Using Phased Microphone Arrays

This Appendix describes how sound sources can be localized and quantified using an array of microphones. In order to provide an intuitive understanding of the technique, we will first give a qualitative description of delay-and-sum beamforming in the time domain. Then, we will see how the different sound sources in a source map can be quantified using a power integration method. Finally, some example applications will be presented, to illustrate the capabilities of acoustic arrays and to help interpretation of the wind turbine noise source maps in Chapter 2.

BEAMFORMING

Consider a compact sound source in a homogeneous stagnant medium with sound speed . The sound field is measured by an array of microphones (Figure 1). For simplicity a linear array is shown here, but the explanation is equally valid for a two-dimensional planar array. Due to the difference in distance r, the sound will arrive slightly earlier at microphone m_1 than at m_2. Furthermore, since the pressure amplitude decreases with distance, the amplitude of the acoustic pressure p will be lower on m_2 than on m_1. Now suppose we know the source is located in a specific plane (e.g. the wing of an aircraft or the rotor of a wind turbine), but the position *within* this plane is unknown. Then the source location can be determined on the basis of the measured signals as follows: first, a scan grid is defined in the plane where the source is expected. For each scan position, the microphone signals are shifted to account for the time the sound needs to travel from the scan position to the microphone. In addition, the amplitudes are adjusted to account for the distance between the scan position and the microphone. Next, the time-shifted, amplitude-adjusted signals from all microphones are summed. This procedure is called delay-and-sum.

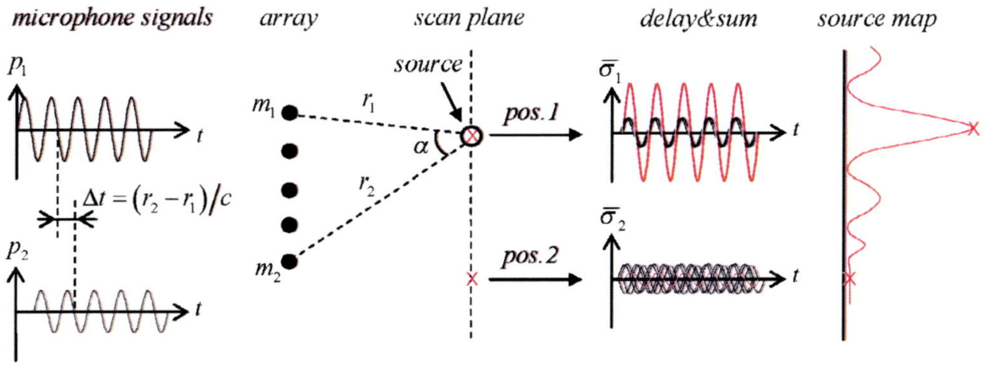

Figure 1. Delay-and-sum beamforming.

If a sound source is present at a given scan position (position 1 in Figure 1), the shifted microphone signals will be in phase and their sum will have a high amplitude. If no source is present (position 2), the shifted signals will have random phase and the amplitude of their sum will be small. Thus, the phased array essentially amplifies the sound from the scan position (or focal point) with respect to sound from other directions. By plotting the (squared) amplitude as a function of scan position, a so-called *source map* is obtained. This source map shows peaks at the source locations. The strength of a source can be determined by dividing the reconstructed peak amplitude in the source map by the number of microphones. In the case of a planar array a two-dimensional source map is obtained and the amplitudes can be indicated by colours (Figure 2).

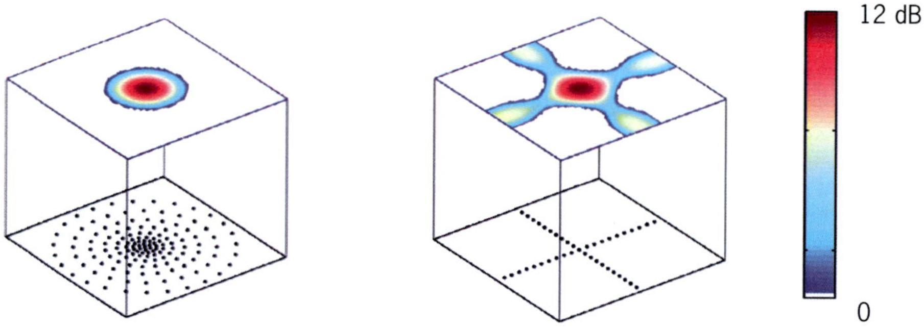

Figure 2. Simulated acoustic source maps of a 2 kHz point source for a random array (left) and a cross array (right). The size of the cube is 1 m^3.

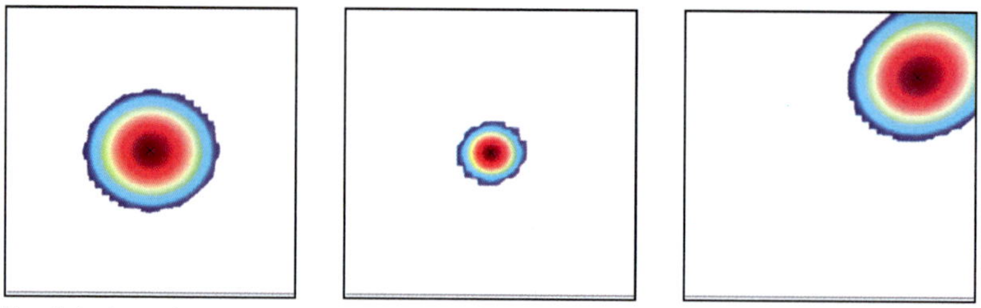

Figure 3. Simulated random-array source maps for 2 kHz (left), 4 kHz (middle), and for an oblique 2-kHz source (right).

The delay-and-sum technique illustrates several aspects of the beamforming methods used in this thesis. First, the main peak in the source map has a certain width, which determines the *resolution* of the array. In order to distinguish two different sources, their distance should be larger than the peak width at 3 dB below the peak level: the narrower the peak, the higher the resolution. The non-zero peak width can be understood if we consider a scan position just next to the actual source position. The time-shifted microphone signals for this position will still be partly in phase, resulting in a non-zero reconstructed amplitude. For a given scan position next to the actual source, the phase error between the shifted signals will decrease with increasing wavelength, resulting in a higher reconstructed amplitude. Thus, the resolution of the array decreases with increasing wavelength, and therefore increases with frequency f (Figure 3). Similarly, it can be deduced that the resolution increases with the opening angle α of the array, so that the peak width is roughly proportional to r/Df, where D is the size of the array. Note that the opening angle, and thus the effective array size, decreases when the source is not in front of the array. Thus, the array resolution decreases for oblique source directions (Figure 3).

Besides the main peak (or main lobe), the source map may also show *side lobes*, local maxima around the source position. These spurious sources are an inevitable consequence of the finite number of microphones in the array, and are difficult to distinguish from real secondary sources. Therefore, an important requirement of the array design is to keep the side lobe levels as low as possible. This can be achieved by the use of 'random arrays', in which the periodicity in microphone spacing is reduced (Figure 2).

In principle, multiple parallel scan planes (or any three-dimensional grid) can be used to estimate the distance to an unknown source. However, it should be noted that for a planar array the depth resolution (i.e., the resolution in the direction perpendicular to the array surface) is much lower than the lateral resolution. As a result, the three-dimensional main lobe of a point source has the shape of a cigar, with its long axis pointing to the array. In many cases this low depth resolution is actually an advantage, since the array results are rather insensitive to small deviations in distance (e.g. for curved wings or rotor blades).

Note that in the above explanation of the delay-and-sum technique no assumptions were made regarding the nature of the source. The method only requires that the pressure disturbance travels with the speed of sound, and that the pressure amplitude is inversely proportional to the distance. This is true for monopole-type sources in the free field, and in the acoustic far field also for dipoles and quadrupoles. If the source has a non-uniform directivity, the reconstructed source strength will be the average value over the solid angle covered by the array. Another requirement for beamforming is that the source radiates sound coherently in the direction of the array. The above explanation also shows that delay-and-sum beamforming is *not* an inverse method in the sense that a source distribution is 'sought' which matches the incoming sound field. The method simply performs an operation on the measured signals and involves no fitting or iterative procedures.

Delay-and-sum beamforming with a phased array is similar to source detection using parabolic or elliptic antennas. An elliptic mirror focuses the sound rays radiated from a potential source location to its focal point, where a microphone is placed. However, phased arrays have a number of important advantages. Whereas antennas must be physically steered to each scan position, so that many time-consuming measurements are needed to obtain a complete source map, a phased array only needs one measurement, and the scanning is done afterwards. The electronic processing also provides the flexibility to include e.g. flow effects, moving sources, or spatial weighting of the microphone signals, whereas for an antenna the time delays and weighting are fixed because they are imposed by the geometry. Furthermore, phased array processing can be performed in the frequency domain, which allows the use of several special data analysis techniques.

QUANTIFICATION OF ARRAY RESULTS

In general we are not only interested in the source *locations*, but also in their *levels*. For well-resolved monopole sources the source level simply corresponds to the peak level in the acoustic source map. However, in practice this situation seldom occurs. Firstly, the main lobes from different sources may overlap or the source may be spatially extended (e.g. trailing edge noise). As a consequence, the peak levels in the source map depend on the extent of the source and on the (frequency-dependent) resolution of the array. Secondly, the array signals may suffer from coherence loss, which typically results in broader main lobes with a reduced level. Thirdly, the levels in the source map may be influenced by side lobes from other sources. To overcome these complications, a *power integration method* can be used. This technique sums the source powers in (part of) the measured source map, and corrects the result with a scaling factor obtained by performing a simulation for a monopole source at the centre of the integration region. The method is schematically explained in Figure 4. An integration contour is defined around the distributed source in the measured source map, and the integrated level is corrected using a (frequency-dependent) calibration function which is determined from a monopole simulation. The integration technique has been applied in many wind tunnel and field tests, and generally yields accurate relative source levels (i.e., differences between configurations). For certain conditions the *absolute* levels may be too low due to, for example, coherence loss effects.

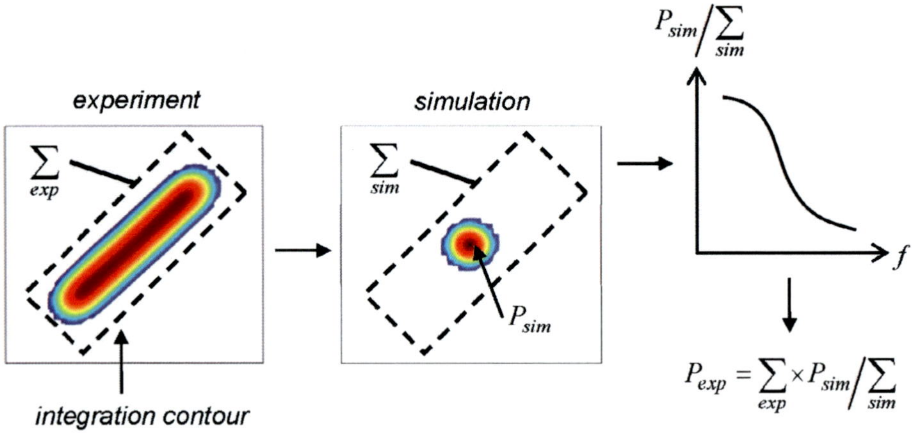

Figure 4. Source power integration method.

EXAMPLE APPLICATIONS

In this section some examples of acoustic array applications will be presented, to illustrate the capabilities and to help interpretation of the wind turbine noise source maps in Chapter 2. We will see several features that are also applicable to the wind turbine noise measurements, including (1) source localization for stationary and moving sources, (2) quantification of array results using the power integration method, and (3) application of the array technique in the wind tunnel and in the field.

Our first example is a (stationary) 1:10 scaled aircraft model in the 8x6 m^2 closed test section of the DNW-LLF wind tunnel (Figure 5). A 1-m diameter acoustic array was mounted flush in the wind tunnel floor. The resulting source maps show that for this particular configuration practically all wing noise is radiated from the leading edge slats.

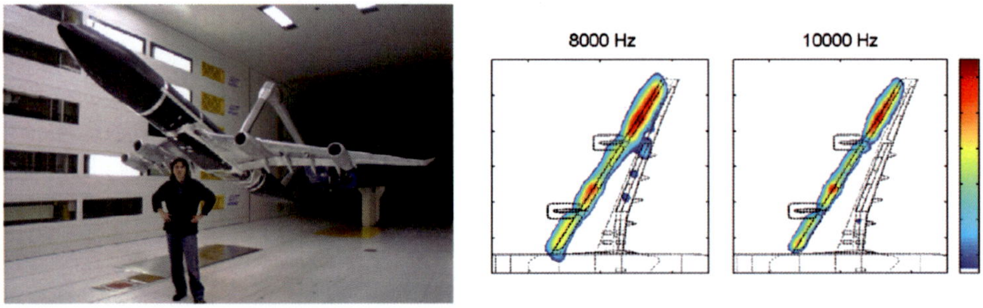

Figure 5. Aircraft model in DNW-LLF wind tunnel and corresponding source maps at model scale frequencies of 8 kHz and 10 kHz.

Our second example is a helicopter model with main rotor and tail rotor in the open jet of the same DNW-LLF wind tunnel (Figure 6). In this case a 4x4 m^2 out-of-flow microphone array was used. Even though main and tail rotor blades are rotating, we can apply a stationary scan plane which coincides with the rotor plane. The resulting source maps (Figure 7) show that for this flight condition sound is radiated from both main and tail rotor, but that their relative importance depends on frequency. This is illustrated in the integrated spectra for both source areas, which confirm that the tail rotor is dominant at low frequencies, and the main rotor at high frequencies.

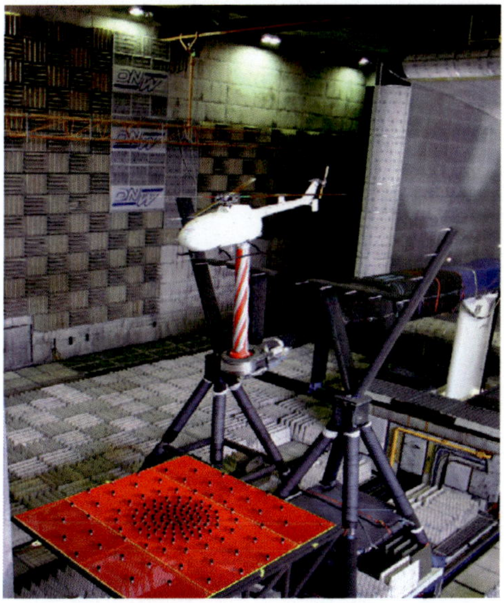

Figure 6. Helicopter model in DNW-LLF wind tunnel, with (red) out-of-flow microphone array.

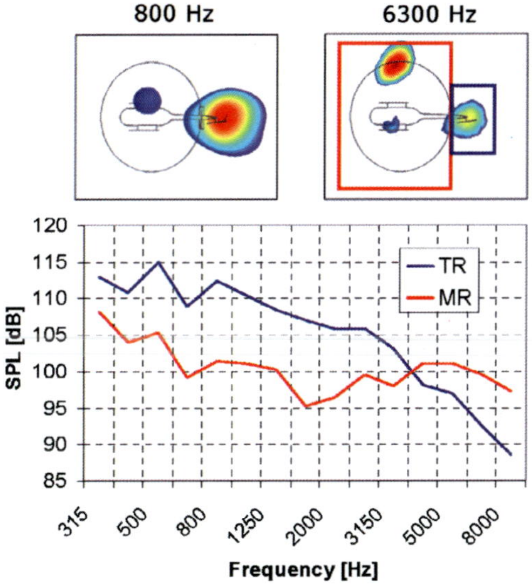

Figure 7. Source maps and integrated spectra for main rotor and tail rotor (level flight). The integration contours are shown in the 6.3 kHz source map.

Figure 8 shows source maps for the same helicopter model, but now for the descent flight condition. The stationary source maps show that in this case most of the noise is generated by the main rotor, when the blades are on the retreating side. Now, using the same measurement data, we can also apply a scan plane which rotates along with the main rotor. The resulting 'rotating' source maps clearly show four equal spots, corresponding to the rotor blades. As expected, the four blades are equally noisy.

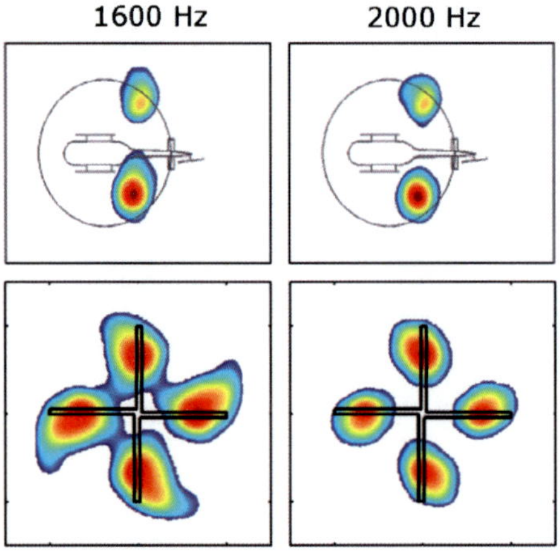

Figure 8. Stationary source maps (top) and corresponding rotating source maps (bottom) for descent flight.

A final example is shown in Figure 9. This is a result of acoustic field measurements on landing aircraft at Amsterdam Airport. A 12-m diameter acoustic array, consisting of 243 microphones, was used. By employing a scan plane moving along with the landing aircraft, the different noise sources were localized. It can be seen that besides the engines, several airframe noise sources are present, including the slats, flaps, and landing gears.

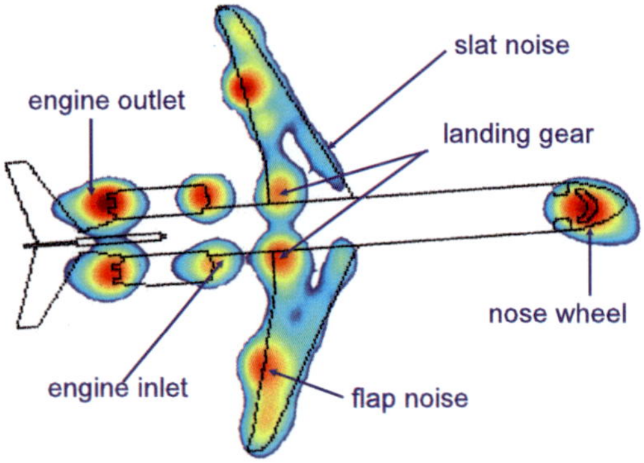

Figure 9. Acoustic source map for landing aircraft.

Appendix 2. Background Noise – Wind Shear and Shelter

Where noise standards for a development are based on the pre-existing background noise levels (the relative method described in Chapter 8) the normal procedure is to measure the wind speed on the development site and the noise level as L_{A90} at residential receptors. The aim of the procedure is to compare calculated turbine noise with pre-existing background noise in order to assess the impact that the turbines will have locally. David Hessler has dealt with measurement techniques and in particular compliance techniques in detail in Chapter 7. This appendix looks briefly at the factors that can influence the presentation of background noise at application stage.

A planning application assessed under any test that relies on background noise measurements may stand or fall on one or two decibels difference in the background noise. So it is important both for developers (so that they can properly plan the development) and for residents (so that they can have adequate protection from noise) that the measurements are not only accurate but that they truly represent the "real" conditions at each location. What we are trying to do is to create a graph that shows turbine noise and background noise both plotted against wind speed so that we can compare the two. To do this, background noise and wind speed are measured and synchronised, usually in 10 minute periods, and plotted against each other on a scatter graph.

The graph in Figure 1 below shows a typical plot of background noise levels. Each marker is a ten minute reading of background noise plotted against a ten minute reading of wind speed measured at 10m height taken at the same time. The black line is a third order best fit polynomial with the characteristic flat S-shape. The curve can be considered as being made up of two separate noise sources each consisting of a set of data. The first is a horizontal line shown in blue consisting of noise that is not wind related (the NWR element). It may be inherent meter noise, streams, road traffic or any other continuous or varying noise other than that associated with the wind. Because it is not wind related the best fit curve for this first set of data must be horizontal. The second noise source, shown in red, produces a set of data of noise of wind in trees and so on that is related to wind speed in a logarithmic relationship (the WG element).

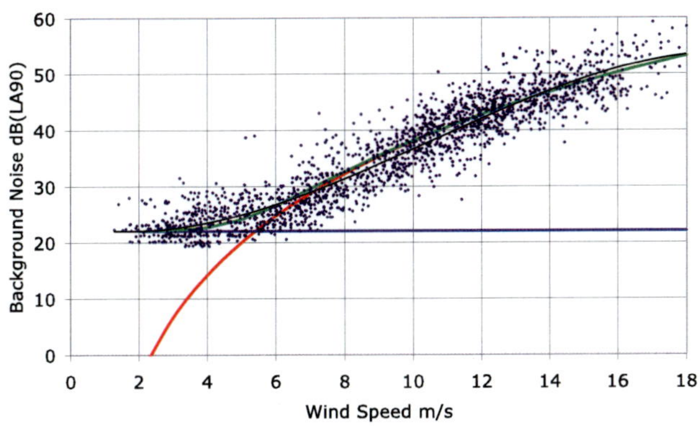

Figure 1. Structure of background noise curve

The figure shows how these two elements go together to produce the typical best fit curve. Because of the way the curve is made up a third order polynomial is the preferred trend line.

The WG element can be described by:

$$L_{WG} = A \times \log(V) + C \tag{1}$$

Where L_{WG} is the 10 minute measured sound pressure level in dB(L_{A90}), V is the wind speed at the noise measurement location and A and C are constants.

The NWR element can be described by:

$$L_N = D \tag{2}$$

Where L_N is the 10 minute measured sound pressure level in dB(L_{A90}) and D is a constant.

When turbine sound power levels are measured according to IEC 61400-11 wind speed is measured at hub height but is reported at 10m high using a standard wind shear equal to a roughness length of 0.05m. So it is a derived 10m wind speed rather than a measured one. It has been common practice in some countries to plot background noise against the measured 10m wind speed rather than the derived 10m wind speed but, as van den Berg pointed out in 2003[1] increasing wind shear results in increasing sound power level output of the turbine for the same 10m wind speed – and so the sound level may be higher than expected.

So if we plot a set of typical background noise data together with the best fit curve we get something like Figure 2 below.

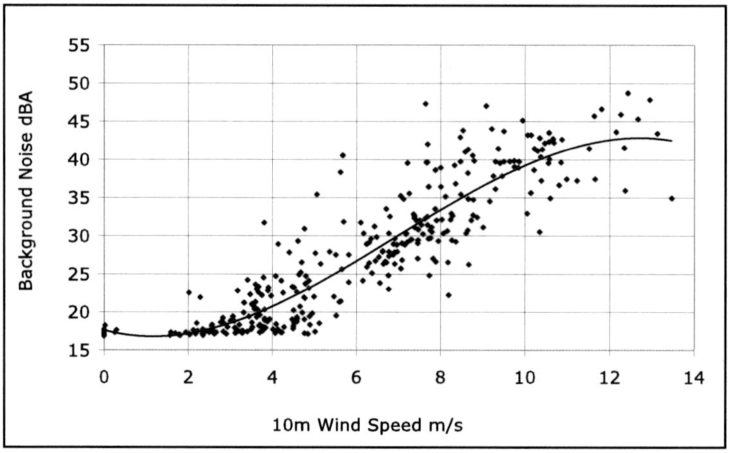

Figure 2. Background noise data with best-fit curve

The next step in the assessment process is to calculate the turbine noise at each wind speed and superimpose it on the background noise graph. If we simply use the manufacturers wind turbine sound power levels then it is implicit that we are using the standard shear with a roughness length of 0.05m. The result is shown in Figure 3 below.

However, if there is excess wind shear – that is to say shear with a roughness length of more than 0.05m – then the turbine noise may be higher than expected at any particular wind speed. What happens is that the turbine noise level curve shifts to the left as shown by the blue curve in Figure 4 below. The effect is to increase the margin of turbine noise over background noise and hence increased wind shear produces noise that has a greater impact on residents.

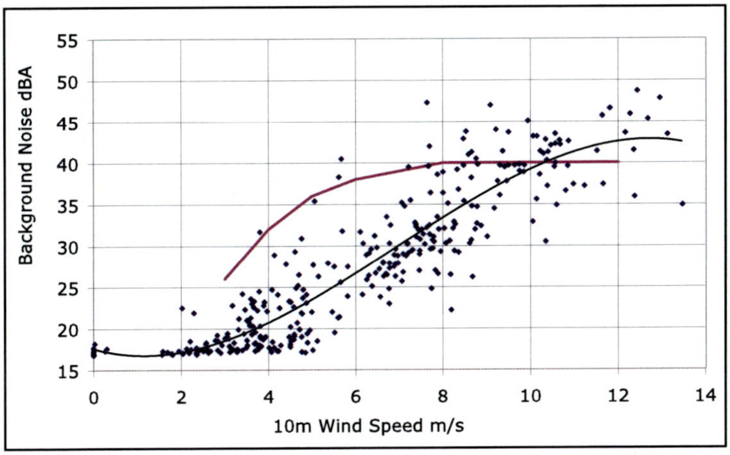

Figure 3. Background noise with turbine noise

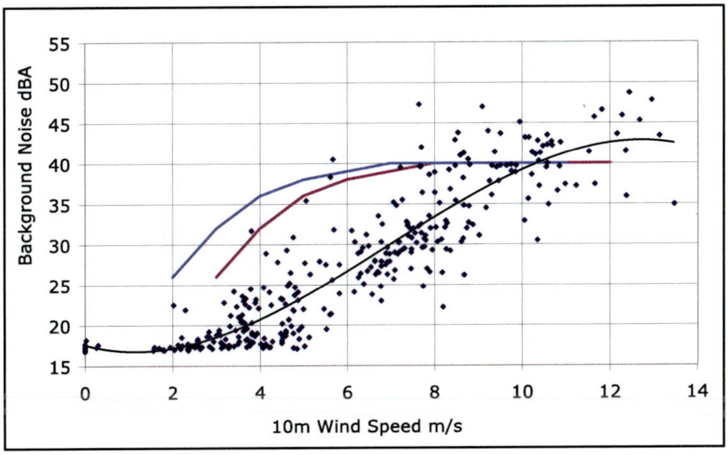

Figure 4. Turbine noise shifted in high wind shear

If, on the other hand, instead of using the measured 10m wind speed for background noise we use the derived 10m wind speed as is used to define turbine sound power level in IEC 61400-11 then (hub height reduced to 10m using a roughness length of 0.05m), in high wind shear situations, instead of the turbine noise level shifting to the left, the turbine noise curve would stay the same but the background noise curve would shift to the right. The result – in terms of the relation between the turbine noise curve and the background noise curve would be the same.

In fact, wind shear does not behave as simply as that. In the Figure 5 we can see night time wind shear plotted against wind speed measured at 10m. In this case the wind shear is shown as the wind shear exponent. A value of the wind shear exponent of about 0.16 is the same as a roughness length of 0.05.

What can be seen here is that the wind shear decreases with increasing wind speed. At a 10m wind speed of 2m/s the average wind shear is 0.4 but this reduces with increasing wind speed to almost the "standard" value of 0.16 (or 0.05 roughness length).

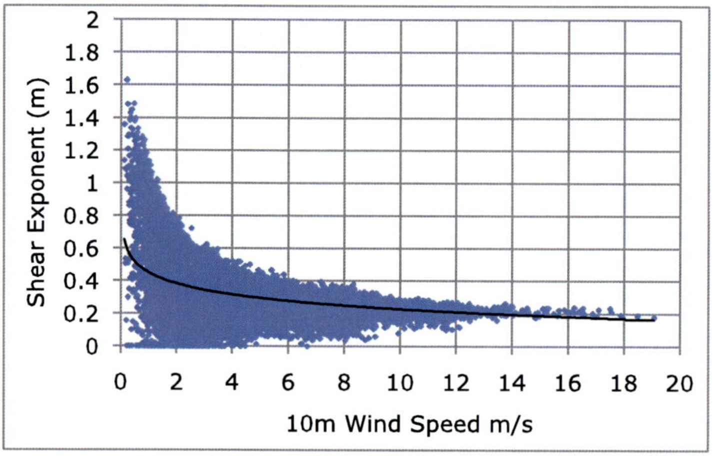

Figure 5. Wind shear reduces with wind speed

So, if we plot the background noise levels against the derived wind speed instead of the measured wind speed, the curve does not just shift to the left but rotates about a point at some high wind speed at which the average wind shear has a roughness values of 0.05. This can be seen in Figure 6.

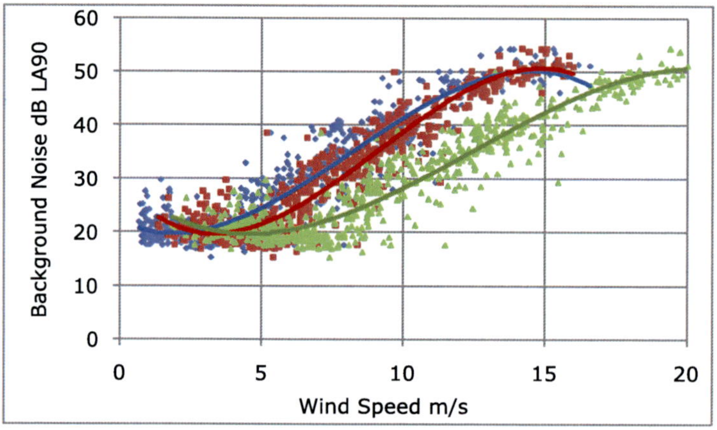

Figure 6. Background noise plotted against wind speed at three different heights

One set of background noise data is plotted here against three different wind speeds. The blue data is plotted against the measured 10m wind speed and the red data is plotted against the 10m wind data derived from hub height using a roughness length of 0.05m. The red line is the blue line shifted to the right at lower wind speeds showing that the wind shear is greater than a roughness length of 0.05m. The shift is less as wind speed increases so that the red curve is effectively the blue curve rotated round a point at a wind speed of about 14m/s where the wind shear is defined by a roughness of 0.05m. The set of green data is the same noise data plotted against the hub height wind speed.

Wind shear occurs mostly in flat areas of the country and particularly flat areas in a large bowl surrounded by hills. In more hilly areas wind shear is not generally as high as in flat areas but there is another feature that causes many residential locations to have lower wind speeds than might be expected compared with that at hub height. That is the fact that houses in hilly areas are often built in sheltered valleys whereas the turbines are more likely to be on the hill. So the effect of shelter from valleys in hilly land is the same in principle as the effect of wind shear in flat land.

This can be demonstrated by taking two noise sensitive locations with similar noise generating objects (trees, shrubs, grass and so on). If the wind speed were measured at the noise measurement locations there would be no difference in the graphs for a sheltered location down in a valley and for an exposed location on a hill. But the wind speed is measured on the exposed wind farm site so the noise levels in the sheltered valley will be less than those further up the hill.

Figure 7 shows the polynomials for two locations where noise has been measured simultaneously. One is exposed on the hill at an elevation only a little below the anemometer mast (the top, black, line) and the other is further down in the valley.

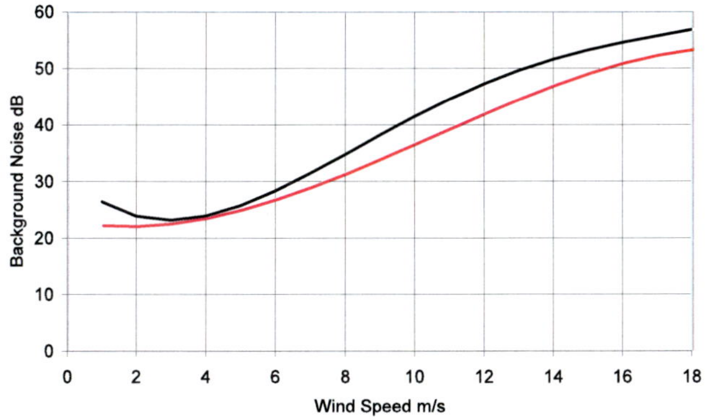

Figure 7. Background noise curves for exposed and sheltered properties

The sheltered location has a similar curve to the exposed location but it is shifted to the right in the same way as the increase in wind shear shifts the curve to the right. The difference is that the curve does not rotate in this case but it shifts by about the same amount at all wind speeds.

It is therefore important that the method of measuring and presenting background noise is the same as that used for measuring and presenting turbine noise so that a like for like comparison can be made. It is also important that the wind speed is measured on a representative part of the wind farm site and that background noise is measured in a location which residents use for relaxation.

REFERENCES

1 G.P.van den Berg. Wind turbines at night: acoustical practice and sound research. *Euronoise* 2003.

Index